Bernd G. Lottermoser

Mine Wastes

Characterization, Treatment, Environmental Impacts

Bernd G. Lottermoser

Mine Wastes

Characterization, Treatment, Environmental Impacts

Second Edition

With 70 Figures and 43 Tables

Author

Bernd G. Lottermoser

School of Earth and Environmental Sciences
James Cook University
P.O. Box 6811
Cairns, Queensland 4870, Australia
E-Mail: bernd.lottermoser@jcu.edu.au

Library of Congress Control Number: 2007924819

ISBN-10 3-540-48629-1 **Springer Berlin Heidelberg New York**
ISBN-13 978-3-540-48629-9 **Springer Berlin Heidelberg New York**

2. Edition 2007
1. Edition 2003

Springer is a part of Springer Science+Business Media
springer.com

Cover design: WMXDesign, Heidelberg
Typesetting: Uwe Imbrock, Stasch · Verlagsservice, Bayreuth (stasch@stasch.com)
Production: Agata Oelschläger

Printed on acid-free paper 30/2132/AO – 5 4 3 2 1 0

Preface

This book is not designed to be an exhaustive work on mine wastes. It aims to serve undergraduate students who wish to gain an overview and an understanding of wastes produced in the *mineral industry*. An introductory textbook addressing the science of such wastes is not available to students despite the importance of the mineral industry as a resource, wealth and job provider. Also, the growing importance of the topics "*mine wastes*", "*mine site pollution*" and "*mine site rehabilitation*" in universities, research organizations and industry requires a textbook suitable for undergraduate students. Until recently, undergraduate earth science courses tended to follow rather classical lines, focused on the teaching of palaeontology, crystallography, mineralogy, petrology, stratigraphy, sedimentology, structural geology, and ore deposit geology. However, today and in the future, earth science teachers and students also need to be familiar with other subject areas. In particular, earth science curriculums need to address land and water degradation as well as rehabilitation issues. These topics are becoming more important to society, and an increasing number of earth science students are pursuing career paths in this sector. Mine site rehabilitation and mine waste science are examples of newly emerging disciplines.

This book has arisen out of teaching mine waste science to undergraduate and graduate science students and the frustration at having no appropriate text which documents the scientific fundamentals of such wastes. There are books which cover the principles and practices of environmental management at mine sites (Hutchison and Ellison 1992; Mulligan 1996) and the environmental impacts of mining (Ripley et al. 1996). There are also a number of books and reports addressing particular mine waste topics such as tailings (Ritcey 1989), sulfide oxidation (Alpers and Blowes 1994; Evangelou 1995), mine waters (Morin and Hutt 1997; Younger et al. 2002; Younger and Robins 2002), acid mine drainage (Skousen and Ziemkiewicz 1996), mine water treatment (Brown et al. 2002), and cyanide-bearing wastes (Mudder et al. 2001). Some of these books and reports, written for researchers or industry practitioners, contain a lot of useful theoretical or practical information. However, a single introductory text explaining the scientific principles of problematic mine wastes is still missing. This book aims to fill this gap and will thereby complement the existing literature. It has been written with undergraduate science, environmental science and engineering students in mind who have already gained a basic knowledge in chemistry and the earth sciences. Details of mineralogical and geochemical aspects have been deliberately omitted from this work as these are already covered by the existing literature. This book will be particularly of use to those students with a preliminary understanding of inorganic chemistry, hydrology, mineralogy, and geochemistry. Postgraduate students working on mine wastes are advised to consult the specialized literature.

I would like to express my appreciation to the many colleagues and students who fuelled my interest in wastes. Of all my colleagues I am most grateful to Associate Professor Paul Ashley (University of New England, Armidale, Australia) whose cooperation over the years has been so enjoyable and most stimulating. The funding and technical support for my research programs and those of my students came over the years from the Australian Research Council, Australian Institute of Nuclear Science and Engineering, Australasian Institute of Mining and Metallurgy Bicentennial Gold '88 Endowment Fund, Environment Australia, Deutsche Forschungsgemeinschaft, Alexander von Humboldt Foundation, University of New England, James Cook University, State Government agencies of New South Wales, South Australia and Queensland, and various private companies. An Alexander von Humboldt Fellowship made this book possible. Special thanks to Johanna for her professional editing, encouragement and understanding. To my family, especially my mother, Gisela and Hella – thank you for being there! To Gisela, thank you for an amazing trip to Greenland (Case Study 4.2). Finally, this book would not have happened at all without the initial suggestion by my father – dieses Buch ist für Dich.

Contents

Glossary

Table G.1. Standard International (SI) system of units

Physical quantity	Basic unit	Symbol
Length	Meter	m
Mass	Kilogram	kg
Time	Second	s
Electric current	Ampere	A
Temperature	Kelvin	K
Amount of substance	Mole	mol

Table G.2. Derived and SI-related units

Physical quantity	Unit	Symbol / Conversion
Force	Newton	$N = kg\ m\ s^{-2}$
Energy, work, heat	Joule	$J = N\ m$
Electric potential	Volt	$V = J\ A^{-1}\ s^{-1}$
Conductance	Siemens	$S = A\ V^{-1}$
Specific conductance	Microsiemens per centimeter	$\mu S\ cm^{-1}$
Area	Hectare	$ha = 10^4\ m^2$
Liquid volume	Liter	$l = 10^{-3}\ m^3$
Solid volume	Cubic meters	m^3
Flow	Liters per second	$l\ s^{-1}$
Weight	Tonne (metric)	$t = 10^3\ kg$
Celsius temperature	Degree Celsius	$°C = K - 273.15$
Radioactivity	Curie	$Ci = 3.7 \times 10^{10}$ disintegrations s^{-1}

Table G.3. Weight-based concentrations

Unit	Symbol
Parts per billion	$ppb = \mu g\ kg^{-1}$
Parts per million	$ppm = mg\ kg^{-1}$
Weight percent	$wt.\% = (kg\ kg^{-1}) \times 100$

Table G.4. Volume-based concentrations

Unit	Symbol
Micrograms per liter	$\mu g\ l^{-1}$
Milligrams per liter	$mg\ l^{-1}$
Volume percent	vol.% = $(l\ l^{-1}) \times 100$

Table G.5. Prefixes to SI and derived SI units

Prefix	Symbol	Value
Mega-	M	10^6
Kilo-	k	10^3
Centi-	c	10^{-2}
Milli-	m	10^{-3}
Micro-	µ	10^{-6}
Nano-	n	10^{-9}
Pico-	p	10^{-12}

Table G.6. Abbreviations

(aq)	Aqueous
B.C.	Before Christ
ca.	Circa
cf.	Compare with
e.g.	For example
etc.	Et cetera
EC	Electrical conductivity
Eh	Oxidation potential relative to the standard hydrogen electrode
i.e.	That is
(g)	Gas
(l)	Liquid
REE	Rare earth elements: La to Lu
(s)	Solid
TDS	Total dissolved solids
UV	Ultraviolet radiation

Chapter 1

Introduction to Mine Wastes

1.1 Scope of the Book

This book focuses on "problematic" solid wastes and waste waters produced and disposed of at modern mine sites. They are problematic because they contain hazardous substances (e.g. heavy metals, metalloids, radioactivity, acids, process chemicals), and require monitoring, treatment, and secure disposal. However, not all mine wastes are problematic wastes and require monitoring or even treatment. Many mine wastes do not contain or release contaminants, are "inert" or "benign", and pose no environmental threat. In fact, some waste rocks, soils or sediments can be used for landform reconstruction, others are valuable resources for road and dam construction, and a few are suitable substrates for vegetation covers and similar rehabilitation measures upon mine closure. Such materials cannot be referred to as wastes by definition as they represent valuable by-products of mining operations.

This books attempts to gather the scientific knowledge on problematic wastes accumulating at modern mine sites. Wastes are also produced at mineral processing plants and smelter sites and include effluents, sludges, leached ore residues, slags, furnace dusts, filter cakes, and smelting residues. Such wastes are not mine wastes by definition as they generally do not accumulate at mine sites. Thus, this book largely focuses on mining wastes. It limits the presentation of mineral processing and metallurgical wastes to those waste types accumulating at or near mine sites. Readers interested in the general areas of mineral processing and metallurgical wastes are advised to consult the relevant literature (e.g. Petruk 1998).

Mine wastes are commonly classified according to their physical and chemical properties and according to their source. Such a classification scheme is followed in this work. The book attempts to cover the major sources of mine wastes including the mining of metal, energy and industrial mineral resources. Wastes of the petroleum industry, in particular, wastes of the oil shale and oil sand industry, have been excluded as they are beyond the scope of this book. The book has been organized into seven chapters which document the different sources and properties of mine wastes. The contents of Chapters Two (Sulfidic Mine Wastes), Three (Mine Water), Four (Tailings), and Five (Cyanidation Wastes of Gold-Silver Ores) are inherent to most metal and/or coal mines. The contents of Chapter Six (Radioactive Wastes of Uranium Ores) are of importance to uranium mining operations, whereas Chapter Seven (Wastes of Phosphate and Potash Ores) discusses topics that are relevant to the fertilizer producing industry.

1. Chapter 1 sets the scene as introduction. It gives important definitions, describes the environmental impacts of mine wastes in human history, presents the nature and scope of waste production in the mining industry, and lists the general resources available to acquire knowledge on mine wastes.
2. Chapter 2 provides an insight into sulfidic mine wastes. Mining of many metal ores and coal exposes and uncovers sulfide minerals to oxidizing conditions. This chapter documents the oxidation and weathering processes of sulfides which cause and influence acid mine drainage. This is followed by discussions of the available tools to predict and to monitor the behaviour of acid generating wastes. The chapter also lists the various technologies available for the control and prevention of sulfide oxidation.
3. Chapter 3 covers the fundamentals of acid mine waters. It explains important processes occuring within such acid waters and documents predictive and monitoring techniques. A documentation of technologies applied for the treatment of acid mine drainage completes the chapter.
4. Chapter 4 addresses the wastes of mineral processing operations (i.e. tailings). The chapter presents the characteristics of tailings solids and liquids. It also gives details on the disposal options of tailings whereby most tailings are stored in engineered structures, so-called "tailings storage facilities" or "tailings dams".
5. Chapter 5 covers the characteristics of cyanide-bearing wastes which are produced during the extraction of gold and silver. The chemistry of cyanide is explained before the use of cyanide in the mineral industry is shown. A documentation of treatment options for cyanidation wastes concludes the chapter.
6. Chapter 6 summarizes radioactive wastes of uranium ores. It provides the mineralogical and geochemical characteristics of uranium ores and gives the principles of radioactivity. The chapter describes uranium mine wastes and the techniques available for their disposal and treatment. The potential hazards and environmental impacts of uranium mining have been discussed in some detail.
7. Chapter 7 describes wastes of the phosphate and potash mining and fertilizer producing industry. Phosphogypsum is the major waste product of fertilizer production. The characteristics, storage and disposal practices, and recycling options of this waste material are documented to some extent.

Sulfidic wastes and acid mine waters have been studied extensively from all scientific angles, and there is a vast literature on the subjects including books, reviews, technical papers, conference proceedings, and web sites. On the other hand, wastes of potash ores have received in comparison only limited attention. Such a disproportionate knowledge has influenced the presentation of this work and is reflected in the length of individual chapters.

Several chapters contain case studies and scientific issues which demonstrate particular aspects of chapter topics in greater detail. Some case studies highlight the successes in handling mine wastes, others point to future opportunities, whereas some document the environmental impacts associated with them. The reasoning behind this is that we have to learn not only from our successes but also from our mistakes in handling mine wastes. Most of all, we have to pursue alternative waste treatment, disposal, use and rehabilitation options.

1.2 Definitions

1.2.1 Mining Activities

Definitions are essential for clear communication especially when discussing technical issues. Therefore, important and relevant terms have been defined in the following sections. Operations of the mining industry include mining, mineral processing, and metallurgical extraction. "*Mining*" is the first operation in the commercial exploitation of a mineral or energy resource. It is defined as the extraction of material from the ground in order to recover one or more component parts of the mined material. "*Mineral processing*" or "*beneficiation*" aims to physically separate and concentrate the ore mineral(s), whereas "*metallurgical extraction*" aims to destroy the crystallographic bonds in the ore mineral in order to recover the sought after element or compound. At mine sites, mining is always associated with mineral processing of some form (e.g. crushing; grinding; gravity, magnetic or electrostatic separation; flotation). It is sometimes accompanied by the metallurgical extraction of commodities such as gold, copper, nickel, uranium or phosphate (e.g. heap leaching; vat leaching; in situ leaching).

All three principal activities of the mining industry – mining, mineral processing, and metallurgical extraction – produce wastes. "*Mine wastes*" are defined herein as solid, liquid or gaseous by-products of mining, mineral processing, and metallurgical extraction. They are unwanted, have no current economic value and accumulate at mine sites.

1.2.2 Metals, Ores and Industrial Minerals

Many mine wastes, especially those of the metal mining industry, contain metals and/or metalloids at elevated concentrations. There is some confusion in the literature over the use of the terms "*metals*", "*metalloids*", "*semi-metals*", "*heavy metals*" and "*base metals*". Metals are defined as those elements which have characteristic chemical and physical properties (e.g. elements with the ability to lose one or more electrons; ability to conduct heat and electricity; ductility; malleability). In contrast, metalloids or semi-metals are elements with metallic and non-metallic properties; that is, arsenic, antimony, bismuth, selenium, and tellurium (e.g. elements with the ability to gain one or more electrons; lower ability to conduct heat and electricity than metals). Heavy metals are those metals with a density greater than 6 g cm^{-3} (i.e. Fe, Cu, Pb, Zn, Sn, Ni, Co, Mo, W, Hg, Cd, In, Tl) (Thornton et al. 1995). The term "heavy metals" is used in this work reluctantly because there are alternative, scientifically rigorous definitions (Hodson 2004). Base metals are those metals used in industry by themselves rather than alloyed with other metals (i.e. Cu, Pb, Zn, Sn).

In most metal ores, the metals are found in chemical combination with other elements forming metal-bearing "*ore minerals*" such as oxides or sulfides. Ore minerals are defined as minerals from which elements can be extracted at a reasonable profit.

In contrast, "*industrial minerals*" are defined as any rock or mineral of economic value excluding metallic ores, mineral fuels, and gemstones. The mineral or rock itself or a compound derived from the mineral or rock has an industrial use. Ore and industrial minerals are commonly intergrown on a microscopic or even sub-microscopic scale with valueless minerals, so-called "*gangue minerals*". The aggregate of ore minerals or industrial minerals and gangue minerals is referred to as "*ore*". Thus, ore is a rock, soil or sediment that contains economically recoverable levels of metals or minerals. Mining results in the extraction of ore/industrial minerals and gangue minerals. Mineral processing enriches the ore/industrial mineral and rejects unwanted gangue minerals. Finally, metallurgical extraction destroys the crystallographic bonds of minerals and rejects unwanted elements.

1.2.3 Mine Wastes

Mining, mineral processing, and metallurgical extraction produce solid, liquid and gaseous wastes. Mine wastes can be further classified as solid mining, processing and metallurgical wastes and mine waters (Table 1.1):

- *Mining wastes.* Mining wastes either do not contain ore minerals, industrial minerals, metals, coal or mineral fuels, or the concentration of the minerals, metals, coal or mineral fuels is subeconomic. For example, the criterion for the separation of waste rock from metalliferous ore and for the classification of materials as economic or subeconomic is the so-called "cut-off grade". It is based on the concentration of the ore element in each unit of mined rock and on the cost of mining that unit. As a result, every mine has a different criterion for separating mining waste from ore. *Mining wastes* include *overburden* and *waste rocks* excavated and mined from surface and underground operations. *Waste rock* is essentially wall rock material removed to access and mine ore (Fig. 1.1). In coal mining, waste rocks are referred to as "*spoils*".

  Mining wastes are heterogeneous geological materials and may consist of sedimentary, metamorphic or igneous rocks, soils, and loose sediments. As a consequence, the particle sizes range from clay size particles to boulder size fragments. The physical and chemical characteristics of mining wastes vary according to their mineralogy and geochemistry, type of mining equipment, particle size of the mined material, and moisture content. The primary sources for these materials are rock, soil, and sediment from surface mining operations, especially open pits, and to a lesser degree rock removed from shafts, haulageways, and underground workings (Hassinger 1997).

  Once the metalliferous ore, coal, industrial minerals or mineral fuels are mined, they are processed to extract the valuable commodity. In contrast, mining wastes are placed in large heaps on the mining lease. Nearly all mining operations generate mining wastes, often in very large amounts.
- *Processing wastes.* Ore is usually treated in a physical process called beneficiation or mineral processing prior to any metallurgical extraction (Fig. 1.2). Mineral processing techniques may include: simple washing of the ore; gravity, magnetic, electrical or optical sorting; and the addition of process chemicals to crushed and sized ore in

Table 1.1. Simplified mining activities whereby a resource is mined, processed and metallurgically treated. Each step of the operation produces solid, gaseous and liquid wastes

Acitivity generating the mine waste	Mine wastes
Open pit mining, underground mining	Mining wastes (e.g. waste rocks, overburden, spoils, mining water, atmospheric emissions)
Mineral processing, coal washing, mineral fuel processing	Processing wastes (e.g. tailings, sludges, mill water, atmospheric emissions)
Pyrometallurgy, hydrometallurgy, electrometallurgy	Metallurgical wastes (e.g. slags, roasted ores, flue dusts, ashes, leached ores, process water, atmospheric emissions)

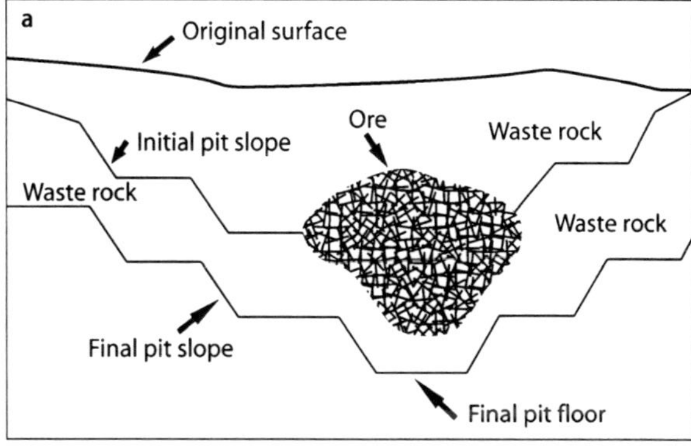

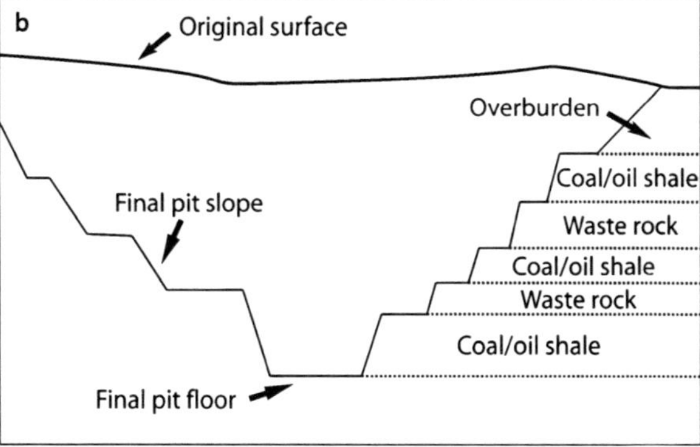

Fig. 1.1. Schematic cross-sections of open pit mines: **a** metal mines; **b** coal and oil shale mines. Waste rocks have to be mined in order to obtain ore, coal or oil shale

order to aid the separation of the sought after minerals from gangue during flotation. These treatment methods result in the production of "*processing wastes*". Processing wastes are defined herein as the portions of the crushed, milled, ground, washed or treated resource deemed too poor to be treated further. The definition thereby includes tailings, sludges and waste water from mineral processing, coal washing, and mineral fuel processing. "*Tailings*" are defined as the processing waste

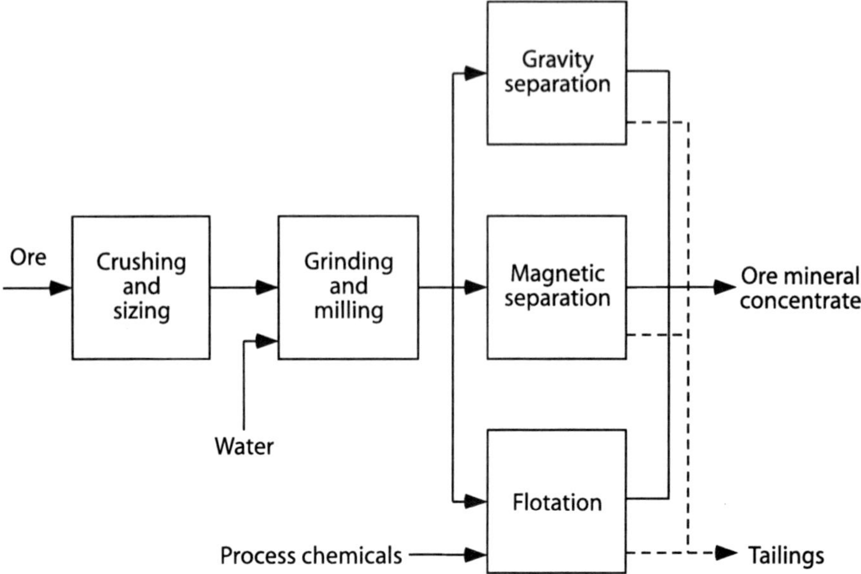

Fig. 1.2. Simplified flow-chart of a mineral processing operation, in which ore is processed to yield an ore mineral concentrate and tailings (after Ripley et al. 1996)

from a mill, washery or concentrator that removed the economic metals, minerals, mineral fuels or coal from the mined resource.

The physical and chemical characteristics of processing wastes vary according to the mineralogy and geochemistry of the treated resource, type of processing technology, particle size of the crushed material, and the type of process chemicals. The particle size for processing wastes can range in size from colloidal size to fairly coarse, gravel size particles. Processing wastes can be used for backfilling mine workings or for reclamation and rehabilitation of mined areas, but an alternative method of disposal must be found for most of them. Usually, this disposal simply involves dumping the wastes at the surface next to the mine workings. Most processing wastes accumulate in solution or as a sediment slurry. These tailings are generally deposited in a tailings dam or pond which has been constructed using mining or processing wastes or other earth materials available on or near the mine site.

- *Metallurgical wastes.* Processing of metal and industrial ores produces an intermediate product, a mineral concentrate, which is the input to extractive metallurgy. Extractive metallurgy is largely based on hydrometallurgy (e.g. Au, U, Al, Cu, Zn, Ni, P) and pyrometallurgy (e.g. Cu, Zn, Ni, Pb, Sn, Fe), and to a lesser degree on electrometallurgy (e.g. Al, Zn) (Ripley et al. 1996; Warhurst 2000). Hydrometallurgy involves the use of solvents to dissolve the element of interest. For example, at gold mines leaching of the ore with a cyanide solution is a common hydrometallurgical process to extract the gold. The process chemical dissolves the gold particles and a dilute, gold-laden solution is produced which is then processed further to recover

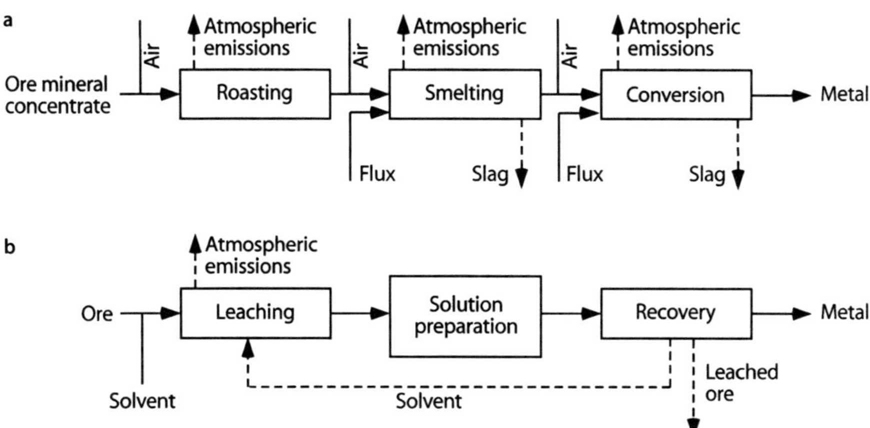

Fig. 1.3. Simplified flow-charts of **a** pyrometallurgical and **b** hydrometallurgical operations, in which ore is treated to yield metals and wastes

the metal. In contrast, pyrometallurgy is based on the breakdown of the crystalline structure of the ore mineral by heat whereas electrometallurgy uses electricity. These metallurgical processes destroy the chemical combination of elements and result in the production of various waste products including atmospheric emissions, flue dust, slag, roasting products, waste water, and leached ore (Fig. 1.3).

"*Metallurgical wastes*" are defined as the residues of the leached or smelted resource deemed too poor to be treated further. At many gold, uranium or phosphate mines hydrometallurgical extraction is performed, and hydrometallurgical wastes accumulate on site. In contrast, electro- and pyrometallurgical processes and their wastes are generally not found at modern mine sites, unless there is cheap fuel or readily available energy for these extractive processes. At many historical metal mines, the ore or ore mineral concentrate was smelted or roasted in order to remove sulfur and to produce a purer marketable product. Consequently, roasted ore, slag, ash, and flue dust are frequently found at historical metal mine sites.

- *Mine waters.* Mining, mineral processing and metallurgical extraction not only involve the removal and processing of rock and the production and disposal of solid wastes, but also the production, use and disposal of mine water. "*Mine water*" originates as ground or meteoric water which undergoes compositional modifications due to mineral-water reactions at mine sites. The term "mine water" is collective and includes any water at a mine site including surface water and subsurface ground water (Morin and Hutt 1997) (Table 1.2).

  Water is needed at a mine site for dust suppression, mineral processing, coal washing, and hydrometallurgical extraction. The term "*mining water*" is used here in a general sense to refer to waters which run off or flow through a portion of a mine site and had contact with any of the mine workings (Table 1.2). "*Mill water*" is water that is used to crush and size the ore. "*Process water*" is water that is used to process the ore using hydrometallurgical extraction techniques. The water commonly contains process chemicals. At some stage of the mining operation, water is unwanted and has no value to the operation. Such mine water is generated and disposed of at various stages during mining, mineral processing or metallurgical extraction. Water of poor quality requires remediation as its uncontrolled discharge, flow, drainage or seepage from the mine site may be associated with the release of heat, suspended solids, bases, acids, and dissolved solids including process chemicals, metals, metalloids, radioactive substances or salts. Such a release could result in a pronounced negative impact on the environment surrounding the mine site.

"*Acid mine drainage*" (AMD) refers to a particular process whereby low pH mine water is formed from the oxidation of sulfide minerals. A number of other terms are also used to describe this process such as "*acid drainage*" or "*acid rock drainage*" (ARD). These latter two terms highlight the fact that there are naturally outcropping sulfide orebodies and sulfidic rocks, which actively weather, oxidize, and cause acidic springs and streams (Furniss et al. 1999; Posey et al. 2000; Munk et al. 2002). In fact, the acid streams draining such ores and rocks can contain high levels of metals and metalloids that exceed water quality standards and result in toxic effects to aquatic life. The use of these terms tries to highlight the fact that AMD occurs naturally and unrelated to mining activities. However, such natural situations are rare compared to those where mining has been directly responsible for the acidification of waters. Therefore, in this work the term AMD is preferred. AMD

Table 1.2. Mine water terminology

Term	Definition
Type of mine water	
Mine water	Any surface water or ground water present at a mine site
Mining water	Water that had contact with any of the mine workings
Mill water	Water that is used to crush and size the ore
Process water	Water that is used to process the ore using hydrometallurgical extraction techniques; it commonly contains process chemicals
Leachate	Mine water that has percolated through or out of solid mine wastes
Effluent	Mining, mill or process water that is discharged into surface waters
Mine drainage water	Surface or ground water that actually or potentially flows from the mine site into surrounding areas
Acid mine drainage (AMD) water	Low pH surface or ground water that formed from the oxidation of sulfide minerals and that actually or potentially flows from the mine site into surrounding areas
Type of process	
Mine seepage	Slow flow of ground water to the surface at pit faces, underground workings, waste dumps, tailings dams, and heap leach piles
Mine drainage	Process of water discharge at a mine
Acid mine drainage (AMD)	Process whereby low pH mine water is formed from the oxidation of sulfide minerals

is still an unfortunate term since AMD impacts more frequently on ground water quality than on the surface drainage from a mine (Bennett and Ritchie 1993). Such impacted ground water has also been named "*acid ground water*" (AG). Finally, the waters generated by the oxidation of sulfide minerals are also referred to by some authors as "*acid sulfate waters*" (ASW).

1.3 Mine Waste Production

The modern mining industry is of considerable importance to the world economy as it provides a great diversity of mineral products for industrial and household consumers (Table 1.3). The consequence of the large size of the mining and mineral processing industry is not only the large volume of materials processed but also the large volume of wastes produced.

Table 1.3. World production of selected non-fuel mineral commodities in 1999 (USGS Mineral Resources Program 2001)

Metals	Production		Industrial minerals	Production	
Antimony	0.122	Mt	Asbestos	1.93	Mt
Arsenic trioxide	38 800	t	Barite	5.66	Mt
Bauxite	127	Mt	Boron minerals	6.37	Mt
Beryl	6 210	t	Cement	1 610	Mt
Chromite	14	Mt	Bentonite	9.82	Mt
Cobalt	29 900	t	Fuller's earth	3.52	Mt
Copper	12.6	Mt	Kaolin	41.6	Mt
Gold	2 540	t	Feldspar	8.98	Mt
Iron ore	990	Mt	Fluorite	4.51	Mt
Lead	3.02	Mt	Graphite	0.685	Mt
Manganese ore	20.4	Mt	Gypsum	106	Mt
Mercury	1 800	t	Lime	141	Mt
Molybdenum	0.123	Mt	Magnesite	10.7	Mt
Nickel	1.12	Mt	Mica	0.304	Mt
Niobium-tantalum concentr.	57 100	t	Peat	27.2	Mt
Platinum-group elements	379	t	Perlite	1.85	Mt
Silver	17 700	t	Phosphate rock	141	Mt
Tin	0.198	Mt	Potash	25.7	Mt
Titanium concentrates	4.17	Mt	Salt	209	Mt
Tungsten	31 000	t	Sand and gravel	107	Mt
Vanadium	42 200	t	Sulfur	57.1	Mt
Zinc	8.04	Mt	Vermiculite	0.534	Mt

The exploitation of mineral resources results in the production of large volumes of waste rocks as they have to be removed to access the resource. Once the resource has been extracted from the Earth, it is processed for its valuable components. These valuable components vary in mass from 100% to a few parts per million of the original resource. For instance, extraction and production of clay, sand and gravel generally do not produce any waste. Operators extract and process the entire mined material. Also, crushing, washing and sizing of rock aggregates generate only minor amounts of unwanted fine-grained slimes and dust particles. These slimes and dust particles can be put to good use as mineral fertilizer. In contrast, exploitation of a metalliferous mineral resource aims to extract only a few percent concentrations of copper, lead or zinc or even parts per million values of gold. Only a very small valuable component is extracted from metalliferous ores during processing and metallurgical extraction. The great majority of the total mined material is gangue which is generally rejected as processing and metallurgical waste. Therefore, mining, mineral processing, and metallurgical extraction result in the production of a high volume of unwanted material.

In general, coal mining and processing generate the largest quantity of waste followed by non-ferrous and ferrous ores and industrial minerals. Waste production varies greatly from nation to nation. For example, more than 4700 Mt of mining waste and 1200 Mt of tailings are stored all over the European Union, most of the mine waste in Finland, Germany, Greece, Ireland, Portugal, Spain, Sweden, and the United Kingdom (BRGM 2000). The production of mine wastes is particularly significant in nations with a major mining industry. In Australia, the mining industry produces 1750 Mt of mine wastes per year. It is by far the largest producer of solid, liquid and gaseous waste and exceeds municipal waste production by at least 450 Mt. Of the 2100 Mt of solid waste generated annually in Australia, 80% is produced by the mining sector (Connor et al. 1995; cited by Boger 1998). In South Africa, over 1100 mines contribute to 72.3% of the country's total solid waste stream, with approximately 25000 ha of land utilized as dumping areas in the form of tailings storage facilities (Maboeta and Rensburg 2003).

The global quantity of non-fuel mineral commodities removed from the Earth's crust each year by mining is now of the order of 3700 Mt (Table 1.3). While the consumption of mineral commodities is well documented, there is no data available on the global production of mine wastes. Therefore, an estimation of the annual quantity of mine waste produced globally has to be based on several assumptions. In 1999, approximately 40 Mt of metals (As, Be, Co, Cr, Cu, Hg, Nb, Ni, Pb, Sb, Sn, Ta, V, W, Zn) were produced worldwide (Table 1.3). Assuming that the average ore grade of the metal deposits was 0.5%, mining, processing and extraction of the ores generated 8000 Mt of solid wastes. Similarly, the production of 2540 t of gold generated about 1250 Mt of solid wastes, assuming an average gold ore grade of 2 ppm. In addition, every year approximately 4500 Mt of coal, 990 Mt of iron ore, 127 Mt of bauxite, and 2500 Mt of industrial minerals are consumed globally. For every tonne of these ores consumed, there will be at least the same amount of solid waste generated (i.e. waste rocks, tailings). Such calculations indicate that approximately 15000 to 20000 Mt of solid mine wastes are being produced annually around the world.

These calculations and statistics represent approximations and can only serve as an indication of the magnitude of waste production. Furthermore, every mine site has its own unique waste because there are compositional differences in the mined ore, and there is a great diversity of applied mining and mineral processing methods. Wastes generated at different mines vary considerably in their properties. While certain scientific principles apply to particular commodities, every mine requires its very own waste characterization, prediction, monitoring, control, and treatment.

Mine wastes represent the greatest proportion of waste produced by industrial activity. In fact, the quantity of solid mine waste and the quantity of Earth's materials moved by fundamental global geological processes are of the same order of magnitude – approximately several thousand million tonnes per year (Fyfe 1981; Förstner 1999). Fundamental global geological processes such as oceanic crust formation, soil erosion, sediment discharge to the oceans, and mountain building naturally move Earth's materials around the Earth's crust and shape our planet. In contrast, mankind extracts material from the Earth during mining and discards most of the extracted crust as waste. As a result, the Earth is getting increasingly shaped by mine wastes rather than by natural geological processes. In addition, metal ores of increasingly lower grades are being exploited, and more wastes are being produced as a result of it. The production of mine wastes may even double within a period of 20 to 30 years (Förstner 1999). Today and in the future, commercial exploitation of a mineral resource is about waste production and waste disposal as well as resource production and provision.

1.4 Mine Wastes: Unwanted By-Products or Valuable Resources?

The term "mine waste" implies that the material has no current economic value and is an unwanted by-product of mining. However, some mine wastes can be useful and this has been recognized since the beginning of mining and smelting. For example, the use of slag in road construction can be traced back to the very early days when the Romans used iron slag as a pavement material for their roads. Also, while wastes of the mineral industry are generally useless at the time of production, they can still be rich in resource ingredients. Unfavourable economics, inefficient processing, technological limitations or mineralogical factors may not have allowed the complete extraction of resource ingredients at the time of mining. In the past, inefficient mineral processing techniques and poor metal recoveries produced wastes with relatively high metal concentrations (Scientific Issue 1.1; Fig. 1.4). In some cases, old tailings and waste rock piles that were considered worthless years ago are now "re-mined", feeding modern mining operations. This approach is widely used in the mining industry.

Hence, changing circumstances may turn a particular waste into a valuable commodity, either because the economic extraction of resource ingredients may now be possible using improved technology, or a market has been found for the previously unwanted material. What may be waste to some miners, can be a very important, useful resource to other mining operations, either now or in the future. Yesterday's waste can become today's resource.

Recycling today's waste is similarly possible. Manganese tailings may be used in agro-forestry, building and construction materials, coatings, resin cast products, glass, ceramics, and glazes (Verlaan and Wiltshire 2000). Tailings can also be suitable fertilizers for golf courses; phosphogypsum can be used in the agricultural and the building industry; clay-rich wastes can improve sandy soils or are the raw material for brick

Scientific Issue 1.1. Historical Base Metal Smelting Slags

The Principles of Smelting

Metals are extracted from metal ores by heating the ore minerals to their melting points. Flux (e.g. limestone, ironstone, ferrous silicate, silica) may be added to the charge of ore and fuel to lower the melting temperature and to decrease the viscosity of the slag. The fuel for the smelting may include coke, coal, firewood, and/or charcoal. Oxygen for the reactions is provided by a blast of compressed air, and temperatures within the interior of the furnace reach over 1000 °C. During the smelting process, the materials in the furnace react and a layered melt is produced. A lighter silicate melt accumulates above a heavier molten metal liquid, present at the bottom of the furnace. The silicate melt (i.e. slag) forms by the combination of the elements and compounds within the ores and fluxes. Once melting of the ore is achieved, the slag is drained off. The slag is dumped, commonly while still liquid, and solidifies and cools into a cohesive mass. Thus, historical smelting slag dumps are largely in the same form as they existed when smelting stopped.

Pyrometallurgy has its origin around 7000 to 6000 B.C. when early humans started to smelt sulfide and oxide ores. The earliest physical evidence for smelting is some slaggy material found in Turkey (Lambert 1997). The rise of early civilizations in Southwest Asia and the Mediterranean region came with the widespread use of and demand for metal tools and products. This in turn led to a steady increase of metal smelting, particularly of lead, which reached peak production during Roman times.

Slags of Lavrion, Greece

The Roman and Greek Empires had mining and smelting operations throughout the Mediterranean region, and ancient smelting slags are common features of these historic sites. The mines and smelters of Lavrion (Greece) are examples of such an historic site. The Lavrion ores have been a source of lead, silver and copper since 3000 B.C., with much of the mining performed by 10000 to 20000 slaves between 600 and 400 B.C. (Jaxel and Gelaude 1986; Wendel 1999). The mining activities not only resulted in over 2000 shafts and over 300 km of underground workings but also in the production of several million tonnes of slag. The slag was dumped next to the mine sites. As the ancient pyrometallurgical processes and metal recovery technologies were relatively inefficient compared to today's standards, the slag dumps are rich in metals and metalloids. Today, the Lavrion smelter slags are famous among mineral collectors for their rare and exotic secondary minerals which have formed during several thousand years of slag weathering.

Potential Recycling

Since the beginning of the industrial age, slag has been considered harmless. It is therefore widely used for sand blasting and to construct roadways and railway beds. However, historical base metal smelting slags are not suitable for such applications as they can contain high levels of potentially toxic elements. Natural weathering can release them into the environment causing contamination of soils, sediments and waters.

The metal concentrations of many historical base metal smelting slags are similar to or even higher than those of geological ore deposits currently being mined for metals. Consequently, the recovery of resource constituents from the slags has been considered. Much of the metals and metalloids are hosted by glass and microcrystalline silicates. This leaves hydrometallurgical processing using acids as the only treatment option.

manufacturing; mine waters can be converted into drinking water (Schwartz and Ploethner 2000; Smit 2000; Varnell et al. 2004); mine water can be used for heating or cooling purposes (Banks et al. 2004; Watzlaf and Ackman 2006); mine drainage sludges can be a resource for pigment (Kirby et al. 1999); and pyritic waste rock can be an excellent soil amendment to neutralize infertile alkaline agricultural soils (Castelo-Branco et al. 1999). If such innovative alternatives to current waste disposal practices are pursued and if wastes are used as raw materials, then waste disposal problems are

Fig. 1.4. Derelict copper-lead-gold-silver smelting works at Chillagoe, Australia. Smelting operations were conducted from 1901 to 1943 and produced 1 Mt of slag. The slag contains wt.% levels of zinc that is principally hosted by glass, olivine and hedenbergite

eliminated. Total resource utilization, where all of the material extracted is put to good use, is a challenging concept for researchers and miners.

1.5 Mining and Environmental Impacts

Major impacts of mining on land can occur before, during and after operation and may include: vegetation clearance; construction of access roads, infrastructure, survey lines, drill sites, and exploration tracks; creation of large voids, piles of wastes, and tailings dams; surface subsidence; excessive use of water; destruction or disturbance of natural habitats or sites of cultural significance; emission of heat, radioactivity, and noise; and the accidental or deliberate release of solid, liquid or gaseous contaminants into the surrounding ecosystems.

An understanding of the long-term release of contaminants requires a solid knowledge of the factors that control such discharge. The major factor that influences contaminant release is the geology of the mined resource (Scientific Issue 1.2). Climate and topography as well as the applied mining and mineral processing activities also play their role in the type and magnitude of contaminant release from a specific mine site or waste repository. The long-term off-site release of contaminants is particularly possible from mining, processing or metallurgical wastes or waste repositories. As a result, the operations of the mining industry have been criticized by the conservation lobby for some time (Scientific Issue 1.3).

Scientific Issue 1.2. Geology and Its Influence on the Environmental Impacts of Mineral Deposits

The Environmental Geology of Mineral Deposits

Mineral deposits are concentrations of metallic or other mineral commodities in the Earth's crust and result from various geological processes (Plumlee 1999). During these processes, the deposits acquire specific geological characteristics, including the amount and type of metals enriched in the deposits, the kind of minerals formed and their grain size, and the type of rocks associated with the deposit. Such fundamental geological aspects of mineral deposits exert important and predictable impacts on the environment (Plumlee 1999; Seal and Hammarstrom 2003). The geology of a deposit may influence, for example, the chemistry of local ground and surface waters and the properties of soils. Also, local soils, sediments and waters can be naturally burdened with trace elements. This is especially the case where weathering and erosion have exposed metallic mineral deposits and have led to the mobilization of trace elements into the environment. At such sites, soils, sediments and waters are naturally enriched in metals and metalloids.

The natural occurrence of elements varies between different ore deposit and rock types. Certain ores and rocks can provide exceptionally high metal and metalloid concentrations to soils, sediments and waters. Different rocks and ores supply different elements. For example, ultramafic igneous rocks and serpentinites and their associated ores can provide high iron, magnesium, chromium and nickel concentrations to overlying soils. Such bedrocks form nutrient-poor, metal-rich soils, and the vegetation has to adapt to such substrates. The flora is so distinct that it is referred to as "serpentine flora". Also, marine and lacustrine sediments including coals may possess high boron and selenium concentrations. Boron enrichment in topsoils overlying such bedrocks can be a widespread constraint to cereal and legume crop and pasture production. By contrast, the selenium abundance in soils and plants may induce acute toxicity in grazing animals.

Thus, rocks and ores with particular element enrichments cause environmental signatures in receiving streams, soils and sediments, and these enrichments may even bring about adverse effects on local and regional ecosystems. The environmental signatures and impacts of mineral deposits may occur naturally. Alternatively, they can be exacerbated or even caused by improper mining and mine waste disposal practices.

The Environmental Geology of Gold Deposits, New Zealand

New Zealand has a number of mineral deposits, including mesothermal and epithermal gold deposits. These deposits possess noticeably different geological characteristics which in turn lead to different impacts on the environment (Craw 2001).

Mesothermal gold deposits are vein-type gold deposits that formed by intermediate-temperature (200–300 °C) hydrothermal solutions in continental collision zones. On the South Island of New Zealand, the mineral deposits are located in cool semiarid or alpine settings and have calcite-bearing host rocks. The abundance of acid-buffering calcite and the lack of water interacting with sulfides control the environmental impacts of these deposits. The calcite buffers any acid generated from the oxidation of sulfides and, the lack of available meteoric water restricts acid generation. As a result, the oxidation of mesothermal gold ores does not result in significant acidification of streams. Yet, these deposit types contain elevated arsenic contents, and this metalloid is mobilized from the deposits at neutral to alkaline pH values (Craw and Pacheco 2002).

Epithermal gold deposits are found on the North Island of New Zealand and represent shallow gold deposits formed by low-temperature (50–200 °C) hydrothermal solutions. These mineral deposit types are characterized by a lack of carbonate and an abundance of pyrite. The deposits occur in a temperate to subtropical climate and hence, unconstrained sulfide oxidation favoured by higher temperatures and abundant meteoric water causes the development of low pH waters from mineralized materials. Consequently, oxidation of these deposits leads to the acidification of streams, and copper, cadmium, lead and zinc are mobilized into the environment (Craw 2001).

Thus, the geology of mineral deposits as well as climate and topography together with mining and mineral processing practices control the release of contaminants from mine sites and mine wastes (Plumlee 1999; Craw 2001). A solid understanding of the environmental geology of mineral deposits is vital to any mining operation, environmental impact assessment or rehabilitation plan. Such knowledge allows the development of effective prediction, prevention and remediation tools necessary for the successful environmental management of these sites.

Scientific Issue 1.3. The Debate on Mining and Its Environmental Impacts

In recent decades the conservation lobby has criticized the operations and actions of the mining industry. Some of the issues raised by conservation groups also relate to the production and environmental impacts of mine wastes. These issues and the likely responses of both parties are presented below.

Issue 1 – Land Disturbance

Critics of the Mining Industry
The mining industry "rapes virgin territory" and leaves permanent scars and massive "footprints" in landscapes. For example, there are more than 32 000 abandoned or inactive underground mines in the western United States alone, and the total area of land having been mined in India is equivalent to one-third of that under agricultural production (Hossner and Hons 1992). The area of land disturbed by mining globally is estimated to increase by approximately 1 million hectares per year (Hossner and Hons 1992).

The Mining Industry
Mining of an area should be regarded as a stage in the sequential use of land. Once mining ceases, the mined and rehabilitated land should be used for industrial, agricultural or recreational activities. Also, mines are local phenomena and account for only a small part of a land area of a country. For example, the mining industry accounts for less than 1% of the total area of South Africa or the United States. In Australia, mining activities occupy only about 0.06% of the land mass. This is less than one big city and less than land used by the defense forces (0.24%), and it is a tiny fraction compared to the vast tracts used for and degraded by agriculture (6%).

Issue 2 – Waste Production

Critics of the Mining Industry
Waste production by the mining industry is enormous. For example, the extent of derelict land covered by waste is estimated to be 100 000 ha in the United Kingdom, 200 000 ha in Malaysia and several million hectares in the United States (Hossner and Hons 1992). The global area covered with mine waste is probably in the order of 100 million hectares containing several hundred thousand million tonnes of mine wastes. Every year another 15 000 to 20 000 Mt of solid mine waste are added to the piles.

The Mining Industry
Mine waste production and waste repositories are a necessary and inevitable adjunct to a mine as are garbage bins and sewage pipes to a dwelling. The modern mining industry plays a leading role in waste management. For example, it is one of the few industries recycling some of its own waste (e.g. the use of slag in cement, the extraction of metals from historical mine wastes) and that of others (e.g. lead batteries, scrap iron). Much of the modern mining industry applies current knowledge and state-of-the-art technology to limit waste production, to recyle wastes, and to reduce the risks of contamination arising from mine wastes.

Issue 3 – Recycling Rather Than Mining

Critics of the Mining Industry
Mining is an unnecessary industrial activity. Recycling is the answer to the ever increasing demand for natural resources.

The Mining Industry
Today's standard of living relies on the supply of natural resources. There will always be a need for mining in addition to recycling and substitution. Moreover, certain mined rocks and minerals are essential and important "environmental minerals" as they are needed to alleviate environmental problems. For example, clays are used as liners and sealants for industrial waste sites, and calcium carbonate minerals in the form of limestone are applied for the remediation of acid waters and acid sulfate soils.

Scientific Issue 1.3. *Continued*

Issue 4 – Pollution

Critics of the Mining Industry
The mining industry is intrinsically "dirty and polluting".

The Mining Industry
The great majority of mine site pollution problems are legacies from the past. Environmental disasters caused by mining are found in developing nations and former or current communist states where economic growth and mining have been or still are enforced at the expense of the environment. In industrialized countries, the mining industry is one of the most environmentally regulated industries.

Issue 5 – Heavy Metal Release

Critics of the Mining Industry
The mining industry releases toxic heavy metals to the environment that negatively affect people's health and impact on ecosystems.

The Mining Industry
Heavy metals have become the "geochemical bogey men" (Hodson 2004). They are synonymous with pollution and toxicity, responsible for all manner of evils in the environmental world. This is nothing but a myth today. Metals are not only vital for our well-being, perform essential functions in our own bodies and are ubiquitous in our environment, but natural concentrations of metals can exceed their man-made levels (Fig. 1.5). Such natural metal contamination can create some problems, especially in communities attuned to the view that natural must be healthy.

Issue 6 – Environmental Damage

Critics of the Mining Industry
The mining industry is the major "environmental vandal" on our planet.

The Mining Industry
Mining's contribution to environmental damage pales by comparison to other human activities. Critics of the mining industry commonly ignore those human activities which cause damage on a continental scale. For example, cities and towns cover significant parts of continents and urban dwellers, including the conservation lobbyists, continue to impact on the environment. They themselves cause the consumption of natural resources, suburban sprawl and associated air, soil and water pollution, climate change, habitat destruction, species extinction, concreting of land, and production of household wastes. Such urban activities not only damage the environment but alter and seal natural landscapes for good whereas mine sites are rehabilitated for other land uses once mining ceases. Is it more acceptable to concrete nature, to flush sewage into rivers and oceans, to dump vast amounts of domestic waste, to create smog, or to cause global and micro climate change? Also, farming and pastoral activities have significantly degraded land masses and have impacted on soils and the aquatic environment. Poor agricultural practices cause salinization, acidification and erosion of soils as well as pesticide and nutrient transfer (N, P) into inland and coastal waters. Such activities even impact on World Heritage Areas and have received surprisingly little attention by the conservation lobby.

1.5.1
Contamination and Pollution

Much of the environmental impacts of mining are associated with the release of harmful elements from mine wastes. Mine wastes pose a problem not just because of their

Fig. 1.5. Noble Island, Australia. Much of the island is naturally enriched in metals and metalloids (i.e. tungsten-tin-arsenic-copper ores). Weathering and erosion lead to the physical and chemical transport of metals and metalloids into the surrounding Great Barrier Reef

sheer volume and aerial extent, but because some of them may impact on local ecosystems. As a result, in many cases mine wastes must be isolated or treated to reduce oxidation, toxicity, erosion or unsightliness and to allow the waste repositories to be used for other purposes after mining ceases. If uncontrolled disposal of mine wastes occurs, it can be associated with increased turbidity in receiving waters or with the release of significant quantities of potentially harmful elements, acidity or radioactivity. These contaminants may spread to the pedosphere, biosphere, atmosphere, and hydrosphere and cause environmental effects. For example, anthropogenic inputs of metals and metalloids to atmospheric, terrestrial and aquatic ecosystems as a result of mining have been estimated to be at several million kilograms per year (Nriagu and Pacyna 1988; Smith and Huyck 1999).

However, it is important to understand that releases of elements or compounds from mine wastes do not necessarily result in damage to the environment. Even if strongly elevated metal and metalloid concentrations are present in mine wastes, the elements may not be readily bioavailable (i.e. available for uptake into the organism) (Williams et al. 1999). Furthermore, even if the elements are bioavailable, they are not necessarily taken up by plants and animals. In cases where the elements are taken up, they do not necessarily lead to toxicity. Many metals are essential for cellular functions and are required by organisms at low concentrations (Smith and Huyck 1999). It is only when these bioavailable concentrations are excessively high that they have a negative impact on the health of the organism and toxicity might be seen. Processes that cause toxicity, disrupt ecological processes, inflict damage to infrastructure, or pose a hazard to human health are referred to as pollution (Thornton et al. 1995). In contrast, contamination refers to processes which do not cause harmful effects (Thornton et al. 1995).

Environmental contamination and pollution as a result of improper mining, smelting and waste disposal practices have occurred and still occur around the world. Problems encountered are as diverse as the emissions from smelters, or the environmental clean-up of collapsed mining ventures which have to be paid for by the taxpayer. This is unacceptable to those of us who believe that technologies can be used to prevent pollution and regulations should be enforced to ensure that the environmental performance of companies is adequate. Regardless of this debate, the challenges for the modern mining industry will remain the same:

- To continue to improve its environmental operations
- To operate in a sustainable manner
- To prove its critics wrong

1.5.2 Historic Mining

Mining has been with us for thousands of years. Even the earliest mining operations during the Copper, Bronze and Iron Ages resulted in the production of gaseous, liquid and solid wastes. In historic times, mine wastes were released into the environment with some of them causing contamination or even pollution on a local or regional scale. Environmental contamination as a result of mining is not new to the industrialized world.

Air contamination as a result of smelting has been detected as far back as 5 000 years ago. Stratigraphic and physicochemical investigations of numerous European peat bogs have confirmed that smelting of sulfide minerals led to metal contamination of the environment (Shotyk et al. 1996; Ernst 1998). For example, the smelting of lead-rich silver ore in Spain by the Romans 2 000 years ago quadrupled the levels of lead in the atmosphere as far away as Greenland (Rosman et al. 1997). Generally, the smelting of sulfide ore in open air furnaces by the Greeks and Romans resulted in a vast area of the Northern Hemisphere being showered with metal-rich dust (Hong et al. 1996; Shotyk et al. 1996; Rosman et al. 1997). Human contamination of the atmosphere with arsenic, antimony, copper, mercury, lead and zinc, at least in the Northern Hemisphere, began well before the Industrial Revolution.

Water and sediment contamination and pollution are similarly not a by-product of industrialization. For example, soil erosion began with clearing of land and primitive agricultural practices 5 000 years ago (Lottermoser et al. 1997a), and metal mining in the northern Harz province of Germany resulted in metal pollution of regional stream sediments as far back as 3 500 years ago (Monna et al. 2000). Similarly, exploitation of the Rio Tinto ores in Spain has caused massive metal contamination of stream and estuary sediments since the Copper Age 5 000 years ago (Leblanc et al. 2000; Davis et al. 2000).

Acid mine drainage resulting from the oxidation of sulfides in mine wastes is a major environmental issue facing the mining industry today. This pollution process has a long history dating back thousands of years when the Rio Tinto mining district of Spain experienced periods of intense mining and the associated production of pyrite-rich wastes and AMD waters. AMD production must have been occurring at least since the first exploitation of the Rio Tinto ores 5 000 years ago, which highlights the long-term nature of AMD.

The knowledge that mining and smelting may lead to environmental impacts is not new to modern science either. The Greek philosopher Theophrastus (ca. 325 B.C.) recognized the oxidation of pyrite, the formation of metal salts and the production of acid. During the Middle Ages, AMD in central Europe was documented by Agricola who wrote the first systematic book on mining and metallurgy. In this 16th century classic, Agricola (1556) also recognized the environmental effects of ubiquitous mining in central Europe and described mining pollution:

> "The fields are devastated by mining operations ... Further, when the ores are washed, the water which has been used poisons the brooks and streams, and either destroys the fish or drives them away. Therefore the inhabitants of these regions, on account of the devastation of their fields, woods, groves, brooks and rivers, find great difficulty in procuring the necessaries of life, and by reason of the destruction of the timber they are forced to greater expense in erecting buildings. Thus it is said, it is clear to all that there is greater detriment from mining than the value of the metals which the mining produces." (Reprinted from Agricola 1556, De re metallica, p. 8. Translated by Hoover HC, Hoover LH, 1950, Dover Publications, New York, with permission of the publisher)

As the scale of mining increased during the Middle Ages, so did the degree of contamination and pollution (Ernst 1998). Coupled with this increase in scale came changes in smelting and processing techniques, including the use of chemicals and the transport of ores and concentrates over greater distances. However, it was not until the Industrial Revolution, with the event of major technological changes including the introduction of blast furnaces, that base metal smelter operations throughout the world became one of the primary sources of metal contamination (Ernst 1998). When this large-scale smelting technique was developed, contamination became even larger in scale. The smelting process released massive amounts of sulfur dioxide and metals into the atmosphere (Fig. 1.6). These activities resulted in ever-increasing environmental impacts which largely went unchecked until the second half of the 19th century (Case Study 1.1). Until then environmental impacts of mining and mineral processing were poorly understood, not regulated, or viewed as secondary in importance to resource extraction and profit maximization. The advent of environmental laws and regulations in the 20th century made the mining industry more accountable and enforced environmental protection (Fig. 1.7).

The concern for the health of miners has evolved in parallel with mining development, particularly in respect to the exposure of humans to mercury and arsenic. Mercury deposits in the Mediterranean were first worked by the Phoenicians, Carthaginians, Etruscans and Romans, who used the ore as a red pigment for paint and cosmetics. To protect local workers and the environment, the Italian mines were closed by the Romans (Ferrara 1999). The mercury was then mined by slaves in occupied Spain.

Mercury has also been used for nearly 3 000 years to concentrate and extract gold and silver from geological ores (Lacerda and Salomons 1997). The use of mercury in gold mining is associated with significant releases of mercury into the environment and with an uptake of mercury by humans during the mining and roasting processes (Lacerda and Salomons 1997). Around 2 100 years ago, Roman authorities were import-

Fig. 1.6. Denuded, bare hills at Queenstown, Australia. During the late 19th and early 20th centuries, smelting operations were conducted at the Mt. Lyell copper-gold mine. The smelting operations combined with timber cutting, frequent bushfires and high annual rainfall resulted in extensive loss of vegetation and considerable soil erosion on the surrounding hills

ing mercury from Spain to be used in gold mining in Italy. Curiously, after less than 100 years, the use of mercury in gold mining was forbidden in mainland Italy and continued in the occupied territories. It is quite possible that this prohibition was already a response to environmental health problems caused by the mercury process (Lacerda and Salomons 1997).

The above mentioned practice to enforce mining operations in occupied territories with no environmental management and no regard for the health of local miners has continued into modern times. For example, the former Soviet Union conducted mining in occupied East Germany from 1946 to 1990. Environmental management of mining and proper waste disposal did not occur, local uranium miners were exposed to deadly radiation levels, and poor regard for the environment left an environmental disaster on a massive scale (Case Study 6.1).

1.5.3 Present-Day Unregulated Mining

Today, mines wastes are produced around the world in nearly all countries. In many developing countries, the exploitation of mineral resources is of considerable importance for economic growth, employment and infrastructure development. In these

Fig. 1.7. Abandoned tin dredge in the dry stream bed of Nettle Creek, Innot Hot Springs, Australia. The bucket dredge was used in the 20th century to extract alluvial cassiterite. During mining, the dredge caused a massive increase in suspended sediment loads and the deterioration of stream water quality. Consequently, the Mining Act Amendment Act 1948 was introduced which is one of the first pieces of Australian legislation specifically concerned with environmental protection. The act required operators to construct settling ponds for turbid mine waters

developing nations and in former communist states the environmental management of municipal, industrial and mine wastes is often unregulated, lax, not enforced or overruled for economic reasons. Strict environmental management of wastes and regulation of mining still remains a luxury of wealthy industrialized nations.

Many of the world's poorest countries and communities are effected by artisan mining and the associated uncontrolled release of mine wastes. Operations are referred to as artisan when the applied mining techniques are primitive and do not employ modern technology. Such small-scale mining has been estimated to account for 15 to 20% of the world's non-fuel mineral production (Kafwembe and Veasey 2001). Artisan mining is highly labour intensive and employs 11.5 to 13 million people worldwide, and up to 100 million people are estimated to depend on small-scale mining for their livelihood (Kafwembe and Veasey 2001). The largely unregulated mining practices and associated uncontrolled release of mine wastes cause environmental harm. One example is the use of mercury in gold mining.

Gold mining and extraction have been with us for over 3 000 years. In the past much of the gold has been exploited either by physically concentrating the gold particles and/or by applying mercury. From the late 19th century onwards, mercury was no longer used since cyanide leaching was invented which allowed large-scale gold mining operations. In the 1970s, the mercury process was reintroduced in developing countries like Brazil, Bolivia, Venezuela, Peru, Ecuador, Colombia, French Guyana, Indone-

Case Study 1.1. Historic Mining in Australia and Its Environmental Impacts

Introduction

Australia has been a major mining nation for more than 150 years. While coal was the first commodity discovered in the late 18th century, the copper finds in South Australia in the early 1840s developed mining in a significant way. This was closely followed by major gold rushes in New South Wales and Victoria in the early 1850s. The development of these resources was associated with the uncontrolled release of gaseous, liquid and solid mine wastes into the environment (Blainey 1991):

"The old gold mining industry usually paid little attention to the environment. Victorians in the 1880s could tell when a new digging had opened up forty miles upstream: the river water downstream quickly changed colour with the clays and gravels that had been overturned upstream ... Stawell, which, in the late 1870s, was the deepest goldfield in Australia, announced its presence to the approaching travellers by the taste of sulfur from the kilns where the gold bearing pyrite was roasted. People did not see Stawell as they approached: they tasted it." (Reprinted from Blainey G (1991) Eye on Australia, p. 186, Schwartz & Wilkinson, with permission of the author).

Environmental Impacts of Historic Mines

In Australia, there are notable examples of large-scale environmental degradation caused by mining and associated smelting operations carried out during the mid to late 19th and early 20th centuries when there were few or no legislative constraints placed on operators. These operations ranged from river dredging tin operations in Queensland to hard rock copper mines in Tasmania. Some of these mining activities still impact on the environment today, even years after the mines closed.

- *Gold mining.* During the 19th century, placer gold was mined and extracted using mercury. It is estimated that several thousand tonnes of mercury were used on Australian goldfields, most of which was lost into the environment during gold recovery. The widespread use of mercury has left a legacy of mercury contamination in the former goldfields, particularly in streams (Bycroft et al. 1982). Today, some of the mercury is entering the foodchain of local streams. In addition, large areas of mined forest and bush land are still without topsoil and display sparse vegetation and stunted regrowth.
- *Base metal mining.* Abandoned copper, lead, zinc and tin mine sites of the late 19th and early 20th centuries are characterized by severely modified (or lack of) vegetation, waste rock heaps, ore stockpiles, tailings dumps, slag and flue residue deposits, disused mining and processing equipment, ruins, and commonly, acid mine drainage. The latter is a consequence of mining activities involving sulfide minerals, especially pyrite. The oxidation of pyrite causes formation of sulfuric acid and the dissolution of many metals (Chapman et al. 1983; Jacobson and Sparksman 1988; Koehnken 1997; Lottermoser et al. 1997b, 1999; Ashley and Lottermoser 1999a,b).
- *Alluvial tin mining.* Alluvial cassiterite was extracted from shallow surface deposits in eastern Australia from the 1870s to the early 1980s. In Tasmania, pyritic overburden sediment had to be removed and this material was dumped next to the mine sites. The pyritic waste heaps generate acid, metal-rich waters which contaminate ground and surface waters and impact on local ecosystems (Jong et al. 2000).
- *Asbestos mining and processing.* Asbestos was mined in Australia from the early 1900s until the 1980s. Asbestos pollution problems are related to dry processing methods and the absence of proper waste management in the pre-1980s. Asbestos fibres in run-off waters and leaching of metals from the waste materials possibly impact on local water quality (Toyer 1981). In addition, uncovered tailings, waste rock dumps and ore stockpiles can be sources of airborne asbestos.

sia, Ghana, and the Philippines. Here, individual artisanal miners use mercury because its application is cheap, reliable and simple (Salomons 1995; Lacerda and Salomons

Case Study 1.1. *Continued*

- *Smelting.* Many historic base metal smelters were constructed in the immediate vicinity of the mines (Ashley and Lottermoser 1999b; Lottermoser 2002). The smelter operations produced metallurgical waste, smelting slags and flue residues, which were dumped close to the smelting works. The large scale smelting techniques caused impacts on human and animal health as well as lead poisoning during smelting operations. The metal particles and sulfur oxides were released during the smelting processes, settled in the surrounding environment, and contaminated and acidified local soils. Today, these polluted areas are clearly visible as they are devoid of vegetation or support only a depauperate flora. Metal poisoning may occur in farm animals eating the contaminated pasture and soil.
- *Riverine discharge of wastes.* Many historic mining operations discharged their mine and process wastes into nearby water courses. These waste management practices of the past continue to cause water and sediment contamination today. The discharge of some mines has led to metal and metalloid contamination of stream, estuary and coastal sediments and can be traced over a distance of hundreds of kilometers.

Early Rehabilitation at Historic Mines

While there are numerous examples of environmental degradation caused by historic mining and smelting activities, there are also early examples of mine site rehabilitation works. For example, Broken Hill is not only famous for its ore and mining history but also for its successful revegetation scheme in the 1930s. The Broken Hill area was cleared of vegetation when base metal mining, settlement and over-grazing began in the late 19th century. Within a few years the area surrounding Broken Hill was bare, entirely denuded of its natural, yet sparse vegetation cover. As a result, the town was plagued by dust storms causing damage to mining equipment and private property. In 1936, a mining company, Zinc Corporation Ltd, initiated a massive revegetation program of the disturbed areas. Belts of European and native vegetation – irrigated with waste water from showers and septic tanks – were established surrounding the city. The revegetation of waste and tailings dumps was similarly pursued to provide surface cover and to reduce wind erosion. This pioneering work not only reduced the frequency and intensity of dust storms and provided local ecosystems for animals and plants, but it also demonstrated that vegetation can be reestablished on mined and degraded lands.

Conclusions

In Australia, historic mining of the late 19th and early 20th centuries occurred while there were no environmental acts, laws and regulations for operators. This has left a legacy of contaminated sites dispersed throughout the continent. Some of these sites have undergone costly environmental clean-up operations, others still await remediation, while a few perhaps should be monitored in perpetuity rather than rehabilitated at extraordinary costs to the taxpayer. While historic mining has produced contaminated sites, soils, sediments and waters and associated impacts on ecosystems today, environmental degradation as a result of mining has also been recognized and addressed by some early miners. These historic rehabilitation efforts largely focused on the revegetation of mined and disturbed land.

1997). However, the unregulated mining practices have caused mercury contamination of rivers such as the Amazon on a massive scale (Case Study 1.2).

1.5.4 Regulation of Modern Mining

The present-day worldwide utilization of mercury by individual miners is a good example of how unregulated mining by non-professionals causes harm to humans and the environment. In contrast, many modern mines particularly in industrialized nations are designed to have minimum environmental impacts outside an area set aside for the mine operation and waste disposal. However, waste discharges into the environment have been allowed to occur and still occur under communist regimes (Fig. 1.8), in developing countries (Fig. 1.9), and also in industrialized nations (Fig. 1.10), and thus even in countries where the mining industry is regulated.

In many countries, mining companies are required to conduct environmental impact assessments prior to the development of proposed mining and mineral processing operations. In preparing such an assessment, operators identify the actions they intend to implement to limit environmental impacts. Acceptance of the proposed actions are subject to the approval of governmental regulatory agencies. These agencies monitor the activities when the facilities are in operation. Nowadays, the environmental aspects of mining are paramount in determining the viability of a modern mining operation, certainly in developed countries (Maxwell and Govindarajalu 1999). Mining companies have to operate under environmental laws and regulations and are required to place multi million dollar bonds to cover all the costs of rehabilitation according to the designated future land use.

Environmental impact assessments and environmental protection are essential parts of a modern mining operation. These aspects become increasingly important

Fig. 1.8. Abandoned slag and waste heaps of copper ores, Eisleben, Germany. Mining in the area occurred for over 800 years, resulting in over 2 000 individual waste heaps, 1 000 km of underground workings and 56 million cubic metere of mine voids. Under the East German communist regime, the extraction of metals from very low grade, uneconomic copper ores was pursued, and large volumes of fly ashes, tailings, and smelting slags were generated. The unconstrained release of wastes into the local environment, especially of atmospheric emissions from smelting works, has caused widespread contamination of streams and lakes with metals and metalloids

Case Study 1.2. Mercury Pollution and Gold Mining in the Brazilian Amazon

Agglutination and Amalgamation

For over 2000 years, mercury has been used to recover gold from alluvial gold ores. Mercury has a strong chemical affiliation with gold, and this chemical relationship is used in the so-called "agglutination" and "amalgamation" processes. The agglutination process is based on the presence of mercury in wooden boxes, so-called "riffles". Gold-bearing sediment is dredged from river beds or taken from river banks. This sediment is channelled down the riffle which has an inclined, cloth- or carpet-lined bottom. The cloth or carpet is impregnated with elemental mercury to make the fine gold particles clump together in the cloth. The sediment-mercury mixture from the riffles is then collected in barrels and treated with mercury. The barrels are stirred, often by hand, to achieve maximum gold amalgamation. In the amalgamation technique mercury dissolves the solid gold in a physical solution called an amalgam. The gold is then recovered by heating the amalgam solution which causes the mercury to evaporate.

Mercury Release to the Amazon Environment

The widespread application of these ancient extraction techniques has resulted in mercury contamination of many streams around the world. Such contamination has been caused by historic gold mining operations in the United States (Leigh 1997; Miller et al. 1998; Mastrine et al. 1999; Ambers and Hygelund 2001; Domagalski 1998, 2001; Thomas et al. 2002; Gray et al. 2002a), Russia (Laperdina 2002), South America (Ogura et al. 2003; Higueras et al. 2004), Asia (Williams et al. 1999), Europe (Covelli et al. 2001), and Australia (Bycroft et al. 1982). Today, artisanal miners use the agglutination and amalgamation techniques to mine and extract alluvial gold ores (James 1994; Kambey et al. 2001; Lacerda 2003; Limbong et al. 2003; Getaneh and Alemayehu 2006).

Since the early 1980s, the Amazon has become the stage of a major gold rush. Several hundred thousand men have been digging for gold in the Brazilian Amazon rainforest. Thousands of tonnes of alluvial gold have been mined from river banks and beds of the Amazon, often dug out by hand without the help of machinery (Cleary and Thornton 1994; Lacerda and Salomons 1997; Lacerda 2003). At all mine sites, mercury has been used to extract gold from the sediment using the agglutination and amalgamation processes.

The use of mercury in gold mining is associated with large losses of mercury waste to the environment. The agglutination process is prone to mercury loss, and much of the mercury used in the agglutination process ends up in the rivers. The elemental mercury after reaching the rivers is transformed partly to the highly toxic methyl-mercury form. A major proportion of the mercury is also lost to the atmosphere through burning of the gold-mercury amalgam or through degassing of metallic mercury from tailings, soils, sediments, and rivers (Pfeiffer et al. 1993; Lacerda and Marins 1997). Nearly 3000 t of mercury have been released into the Amazon environment in the last 15 years. The annual emission of mercury to the atmosphere in the Brazilian Amazon has been estimated to be greater than the total annual emission of mercury from industry in the United Kingdom.

Uncontrolled gold mining in the Amazon not only results in mercury contamination of air, soils, sediments, rivers, fish and plants but also in the destruction of rainforests, increased erosion of riverbanks, and the exposure of miners, gold dealers, fishermen and residents to toxic mercury concentrations. The miners show symptoms of mercury poisoning also known as Mad Hatter's or Minamatta Disease. They burn off the gold-mercury amalgam in an open pan or closed retort to distill the mercury and to concentrate the gold. Roasting of the amalgam commonly takes place in a hut and on the same fire used to cook food. Miners – eager to see how much gold will remain once the mercury is burned off – will stand directly over the amalgam as it is burned. The gold miners end up inhaling the mercury vapor and as a result, many have been poisoned.

as waste production in the mining industry is significant in volume and diverse in composition when compared to other industries. However, the demands on the mining industry are the same as they are for all industries producing waste:

- To reduce and to recycle waste;
- To ensure that there are no or minimal impacts from wastes on the environment and humans;
- To understand the composition, properties, behaviour and impacts of wastes.

1.6 Rehabilitation of Mine Wastes and Mine Sites

Mining creates wastes and disturbs proportions of land and areas of existing vegetation and fauna. Mining activities may also cause distinct changes in topography, hydrology and stability of a landscape. Once mining ceases, the mined land and its waste repositories need to be rehabilitated. Rehabilitation of mine sites is an integral part of mine planning, development and final closure. This process does not start with mine closure. In industrialized nations, the operator of a mine is required to undertake monitoring as well as progressive rehabilitation of areas of the mine site wherever possible. The latter generally involves revegetation and contouring but operational constraints allow in most cases little ongoing rehabilitation work.

There are several issues which must be addressed in the successful rehabilitation of a mine site. Some rehabilitation issues are common to almost all mines regardless of the type of resource extracted or whether they have used open pit or underground mining methods. These general aspects of mine site rehabilitation include:

Fig. 1.9. The Ertsberg open pit, Irian Jaya, Indonesia. The Grasberg-Ertsberg mine – located at about 4 000 m among the jagged alpine peaks of the Jayawijaya mountain range – is one of the world's largest mining operations. Approximately 0.24 Mt of copper ore are mined every day, and 0.2 Mt of tailings are dumped into the Ajkwa River system every day, causing increased sedimentation on the coastal floodplains

- *Removal of mine facilities.* All mine facilities such as crushers and processing plants need to be dismantled and removed.
- *Sealing and securing of mine workings.* Underground workings such as shafts and adits are sealed at the surface, and open pits may need to be fenced.
- *Ensuring long-term stability of waste repositories.* Waste rocks and tailings must be contained within repositories which have to remain stable in the long-term and prevent migration of contaminants into the environment. The safe long-term isolation of problematic mine wastes represents one of the most challenging tasks facing the mining industry.
- *Modeling future water quality and quantity in pit lakes.* Many surface mining operations create voids which may fill with water once mining ceases. Such open pits need to be modelled for future water quality and quantity.
- *Modeling future water quality in underground workings and aquifers.* Similar to open pits, the closure of an underground mine leads to mine flooding. Deep saline ground waters may rise to shallow levels or contaminants may be leached from the workings into shallow aquifers. Therefore, an assessment of the potential contamination of shallow ground waters is needed.
- *Construction of suitable landforms.* Artifical land forms which will remain after mine closure such as waste rock dumps must be shaped to reduce wind and water erosion. Also, the topography of the mine site and waste repositories needs to be sculptured to create adequate drainage and to resemble the surrounding landform.

Fig. 1.10. The copper and lead smelter stacks at Mt. Isa, Australia. Until recently, the sulfurous plume was dispersed over the town and across 100 000 km^2 northwest of Mt. Isa. Over the years, the emissions resulted in the acidification of local soils and killed or reduced vegetation in the vicinity of the smelters. The recent installation of a sulfuric acid plant has reduced sulfur emissions

- *Development of a suitable plant growth medium.* A suitable plant growth medium needs to be developed for the covering of waste repositories and for the revegetation of mined/disturbed areas. This can be achieved through the selective handling of soil and waste. Many mine wastes are structureless, prone to crusting, and low in organic matter and essential plant nutrients (P, N, K), have low water-holding capacity, and contain contaminants such as salts, metals, metalloids, acid, and radionuclides. If the waste is left uncovered, few mine wastes can become colonized by plants.
- *Establishment of a vegetation cover.* Suitable vegetation and soil amendments must be chosen for the local conditions. Planting stock needs to be propagated in a nursery on the mine site, and local species have to be chosen for the vegetation cover.
- *Addressing generic mine waste issues.* Every mine produces its very own unique waste, and this waste requires its very own characterization, prediction, monitoring, treatment, and secure disposal. For example, sulfidic waste rock dumps require covers to prevent sulfide oxidation, and acid mine waters or cyanide-bearing waters need treatment. Hence, mine waste issues are generic to individual mine sites depending on the waste characteristics, the local mining methods, and the hydrology, geology, meteorology and topography of the area. These generic waste issues are presented in the following chapters.

The construction of post-mining landforms including the shaping of waste repositories, the development of suitable plant growth media, and the establishment of vegetation on waste repositories are all integral parts of mine waste disposal. These topics also represent the final aspects of mine waste management. Such detailed soil science and botanical aspects of mine wastes and mine sites are beyond the scope of this book. In addition, these aspects have already been addressed to some degree by the specialized soil science, plant nutrition and botany literature, and interested readers are directed to these works (e.g. Loch and Jasper 2000).

Rehabilitation of mined land does not imply that the mine site and its waste repositories are to be converted to a pristine wilderness, which may or may not have existed prior to mining. Mine site rehabilitation returns mined land to future land use. The future land use of a waste repository and the entire mine site is highly site specific. In sparsely populated areas, mine waste repositories and mine sites may be rehabilitated to a standard which may allow only limited grazing. In densely populated areas, rehabilitated mine waste dumps have become centres of social amenity such as parklands, football fields, golf courses, and even artificial ski slopes. Open pits may be used as: water storage facilities; wetland/wildlife habitats; aquaculture ponds for fish and crustaceans; recreational lakes; heritage sites; engineered solar ponds to capture heat for electricity generation, heating, or desalinisation and distillation of water; or as repositories for mining, industrial or domestic waste. Underground mines are used as storage facilities, archives, and concert halls, and in Europe some radon emitting mines become part of health spas.

Whatever the future land use of a former mine site or a waste repository may be, mining has to be regarded as a stage in the sequential use of land. Therefore, rehabilitation of a mine should aim to return mined land in such a manner that it is consistent with intended future land use. The goals for mine site rehabilitation are based on the anticipated post-mining use of the area, and this process includes the rehabilitation of mine waste repositories. However, some historic mines, particularly those with AMD, can only be monitored and managed in order to restrict contamination to the mined area. In these cases, collection and treatment of AMD waters represents the only viable and cheapest option. Finally, mine site rehabilitation efforts should not be evaluated and awarded too quickly. In many cases the success of mine site rehabilitation can only be judged well after mining has ceased.

1.7 Sources of Information

There are considerable and excellent resources available for the study of mine wastes and other related topics such as surface reclamation of wastes (i.e. revegetation, landform design), environmental impacts of mine wastes, and mine site rehabilitation. Several organizations are instrumental in supplying a number of publications including journals, reports, conference proceedings, workshop abstracts, databases, and textbooks. These publications and other important resources including web sites are listed in Tables 1.4 and 1.5.

1.8 Summary

All human activities produce wastes and mining is no exception. Mine wastes are liquid, solid and gaseous by-products of mining, mineral processing, and metallurgical extraction. They are unwanted and have no current economic value. While some mine wastes are benign and pose no environmental threat, and others can be used for the rehabilitation of mine sites, many mine wastes are problematic wastes as they contain contaminants which may impact on ecosystems.

Mineral resources contain valuable components which vary in mass from 100% to a few parts per million of the original geological resource. For example, in the metal mining industry only a small valuable component is extracted from the originally mined ore during processing and metallurgical extraction. The great majority of the total mined material is gangue which is rejected as mining, processing and metallurgical waste.

Mine wastes present the highest proportion of solid waste produced by industrial activity, with approximately 15 000 to 20 000 Mt being produced annually. The mining industry is the most significant industrial producer of solid, liquid and gaseous wastes. Every mine thereby produces its own unique waste because there are compositional differences in the mined ore and there is a great diversity of applied mining and mineral processing methods.

Changing circumstances may turn a particular waste into a valuable commodity, either because the economic extraction of resource ingredients may now be possible using improved technology, or a market has been found for the previously unwanted material. Yesterday's waste can become today's resource. What may be mine waste to some, can be a very important, useful resource to others, either now or in the future.

Table 1.4. Resource materials covering aspects of mine wastes

Textbooks and reports
Alpers CN, Blowes DW (1994) Environmental geochemistry of sulfide oxidation. American Chemical Society, Washington (Symposium Series 550)
Azcue JM (1998) Environmental impacts of mining activities: Emphasis on mitigation and remedial measures. Springer, Berlin Heidelberg New York
Brown M, Barley B, Wood H (2002) Minewater treatment: Technology, application and policy. International Water Association Publishing
Environment Australia (1995–2002) Best practice in environmental management in mining booklets. Environment Australia, Canberra
Evangelou VP (1995) Pyrite oxidation and its control. CRC Press, Boca Raton
Filipek FH, Plumlee GS (1999) The environmental geochemistry of mineral deposits. Part B: Case studies and research topics. Society of Economic Geologists, Littleton (Reviews in economic geology, vol 6b)
Geller W, Klapper H, Salomons W (1998) Acidic mining lakes: Acid mine drainage, limnology and reclamation. Springer, Berlin Heidelberg New York
Hutchison I, Ellison R (1992) Mine waste management. Lewis Publishers, Boca Raton
Jambor JL, Blowes DW (1994) The environmental geochemistry of sulfide mine-wastes. Mineralogical Association of Canada, Nepean (Short course handbook, vol 22)
Jambor JL, Blowes DW, Ritchie AIM (2003) Environmental aspects of mine wastes. Mineralogical Association of Canada, Nepean (Short course handbook, vol 31)
Lacerda de LD, Salomons W (1997) Mercury from gold and silver mining: a chemical time bomb? Springer, Berlin Heidelberg New York
Marcus JJ (1997) Mining environmental handbook: Effects of mining on the environment and American environmental controls on mining. Imperial College Press, London
Morin KA, Hutt NM (1997) Environmental geochemistry of minesite drainage. MDAG Publishing, Vancouver
Mudder T, Botz M, Smith ACS (2001) Chemistry and treatment of cyanidation wastes. The Mining Journal Ltd., London
Mulligan DR (1996) Environmental management in the Australian minerals and energy industries: Principles and practices. University of New South Wales Press, Sydney
Plumlee GS, Logsdon MS (1999) The environmental geochemistry of mineral deposits. Part A: Processes, techniques and health issues. Society of Economic Geologists, Littleton (Reviews in economic geology, vol 6a)
Ripley EA, Redmann RE, Crowder AA (1996) Environmental effects of mining. St Lucie Press, Delray Beach
Ritcey GM (1989) Tailings management problems and solutions in the mining industry. Elsevier, Amsterdam
Salomons W, Förstner U (1988) Chemistry and biology of solid waste, dredged materials and mine tailings. Springer, Berlin Heidelberg New York
Sengupta M (1993) Environmental impacts of mining: monitoring, restoration and control. Lewis Publishers, Boca Raton
Skousen JG, Ziemkiewicz PF (compilers) (1996) Acid mine drainage control and prevention, 2nd edn. West Virginia University, National Mine Land Rehabilitation Center, Morgantown
Skousen J, Rose A, Geidel G, Foreman J, Evans R, Hellier W (1998) A handbook of technologies for avoidance and remediation of acid mine drainage. Acid Drainage Technology Initiative (ADTI), West Virginia University, National Mine Land Rehabilitation Center, Morgantown

Table 1.4. *Continued*

Textbooks and reports
Warhurst A, Noronha L (1999) Environmental policy in mining. Lewis Publishers, Boca Raton
Younger PL, Robins NS (2002) Mine water hydrogeology and geochemistry. Geological Society, London
Younger PL, Banwart SA, Hedin RS (2002) Mine water: hydrology, pollution, remediation. Kluwer Academic Publishers, Dordrecht
Journals
American Mineralogist
Applied Geochemistry
Ecological Engineering
Engineering and Environmental Geoscience
Environmental Geochemistry and Health
Environmental Geology
Environmental Science and Technology
Geochemistry: Exploration, Environment, Analysis
Geochimica et Cosmochimica Acta
International Journal of Surface Mining, Reclamation and Environment
Journal of Contaminant Hydrology
Journal of Environmental Quality
Journal of Geochemical Exploration
Mine Water and the Environment
Mineralogical Magazine
Mining Environmental Management
Science of the Total Environment
Water, Air, and Soil Pollution

Mine wastes have been with us since the beginning of mining. Thus, anthropogenic contamination of air, water, sediments and soils by mine wastes is as old as people's abilities to mine, smelt and process ores. The impact of mine wastes on the environment has been recognized for more than 2 000 years, well before 20th century organizations took on the issue. Also, the detrimental effects of mine wastes on the health of workers have been of concern to authorities and miners for over 2 000 years.

In many industrialized countries, the majority of mining environmental problems are legacies from the past. Today, improper disposal practices of wastes continue to occur in developing nations and communist states where unregulated mining and uncontrolled release of mine wastes cause environmental harm.

Rehabilitation of mine sites is an integral part of modern mine planning, development and mine closure. Several issues have to be addressed in the successful rehabilitation of a mine site including: removal of all mine facilities; sealing and securing of mine workings; ensuring long-term stability of waste repositories; modeling of future

Table 1.5. Web sites covering general aspects of mine wastes

Organization	Web address and description
United Nations Environmental Programme (UNEP) Mineral Resources Forum	*http://www.mineralresourcesforum.org* Information, publications and case studies on mining and the environment
International Council on Mining and Metals (ICMM)	*http://www.icmm.com* Publications, factsheets, newsletters
US Office of Surface Mining	*http://www.osmre.gov* Protection of the environment during coal mining and reclamation of mined lands
US Bureau of Land Management – Abandoned Mine Lands	*http://www.blm.gov/aml/* Abandoned mine reclamation
US Geological Survey (USGS) Abandoned Mine Lands	*http://amli.usgs.gov/amli* Data, reports, and pictures on abandoned mines
US Environmental Protection Agency (EPA)	*http://www.epa.gov/superfund/programs/aml/tech/index.htm* *http://www.epa.gov/ORD/NRMRL/std/mtb/mwt/* *http://www.epa.gov/osw* Information on abandoned mine lands and information on mining and mineral processing wastes
American Society for Surface Mining and Reclamation (ASSMR)	*http://ces.ca.uky.edu/assmr* Annual conference proceedings on various aspects of minesite rehabilitation
Department of Industry, Tourism and Resources, Australia	*http://www.deh.gov.au/settlements/industry/minerals/index.html* Training kits and electronic booklets on best practice environmental management in mining
Mining Environment Database, Laurentian University, Canada	*http://www.laurentian.ca/library/medb/medlib_e.php* Database on acid mine drainage, abandoned mines, and reclamation
Mining Information Service EnviroMine	*http://technology/infomine.com/enviromine* Information network on mining and the environment

water quality and quantity in pit lakes; modeling future water quality in underground workings and aquifers; construction of suitable landforms; development of a suitable plant growth medium; establishment of a vegetation cover; and addressing generic mine waste issues. Every mine produces its very own unique waste, and this waste requires its very own characterization, prediction, monitoring, treatment, and secure disposal.

Chapter 2

Sulfidic Mine Wastes

2.1 Introduction

Sulfide minerals are common minor constituents of the Earth's crust. In some geological environments, sulfides constitute a major proportion of rocks. In particular, metallic ore deposits (Cu, Pb, Zn, Au, Ni, U, Fe), phosphate ores, coal seams, oil shales, and mineral sands may contain abundant sulfides. Mining of these resources can expose the sulfides to an oxygenated environment. In fact, large volumes of sulfide minerals can be exposed in: tailings dams; waste rock dumps; coal spoil heaps; heap leach piles; run-of-mine and low-grade ore stockpiles; waste repository embankments; open pit floors and faces; underground workings; haul roads; road cuts; quarries; and other rock excavations. When the sulfides are exposed to the atmosphere or oxygenated ground water, the sulfides will oxidize to produce an acid water laden with sulfate, heavy metals and metalloids. The mineral pyrite (FeS_2) tends to be the most common sulfide mineral present. The weathering of this mineral at mine sites causes the largest, and most testing, environmental problem facing the industry today – *acid mine drainage* (AMD).

This chapter documents the weathering processes occurring in sulfidic wastes. An understanding of the complex chemical reactions within sulfidic wastes is essential as the reactions can cause and influence AMD. Discussions of the various techniques used to predict and monitor such acid generating wastes follow. A documentation of environmental impacts of sulfidic wastes and a review of the technologies available for the control and prevention of sulfide oxidation complete the chapter.

2.2 Weathering of Sulfidic Mine Wastes

Sulfidic mine wastes are in most cases polymineralic aggregates. The aggregates contain, apart from sulfides, a wide range of possible minerals including silicates, oxides, hydroxides, phosphates, halides, and carbonates. Silicates are the most common gangue minerals, and the sulfides may represent ore or gangue phases. Thus, the mineralogy of sulfidic wastes and ores is highly heterogeneous and deposit specific.

When mining exposes sulfidic materials to an oxidizing environment, the materials become chemically unstable. A series of complex chemical weathering reactions are spontaneously initiated. This occurs because the mineral assemblages contained in the waste are not in equilibrium with the oxidizing environment. Weathering of the minerals proceeds with the help of atmospheric gases, meteoric water and microorganisms.

The chemical weathering of an individual mineral within a polymineralic aggregate can be classified as an acid producing (i.e. generation of H^+), acid buffering (i.e. consumption of H^+), or non-acid generating or consuming reaction (i.e. no generation or consumption of H^+). For example, the degradation of pyrite is an acid producing reaction, whereas the weathering of calcite is acid buffering, and the dissolution of quartz does not consume or generate any acid. The balance of all chemical reactions, occurring within a particular waste at any time, will determine whether the material will "turn acid" and produce AMD.

2.3 Acid Producing Reactions

2.3.1 Pyrite

Sulfides are stable under strongly reducing conditions. Exposure of these minerals to oxidizing conditions will destabilize them, and the sulfides will be destroyed via various oxidation mechanisms. Pyrite is the most abundant of the sulfide minerals, occurs in nearly all types of geological environments, and is commonly associated with coal and metal ore deposits. Thus, pyrite oxidation has been studied extensively from all scientific angles, and there is a vast literature on the subject (e.g. Luther 1987; Evangelou 1995; Evangelou and Zhang 1995; Keith and Vaughan 2000). In contrast, the oxidation of other sulfides such as galena, sphalerite and chalcopyrite has received in comparison only limited attention.

Pyrite oxidation takes place when the mineral is exposed to oxygen (Rimstidt and Vaughan 2003). Oxidation which occurs in the presence of microorganisms is known as biotic. Pyrite oxidation may also occur without microorganisms as an abiotic chemical oxidation process. Biotic and abiotic degradation can be caused by oxygen (i.e. direct oxidation) or by oxygen and iron (i.e. indirect oxidation) (Evangelou and Zhang 1995). Iron, both in its divalent and trivalent state, plays a central role in the indirect oxidation of pyrite. These different pyrite oxidation mechanisms can be summarized as:

1. Oxidation by oxygen (abiotic direct oxidation)
2. Oxidation by oxygen in the presence of microorganisms (biotic direct oxidation)
3. Oxidation by oxygen and iron (abiotic indirect oxidation)
4. Oxidation by oxygen and iron in the presence of microorganisms (biotic indirect oxidation)

Stoichiometric chemical reactions are commonly used to describe these different oxidation mechanisms. In the abiotic and biotic *direct oxidation* processes (mechanisms 1 and 2), oxygen directly oxidizes pyrite:

$$FeS_{2(s)} + 7/2\,O_{2(g)} + H_2O_{(l)} \rightarrow Fe^{2+}_{(aq)} + 2\,SO^{2-}_{4(aq)} + 2\,H^{+}_{(aq)} + \text{energy} \qquad (2.1)$$

It is generally accepted, however, that pyrite oxidation is primarily accomplished by *indirect oxidation* (mechanisms 3 and 4). The indirect oxidation of pyrite involves the chemical oxidation of pyrite by oxygen and ferric iron (Fe^{3+}), which occurs in three interconnected steps. The following chemical equations show the generally accepted sequence for such indirect oxidation of pyrite:

$$4\,FeS_{2(s)} + 14\,O_{2(g)} + 4\,H_2O_{(l)} \rightarrow 4\,FeSO_{4(aq)} + 4\,H_2SO_{4(aq)} + \text{energy} \quad (2.2)$$

or,

$$FeS_{2(s)} + 7/2\,O_{2(g)} + H_2O_{(l)} \rightarrow Fe^{2+}_{(aq)} + 2\,SO^{2-}_{4(aq)} + 2\,H^{+}_{(aq)} + \text{energy}$$

$$4\,FeSO_{4(aq)} + O_{2(g)} + 2\,H_2SO_{4(aq)} \rightarrow 2\,Fe_2(SO_4)_{3(aq)} + 2\,H_2O_{(l)} + \text{energy} \quad (2.3)$$

or,

$$Fe^{2+}_{(aq)} + 1/4\,O_{2(g)} + H^{+}_{(aq)} \rightarrow Fe^{3+}_{(aq)} + 1/2\,H_2O_{(l)} + \text{energy}$$

$$FeS_{2(s)} + Fe_2(SO_4)_{3(aq)} + 2\,H_2O_{(l)} + 3\,O_{2(g)} \rightarrow 3\,FeSO_{4(aq)} + 2\,H_2SO_{4(aq)} + \text{energy} \quad (2.4)$$

or,

$$FeS_{2(s)} + 14\,Fe^{3+}_{(aq)} + 8\,H_2O_{(l)} \rightarrow 15\,Fe^{2+}_{(aq)} + 2\,SO^{2-}_{4(aq)} + 16\,H^{+}_{(aq)} + \text{energy}$$

Reactions 2.2 to 2.4 release energy. Indirect pyrite oxidation is exothermic. In the initial step (Reaction 2.2), pyrite is oxidized by oxygen to produce dissolved ferrous iron (Fe^{2+}), sulfate and hydrogen ions. The dissolved iron sulfate ions cause an increase in the total dissolved solids of the water. The release of hydrogen ions with the sulfate anions results in an acidic solution unless other reactions occur to neutralize the hydrogen ions. The second step (Reaction 2.3) represents the oxidation of ferrous iron (Fe^{2+}) to ferric iron (Fe^{3+}) by oxygen and occurs at a low pH. In the third reaction (Reaction 2.4) pyrite is oxidized with the help of Fe^{3+} generated in Reaction 2.3. Thus, Fe^{3+} acts as the oxidizing agent of pyrite. The oxidation of pyrite by Fe^{3+} in turn generates more Fe^{2+}. This Fe^{2+} can then be oxidized to Fe^{3+} by oxygen via Reaction 2.3. The Fe^{3+} in turn oxidizes pyrite via Reaction 2.4, which in turn produces more Fe^{2+}, and so on. Reactions 2.3 and 2.4 form a continuing cycle of Fe^{2+} conversion to Fe^{3+} and subsequent oxidation of pyrite by Fe^{3+} to produce Fe^{2+} (Fig. 2.1). This cyclic propagation of

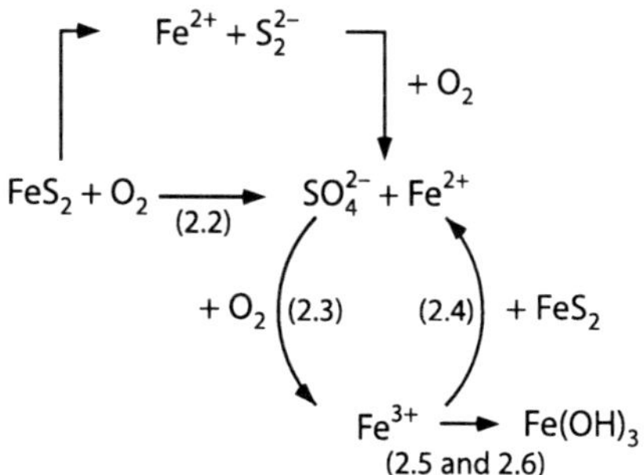

Fig. 2.1. Simplified diagram illustrating the reaction pathways for pyrite oxidation (after Banks et al. 1997). Numbers 2.2 to 2.6 refer to Reactions 2.2 to 2.6 in the text

pyrite oxidation by Fe^{3+} continues until the supply of pyrite or Fe^{3+} to the reaction system is exhausted. While oxygen is not required for the Reaction 2.4 to occur, it is still needed to convert Fe^{2+} to Fe^{3+}.

The abundance of the oxidizing agent Fe^{3+} is influenced by the pH of the weathering solution. The solubility of Fe^{3+} is very low in neutral and alkaline waters. Hence, the concentrations of Fe^{3+} are very low in these solutions, and pyrite oxidation by Fe^{3+} in neutral to alkaline waters is insignificant. Also, the concentration of dissolved Fe^{3+} decreases with increasing pH as Fe^{3+} solubility is limited by the precipitation of ferric hydroxides ($Fe(OH)_3$) and oxyhydroxides (FeOOH). In other words, if the pH increases to more than approximately 3 because of partial neutralization, for example, by carbonate minerals, then the following reactions will occur:

$$Fe^{3+}_{(aq)} + 3\,H_2O_{(l)} \leftrightarrow Fe(OH)_{3(s)} + 3\,H^+_{(aq)} \tag{2.5}$$

$$Fe^{3+}_{(aq)} + 2\,H_2O_{(l)} \leftrightarrow FeOOH_{(s)} + 3\,H^+_{(aq)} \tag{2.6}$$

The precipitation of dissolved Fe^{3+} (Reactions 2.5, 2.6) provides significant acidity to the solution by the release of hydrogen ions into water. This reaction lowers the pH and allows more Fe^{3+} to stay in solution. The Fe^{3+} is then involved in the oxidation of pyrite (Reaction 2.4) which results in a further reduction in pH.

The chemical precipitation of iron hydroxides in Reactions 2.5 and 2.6 is termed "*hydrolysis*". Hydrolysis is the chemical process whereby water molecules react with dissolved cations; the cations become bonded to the hydroxy group and hydrogen ions are released. Consequently, hydrolysis results in the production of hydrogen ions, thereby causing the pH to fall. As mentioned above, the hydrolysis reaction of iron is controlled by pH. Under acid conditions of less than about pH 3, Fe^{3+} remains in solution. At higher pH values, precipitation of Fe^{3+} hydroxides occurs. Such a precipitate is commonly observed as the familiar reddish-yellow to yellowish-brown stain, coating, slimy sludge, gelatinous flocculant and precipitate in AMD affected streams and seepage areas (Sec. 3.5.7).

The Reactions 2.2 to 2.6 show that in the presence of molecular oxygen, Fe^{2+} and S^{2-} in pyrite are oxidized by oxygen to produce solid iron hydroxides and oxyhydroxides as well as dissolved sulfate and hydrogen ions. Clearly, oxygen and Fe^{3+} are the major oxidants of pyrite (Singer and Stumm 1970; Evangelou 1998). The oxidation of pyrite continues indefinitely unless one of the vital ingredients of pyrite oxidation is removed (i.e. Fe^{3+}, oxygen or pyrite), or the pH of the weathering solution is significantly raised.

The reaction pathways of pyrite (Reactions 2.2 to 2.6) have also been referred to as the AMD engine (Fig. 2.2). Pyrite, Fe^{3+} and oxygen represent the fuel, oxygen is also the starter engine, and Fe^{3+} hydroxides, sulfuric acid and heat come out of the exhaust pipe of the sulfidic waste. Such a simplified model of indirect oxidation of pyrite (Reactions 2.2 to 2.6) can be summarized by one overall chemical reaction:

$$FeS_{2(s)} + 15/4\,O_{2(aq)} + 7/2\,H_2O_{(l)} \rightarrow Fe(OH)_{3(s)} + 2\,H_2SO_{4(aq)} + \text{energy} \tag{2.7}$$

The above reaction describes the weathering of pyrite, highlights the need for water and oxygen, and illustrates the production of acid and iron hydroxide. However,

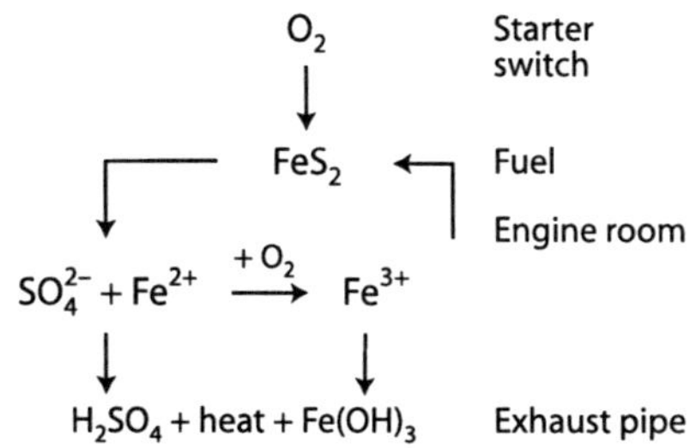

Fig. 2.2. The self-sustaining, cyclic destruction of pyrite simplified as the "AMD engine". The oxidation of pyrite is initiated through oxygen ("starter switch"). Pyrite, oxygen and iron ("fuel") combust in the waste ("engine room"), and release Fe^{3+} hydroxides, sulfuric acid and heat into mine waters ("exhaust pipe")

there is little consensus in the literature on the precise reaction mechanisms describing the chemical oxidation of pyrite. Also, the chemical Equations 2.2 to 2.7 are gross oversimplifications since: (*a*) the reactions do not explain that the Fe^{3+} hydroxides and sulfates are fictious, idealized solid phases; (*b*) they do not illustrate the range of iron hydroxide, oxyhydroxide and oxyhydroxysulfate minerals formed during pyrite oxidation; (*c*) they do not reflect the slow oxidation of Fe^{2+} in acid waters; (*d*) they disregard adsorption, desorption and neutralization reactions; (*e*) they disregard supersaturation of waters with iron and sulfate; (*f*) they do not consider the precipitation of elemental sulfur (S^0) and the formation of sulfite (SO_3^{2-}; S: 4+), thiosulfate ($S_2O_3^{2-}$; S: 2+), and polythionates ($S_nO_6^{2-}$) ions; and (*g*) they do not describe the rate or speed (i.e. kinetics) of pyrite oxidation (Ritchie 1994b; Nordstrom and Alpers 1999a). Hence, the above reaction paths (Reactions 2.1 to 2.7) represent only approximations for actual field conditions.

How quickly pyrite weathers is influenced by its mineralogical properties and by external chemical, physical and biological factors. Mineralogical properties include the particle size, porosity, surface area, crystallography, and trace element content of pyrite. External factors are the presence of other sulfides, the presence or absence of microorganisms, as well as the oxygen and carbon dioxide concentration, temperature, pH and Fe^{2+}/Fe^{3+} ratio of the weathering solution. Therefore, the rate of pyrite oxidation (i.e. the weathering kinetics of pyrite) is influenced by the following factors:

- *Pyrite particle size, porosity and surface area.* The oxidation reactions occur on the surfaces of pyrite particles. Small particle sizes and large surface areas increase the reactivity of pyrite, and maximum oxidation of the pyrite surface occurs along pits, cracks, pores, and solid and liquid inclusions. For example, pyrite grains are exceptionally small in diameter in so-called "*framboidal pyrite*". Framboidal pyrite refers to small-grained pyrite crystals with a grain size less than one micron. The grains are dispersed in the matrix or agglomerated to form a small spherical mass, typically several tens of micron in diameter. Such framboidal pyrite is more reactive than other pyrite morphologies – cubic pyrite crystals or coarse pyrite nodules – because of the greater surface area and porosity per volume of framboidal pyrite. Thus, pyrite oxidation is a surface controlled reaction (Evangelou 1995; Rose and Cravotta 1999).

  Mining, crushing and milling of pyrite-bearing rock to fine particle sizes, for the purpose of metal extraction, vastly increase the pyrite surface area and potentially expose more pyrite to oxidation and weathering. However, crushing and milling of pyritic materials do not necessarily increase the oxidation rate of pyrite in waste rock dumps. This is because coarse-grained pyritic wastes have more pore space and allow greater oxygen movement into the wastes. Consequently, acid generation in coarse-grained wastes may occur to a greater depth than in fine-grained wastes.

- *Pyrite crystallography.* Poorly crystalline pyrites or pyrites with structural defects have an imperfect or distorted crystal lattice. This leads to physical stress in the crystal structure which makes the mineral more susceptible to chemical attack (Hutchison and Ellison 1992; Rose and Cravotta 1999).
- *Trace element substitution.* Trace elements can be present in pyrite in the form of minute mineral inclusions and as chemical impurities in the crystal lattice (Table 2.1). This puts strain on the crystal structure and diminishes the sulfide's resistance to oxidation. For instance, the occurrence of arsenic in pyrite greatly decreases the resistance of pyrite to oxidation (Hutchison and Ellison 1992; Plumlee 1999).
- *Presence of other sulfides.* Sulfidic wastes commonly contain sulfides other than pyrite. If there is direct physical contact between at least two different sulfide minerals, electrons move between the sulfides and a galvanic cell is formed. During weathering the sulfide mineral with the highest electrode potential is galvanically protected from oxidation, while the mineral with the lowest electrode potential is weathered more strongly. Selective oxidation of sulfide minerals occurs as one sulfide mineral is preferentially leached over another (Evangelou and Zhang 1995; Evangelou 1995; Nordstrom and Alpers 1999a; Kwong et al. 2003; Abraitis et al. 2004). This galvanic protection process is the same as that for galvanized iron. The more electroconductive sulfide oxidizes at a slower rate than it would when not in contact with another sulfide. For example, among the three common sulfide minerals – pyrite, galena and sphalerite – pyrite has the highest electrode potential followed by galena and then sphalerite (Sato 1992). If these minerals are in contact with each other, sphalerite will be preferentially weathered and oxidation of pyrite is reduced. Hence, pyrite in direct contact with other sulfides does not react as vigorously as it does in isolation (Cruz et al. 2001a). Also, the oxidative dissolution of pyrite can be delayed, while other sulfides are preferentially oxidized (Kwong et al. 2003).
- *Temperature of the waste.* The oxidation of pyrite is exothermic and generates heat as shown by the above equations. Such elevated temperatures are also advantageous to the growth of thermophilic bacteria. These bacteria use some of the released energy for their metabolic processes. However, most of the energy is released as heat and within the physical confines of waste dumps and tailings dams, there is little dissipation of the heat due to the abundance of gangue minerals with poor heat conductivity. Thus, the pyritic waste gets warmer. Pyrite oxidation occurs faster as its oxidation rate nearly doubles with each 10 °C increase in temperature (Smith et al. 1992) (Scientific Issue 2.1).
- *Microbiological activity (bacteria, archaea, fungi, algae, yeasts, and protozoa).* AMD environments commonly contain an abundance of microorganisms. Some of these microorganisms thrive under aerobic or anaerobic conditions and favour acid or neutral pH regimes. Bacteria isolated from AMD environments are diverse and include *Acidithiobacillus thiooxidans* (previously *Thiobacillus;* Kelly and Wood 2000), *Acidithiobacillus ferrooxidans* (previously *Thiobacillus*), *Leptospirillum ferrooxidans*,

Table 2.1. Sulfide minerals and their chemical formula. The ability of sulfides to contain minor and trace element constituents in the form of cation substitutiuons is illustrated for common sulfides (after Vaughan and Craig 1978). However, some of these elements may be present as small inclusions in the host sulfides

Mineral name	Chemical formula	Minor and trace element substitution
Arsenopyrite	$FeAsS$	
Bornite	Cu_3FeS_4	
Chalcocite	Cu_2S	
Chalcopyrite	$CuFeS_2$	Ag, As, Bi, Cd, Co, Cr, In, Mn, Mo, Ni, Pb, Sb, Se, Sn, Ti, V, Zn
Cinnabar	HgS	
Cobaltite	$CoAsS$	
Covellite	CuS	
Cubanite	$CuFe_2S_3$	
Enargite	Cu_3AsS_4	
Galena	PbS	Ag, As, Bi, Cd, Cu, Fe, Hg, Mn, Ni, Sb, Se, Sn, Tl, Zn
Mackinawite	$(Fe,Ni)_9S_8$	
Marcasite	FeS_2	As, Hg, Se, Sn, Ti, Tl, Pb, V
Melnikovite	Fe_3S_4	
Millerite	NiS	
Molybdenite	MoS_2	
Orpiment	As_2S_3	
Pentlandite	$(Ni,Fe)_9S_8$	
Pyrite	FeS_2	Ag, As, Au, Bi, Cd, Co, Ga, Ge, Hg, In, Mo, Ni, Pb, Sb, Se, Sn, Ti, Tl, V
Pyrrhotite	$Fe_{1-x}S$	Ag, As, Co, Cr, Cu, Mo, Ni, Pb, Se, Sn, V, Zn
Realgar	AsS	
Stibnite	Sb_2S_3	
Sphalerite	ZnS	Ag, As, Ba, Cu, Cd, Co, Cr, Fe, Ga, Ge, Hg, In, Mn, Mo, Ni, Sb, Se, Sn, Tl, V
Tennantite	$(Cu,Fe)_{12}As_4S_{13}$	
Tetrahedrite	$(Cu,Fe)_{12}Sb_4S_{13}$	
Violarite	$FeNi_2S_4$	

and *Thiobacillus thioparus* (e.g. Gould et al. 1994; Ledin and Pedersen 1996; Johnson 1998a,b; Schrenk et al. 1998; Blowes et al. 1998; Fowler et al. 1999; Schippers and Sand 1999; Bond et al. 2000; Gould and Kapoor 2003; Hallberg and Johnson 2005; Gleisner et al. 2006). Certain bacteria grow particularly well in pH 2 to 3 environments. These acidophilic (i.e. acid loving) bacteria participate in the conversion of Fe^{2+} to Fe^{3+} and the oxidation of sulfur and sulfur compounds. They utilize the oxidation of the metal component (i.e. predominantly Fe) and sulfur compounds to obtain energy for their growth. Consequently, some bacteria significantly accelerate the rate of Fe^{2+} oxidation to Fe^{3+}. In fact, these bacteria accelerate the rate of Fe^{2+} oxidation, which is relatively slow under abiotic, acid (pH < 4) conditions (Reaction 2.3), by a factor of hundreds to as much as one million times (Singer and Stumm 1970). In turn, the

Scientific Issue 2.1. Pyrite Oxidation in Permafrost Regions

Permafrost

Mining of sulfide ores and production of sulfidic wastes occur around the world including permafrost areas. Permafrost refers to permanently frozen rock, soil or sediment. Permafrost areas are divided into continuous and discontinuous permafrost regions whereby continuous permafrost is prevalent in the Arctic and Antarctic. Discontinous permafrost occurs at lower latitudes where patches of permafrost alternate with unfrozen ground. In permafrost environments, the ground temperature fluctuates causing the surface layer to thaw annually. This active zone undergoes freeze-thaw cycles which promote frost wedging and frost heaving. The thickness of the active zone tends to be greater in discontinuous permafrost areas than in continuous permafrost environments. While lower plants and an organic layer may be established in the active thawing zone, an arid cold climate generally discourages the development of vegetation.

Pyrite oxidation at low temperatures

The rate of pyrite oxidation is influenced by the temperature of the environment and microbological activity. Temperatures around 30 °C favour a faster oxidation rate of pyrite and the proliferation of iron and sulfur compound oxidizing bacteria. Hence, it may be suggested that simple freezing of sulfidic wastes to less than 0 °C would kill the microorganisms and slow down the oxidation reactions to negligible speeds.

However, studies have shown that sulfide oxidation and AMD generation are still prevalent in permafrost environments (MEND 1993a, 1997a). Here freeze-thaw cycles lead to the annual thawing of the upper permafrost layer. Atmospheric oxygen gas and water infiltrate into the active thawing zone and initiate oxidation of sulfidic wastes. Also, in continuous permafrost environments simple freezing of unsaturated sulfidic wastes does not stop the transport of oxygen into the waste since there are still enough pore spaces for the atmospheric oxygen to enter the waste material. In fact, the flux of oxygen into sulfidic waste is only slightly decreased (MEND 1997a). In addition, sulfide oxidizing bacteria still occur in permafrost regions and actively oxidize sulfides at an annual mean temperature of minus 15.6 °C (Elberling et al. 2000). Freezing of unsaturated sulfidic wastes to temperatures below 0 °C is not sufficient to stop sulfide oxidation. Therefore, in permafrost regions the rate of sulfide oxidation is slowed down but not necessarily reduced to negligible levels. Such changes are not significant enough to prevent AMD generation (MEND 1997a).

Tailings disposal in permafrost regions

Waste disposal and management practices can still take advantage of the low temperatures and permafrost conditions. Sulfidic tailings are commonly placed under water into a tailings dam. In continuous permafrost regions, the permanent freezing of water saturated tailings encapsulates the waste in ice. The ice layer acts as a surface barrier and reduces the transport of atmospheric oxygen into the waste (MEND 1997a). Furthermore, any frozen waste has a very low hydraulic conductivity which limits the ingress of water into the waste. However, a small but significant percentage of pore water remains unfrozen to about minus 5 °C or colder, particularly around smaller particles such as tailings (MEND 1993a). These pore waters are invariably saline, freeze only at temperatures well below 0 °C, and are able to participate in sulfide oxidation reactions. Also, ice scouring and the resultant disturbance of deposited tailings may allow oxygen transport into the waste causing sulfide oxidation (MEND 1997a). Thus, while the sulfide oxidation rate is reduced in frozen sulfidic wastes, permafrost cannot provide an absolute control on sulfide oxidation and AMD generation.

increased concentrations of Fe^{3+} oxidize the pyrite and accelerate acid formation. A so-called "self-perpetuating" or "autocatalytic" reaction develops whereby the bacteria serve as a reaction catalyst for Fe^{2+} oxidation (Reaction 2.3). Iron oxidizing bacteria such as *Acidithiobacillus ferrooxidans* and *Leptospirillum ferrooxidans* oxidize Fe^{2+} to Fe^{3+} whereas sulfur oxidizing thiobacteria such as *Acidithiobacillus thiooxidans* oxidize sulfides and other sulfur compounds. These aerobic bacteria speed up the chemical oxidation rate of Fe^{2+} and sulfur compounds when molecular oxygen is present.

Despite much research on microbiological oxidation of pyrite and especially on the role of *Acidithiobacillus ferrooxidans*, it has been argued that abiotic chemical oxidation of pyrite is more dominant than biotic oxidation and that 95% of bacteria associated with AMD are not *Acidithiobacillus ferrooxidans* (Ritchie 1994a; Morin and Hutt 1997). Indeed, it has been suggested that the microbial ecology rather than a particular individual microorganism is the catalyst of pyrite oxidation and responsible for extreme AMD conditions (Lopez-Archilla et al. 1993; Ritchie 1994a). Also, biological parameters – such as population density of the bacteria, rate of bacterial growth, and supply of nutrients – influence the growth and abundance of the acidophilic bacteria and hence, the rate of pyrite oxidation. Moreover, bacteria are ubiquitous, and the presence of a bacterial population in sulfidic wastes may only indicate a favourable environment for microbial growth (Ritchie 1994a). Thus, the exact role of individual bacteria and other microorganisms in sulfide oxidation is a controversy for some. Also, our knowledge of the microbriology of neutral mine waters is incomplete. A comprehensive understanding of microbial processes in mine waters may enable the development of technologies that may prevent sulfide oxidation and AMD formation (Hallberg and Johnson 2005).

- *Oxygen concentration in the gas and water phase.* Oxidation of pyrite may occur in the atmosphere or in water. A significant correlation exists between the oxidation rate of pyrite and the oxidation concentration of the medium in which oxidation takes place. Generally, the oxidation rate increases with higher oxygen concentrations. Oxygen is essential for the oxidation of sulfides and Ritchie (1994a) considers that the transport of oxygen to the oxidation sites is the rate limiting process in dumps and tailings deposits. If the oxidation takes place in water or in saturated pores under cover, the reactivity of pyrite is greatly affected by the concentration and rate of transport of oxygen in water. The concentration of dissolved oxygen in water is partly temperature-dependent and can vary from 0 mg l^{-1} to a maximum of 8 mg l^{-1} at 25 °C. Such a concentration is significantly less than the oxygen concentration in the atmosphere (21 vol.% or 286 mg l^{-1} of O_2 at 25 °C) (Langmuir 1997). As a result, the oxidation of pyrite in oxygenated water is much slower than the oxidation of pyrite in the atmosphere.

  Changes in oxygen concentrations also influence the occurrence of aerobic iron and sulfur oxidizing bacteria which require oxygen for their survival. Above the water table, abundant atmospheric oxygen is available and oxidation rates are usually catalyzed by aerobic bacteria like *Acidithiobacillus ferrooxidans*. In contrast, oxidation rates in water saturated waste or below the water table are much slower because ground water generally has low dissolved oxygen concentrations and hence lacks catalyzing aerobic bacteria. In extreme cases such as flooded mine workings with no dissolved oxygen, the lack of dissolved oxygen and the absence of aerobic bacteria can reduce pyrite oxidation to negligible rates.
- *Carbon dioxide concentration in the gas and water phase.* Sulfide oxidizing anaerobic bacteria use carbon dioxide as their sole source of carbon in order to build up organic material for their maintenance and growth (Ledin and Pedersen 1996). Carbon dioxide is produced in sulfidic waste rock dumps as a result of carbonate disso-

lution and subsequent release of carbon dioxide into pore spaces. Thus, elevated concentrations of carbon dioxide in the pore space of waste rock dumps have been reported to increase the oxidation of pyrite as the heightened concentrations favour the growth of sulfide oxidizing anaerobic bacteria (Ritchie 1994a).

- *pH of the solution in contact with pyrite.* Acid conditions prevail in microscopic environments surrounding pyrite grains. However, the exact pH of a solution in contact with an oxidizing pyrite surface is unknown since current technologies are unable to measure the pH conditions at a submicroscopic level. The pH value of the solution in contact with pyrite influences the rate of pyrite oxidation. Under low to neutral pH conditions, Fe^{3+} acts as the oxidant of pyrite (i.e. indirect oxidation). The Fe^{3+} concentration is pH dependent. As a consequence, the oxidation rate of pyrite in Fe^{3+} saturated solutions is pH dependent. Significant dissolved concentrations of Fe^{3+} only occur at low pH values because the Fe^{3+} concentration in solution is controlled by the precipitation of insoluble Fe^{3+} precipitates (Reactions 2.5, 2.6). At pH values greater than 3, Fe^{3+} will precipitate and the oxidizing agent is removed from solution (Rose and Cravotta 1999; Ficklin and Mosier 1999). When the pH value falls below 3, sulfide oxidation becomes markedly faster.

  Furthermore, the activity of some microorganisms is pH dependent with optimal conditions for *Acidithiobacillus ferrooxidans* and *Acidithiobacillus thiooxidans* below pH 3 (i.e. they are acidophilic), and for *Thiobacillus thioparus* in the neutral pH range (i.e. they are neutrophilic) (Blowes et al. 1998). Thus, low pH conditions favour the activity of acidophilic sulfide oxidizing bacteria. Once pyrite oxidation and acid production have begun, the low pH conditions allow the proliferation of acidophilic microorganisms which further accelerate the pyrite oxidation rate (Hallberg and Johnson). On the other hand, an increase in pH to more neutral values greatly affects the occurrence of iron oxidizing acidophilic bacteria. They do not contribute significantly to the oxidation process under neutral to alkaline conditions.
- *Abundance of water.* Some researchers consider water to be an essential factor and reactant in the oxidation of pyrite (Rose and Cravotta 1999; Evangelou and Zhang 1995); others consider water as a reaction medium (Stumm and Morgan 1995). Whatever the role of water in sulfide oxidation, water is an important transport medium, and alternate wetting and drying of sulfides accelerate the oxidation process. Oxidation products can be dissolved and removed by the wetting, leaving a fresh pyrite surface exposed for further oxidation.
- *Fe^{2+}/Fe^{3+} ratio in the solution.* The most efficient oxidant for pyrite is dissolved Fe^{3+} and not oxygen, because Fe^{3+} oxidizes pyrite more rapidly than oxygen (Luther 1987). Therefore, the amount of Fe^{3+} produced (Reaction 2.3) controls how much pyrite can be destroyed (Reaction 2.4). As a result, the oxidation of Fe^{2+} to Fe^{3+} by dissolved oxygen is considered to be the rate limiting step in the indirect abiotic oxidation of pyrite (Singer and Stumm 1970). The precipitation of dissolved Fe^{3+} (Reactions 2.5, 2.6) places a limit on available dissolved Fe^{3+} and on the rate of pyrite oxidation (Reaction 2.4).

The rate of pyrite oxidation (i.e. its destruction over a given time period) varies depending on the above parameters. The rapid destruction of pyrite can potentially generate large amounts of acid and mobilize large amounts of metals and metalloids. Consequently, AMD generation and its impact on the environment can be severe. Alternatively, if the rate of pyrite oxidation is very slow, the production of acidity and dissolved contaminants occurs over an extended period of time, and AMD generation is negligible.

2.3.2 Other Sulfides

Pyrite is the dominant metal sulfide mineral in many ore deposits and as such plays a key role in the formation of AMD. However, other sulfide minerals commonly occur with pyrite, and their oxidation also influences the chemistry of mine waters. The weathering of these sulfides may occur via direct or indirect oxidation with the help of oxygen, iron and bacteria (Romano et al. 2001). The oxidation mechanisms of sulfides are analogous to those of pyrite but the reaction rates may be very different (Rimstidt et al. 1994; Domvile et al. 1994; Nicholson and Scharer 1994; Janzen et al. 2000; Keith and Vaughan 2000). Factors which influence the oxidation rate of pyrite such as trace element substitutions may or may not influence the oxidation rate of other sulfides (Jambor 1994; Janzen et al. 2000).

The weathering of various sulfides has been evaluated through laboratory experiments and field studies (Rimstidt et al. 1994; Jambor 1994; Domvile et al. 1994; Schmiermund 2000; Janzen et al. 2000; Jennings et al. 2000; Belzile et al. 2004; Yunmei et al. 2004; Lengke and Tempel 2005; Goh et al. 2006; Walker et al. 2006). The principal conclusion is that sulfide minerals differ in their acid production, reaction rate and degree of recalcitrance to weathering. Different sulfide minerals have different weathering behaviours. Pyrite, marcasite (FeS_2), pyrrhotite ($Fe_{1-x}S$) and mackinawite ($(Fe,Ni)_9S_8$) appear to be the most reactive sulfides and their oxidation generates low pH waters. Other sulfides such as covellite (CuS), millerite (NiS) and galena (PbS) are generally far less reactive than pyrite. This is partly due to: (*a*) the greater stability of their crystal structure; (*b*) the lack of iron released; and (*c*) the formation of low solubility minerals such as cerussite ($PbCO_3$) or anglesite ($PbSO_4$), which may encapsulate sulfides like galena preventing further oxidation (Lin 1997; Plumlee 1999). In contrast, the persistence of minerals such as cinnabar (HgS) and molybdenite (MoS_2) in oxic environments indicates that they weather very slowly under aerobic conditions (Plumlee 1999). These sulfides are most resistant to oxidation and do not generate acidity.

The presence of iron in sulfide minerals or in waters in contact with sulfides appears to be important for sulfide oxidation. Indeed, the amount of iron sulfides present in an assemblage strongly influences whether and how much acid is generated during weathering (Plumlee 1999). Sulfidic wastes with high percentages of iron sulfides (e.g. pyrite, marcasite, pyrrhotite), or sulfides having iron as a major constituent (e.g. chalcopyrite, Fe-rich sphalerite), generate significantly more acidity than wastes with low percentages of iron sulfides or sulfides containing little iron (e.g. galena, Fe-poor sphalerite). Moreover, the release of Fe^{2+} by the oxidation of Fe^{2+}-bearing sulfides is important as Fe^{2+} may be oxidized to Fe^{3+} which in turn can be hydrolyzed generating acidity (Boon et al. 1998; Munroe et al. 1999). Hence, sulfide minerals which do not contain iron in their crystal lattice (e.g. covellite, galena or iron-poor sphalerite) do not have the capacity to generate significant amounts of acid (Plumlee 1999).

The reason is that Fe^{3+} is not available as the important oxidant. Consequently, iron hydrolysis, which would generate additional acidity, cannot occur.

The metal/sulfur ratio in sulfides influences how much sulfuric acid is liberated by oxidation. For example, pyrite and marcasite have a metal/sulfur ratio of 1:2 and are more sulfur-rich than galena and sphalerite which have a metal/sulfur ratio of 1:1. Consequently, pyrite and marcasite produce more acid per mole of mineral. Sulfide minerals commonly contain minor and trace elements as small solid and liquid inclusions, adsorbed films, or substitutions for major metal cations in the crystal lattice (Table 2.1). These elements are liberated and potentially mobilized during the breakdown of the host mineral. Therefore, major amounts of sulfate and metals, as well as trace amounts of other metals and metalloids are released from oxidizing sulfides.

The stability, reaction rate, and acid generating capacity vary greatly among sulfides. Sulfides like pyrite and pyrrhotite readily oxidize and generate acid, whereby

Table 2.2. Examples of simplified acid producing reactions in sulfidic wastes

Mineral undergoing weathering	Chemical reaction
1. Complete oxidation of Fe-rich sulfides	
Pyrite and marcasite	$FeS_{2(s)} + 15/4\,O_{2(g)} + 7/2\,H_2O_{(l)} \rightarrow Fe(OH)_{3(s)} + 2\,SO^{2-}_{4(aq)} + 4\,H^+_{(aq)}$
Pyrrhotite	$Fe_{0.9}S_{(s)} + 2.175\,O_{2(g)} + 2.35\,H_2O_{(l)} \rightarrow 0.9\,Fe(OH)_{3(s)} + SO^{2-}_{4(aq)} + 2\,H^+_{(aq)}$
Chalcopyrite	$CuFeS_{2(s)} + 15/4\,O_{2(g)} + 7/2\,H_2O_{(l)} \rightarrow Fe(OH)_{3(s)} + 2\,SO^{2-}_{4(aq)} + Cu^{2+}_{(aq)} + 4\,H^+_{(aq)}$
Bornite	$Cu_3FeS_{4(s)} + 31/4\,O_{2(g)} + 7/2\,H_2O_{(l)} \rightarrow Fe(OH)_{3(s)} + 4\,SO^{2-}_{4(aq)} + 3\,Cu^{2+}_{(aq)} + 4\,H^+_{(aq)}$
Arsenopyrite	$FeAsS_{(s)} + 7/2\,O_{2(g)} + 3\,H_2O_{(l)} \rightarrow FeAsO_4 \cdot 2\,H_2O_{(s)} + SO^{2-}_{4(aq)} + 2\,H^+_{(aq)}$
Fe-rich sphalerite	$(Zn,Fe)S_{(s)} + 3\,O_{2(g)} + H_2O_{(l)} \rightarrow Fe(OH)_{3(s)} + SO^{2-}_{4(aq)} + Zn^{2+}_{(aq)} + 2\,H^+_{(aq)}$
2. Precipitation of Fe^{3+} and Al^{3+} hydroxides	
Iron hydroxides	$Fe^{3+}_{(aq)} + 3\,H_2O_{(l)} \leftrightarrow Fe(OH)_{3(s)} + 3\,H^+_{(aq)}$
Aluminium hydroxides	$Al^{3+}_{(aq)} + 3\,H_2O_{(l)} \leftrightarrow Al(OH)_{3(s)} + 3\,H^+_{(aq)}$
3. Dissolution of secondary minerals (Fe^{2+}, Mn^2, Fe^{3+}, and Al^{3+} sulfate and hydroxysulfate salts)	
Halotrichite	$FeAl_2(SO_4)_4 \cdot 22\,H_2O_{(s)} + 0.25\,O_{2(g)} \rightarrow Fe(OH)_{3(s)} + 2\,Al(OH)_{3(s)} + 13.5\,H_2O_{(l)} + 4\,SO^{2-}_{4(aq)} + 8\,H^+_{(aq)}$
Römerite	$Fe_3(SO_4)_4 \cdot 14\,H_2O_{(s)} \leftrightarrow 2\,Fe(OH)_{3(s)} + Fe^{2+}_{(aq)} + 8\,H_2O_{(l)} + 4\,SO^{2-}_{4(aq)} + 6\,H^+_{(aq)}$
Coquimbite	$Fe_2(SO_4)_3 \cdot 9\,H_2O_{(s)} \rightarrow 2\,Fe(OH)_{3(s)} + 3\,H_2O_{(l)} + 3\,SO^{2-}_{4(aq)} + 6\,H^+_{(aq)}$
Melanterite	$FeSO_4 \cdot 7\,H_2O_{(s)} + 0.25\,O_2 \rightarrow Fe(OH)_{3(s)} + 4.5\,H_2O_{(l)} + SO^{2-}_{4(aq)} + 2\,H^+_{(aq)}$
Jurbanite	$Al(SO_4)(OH) \cdot 5\,H_2O_{(s)} \rightarrow Al(OH)_{3(s)} + 3\,H_2O_{(l)} + SO^{2-}_{4(aq)} + H^+_{(aq)}$
Jarosite	$KFe_3(SO_4)_2(OH)_{6(s)} + 3\,H_2O_{(l)} \rightarrow K^+_{(aq)} + 3\,Fe(OH)_{3(s)} + 2\,SO^{2-}_{4(aq)} + 3\,H^+_{(aq)}$
Alunite	$KAl_3(SO_4)_2(OH)_{6(s)} + 3\,H_2O_{(l)} \rightarrow K^+_{(aq)} + 3\,Al(OH)_{3(s)} + 2\,SO^{2-}_{4(aq)} + 3\,H^+_{(aq)}$

pyrite generates more acid than pyrrhotite. Other sulfides like cinnabar oxidize very slowly and do not generate acid. Regardles of the oxidation rate and the acid generating capacity, weathering of sulfides contributes contaminants to mine waters. Even the relatively slow oxidation of arsenopyrite (FeAsS) can still release significant amounts of arsenic to mine waters (Craw et al. 1999; Yunmei et al. 2004).

2.3.3 Other Minerals

While some sulfides can produce significant amounts of acid and other sulfides do not, there are non-sulfide minerals whose weathering or precipitation will also release hydrogen ions (Plumlee 1999). Firstly, the precipitation of Fe^{3+} hydroxides and aluminium hydroxides generates acid (Table 2.2). Secondly, the dissolution of soluble Fe^{2+}, Mn^{2+}, Fe^{3+} and Al^{3+} sulfate salts such as jarosite ($KFe_3(SO_4)_2(OH)_6$), alunite ($KAl_3(SO_4)_2(OH)_6$), halotrichite ($FeAl_2(SO_4)_4 \cdot 22\,H_2O$), and coquimbite ($Fe_2(SO_4)_3 \cdot 9\,H_2O$) releases hydrogen ions (Table 2.2). Soluble Fe^{2+} sulfate salts are particularly common in sulfidic wastes and a source of indirect acidity. For example, the dissolution of melanterite ($FeSO_4 \cdot 7\,H_2O$) results in the release of Fe^{2+} which can be oxidized to Fe^{3+}. This Fe^{3+} may precipitate as ferric hydroxide ($Fe(OH)_3$) and generate hydrogen ions, or it may oxidize any pyrite present (Table 2.2). In general, increased hydrogen concentrations and acid production in mine wastes can be the result of:

- oxidation of Fe-rich sulfides;
- precipitation of Fe^{3+} and Al^{3+} hydroxides; and
- dissolution of soluble Fe^{2+}, Mn^{2+}, Fe^{3+} and Al^{3+} sulfate salts.

2.4 Acid Buffering Reactions

The oxidation of pyrite, the precipitation of iron and aluminium hydroxides, and the dissolution of some secondary minerals release hydrogen to solution. These processes increase the solution's acidity unless the hydrogen is consumed through buffering reactions. Much of the buffering of the generated acidity is achieved through the reaction of the acid solution with rock-forming minerals in the sulfidic wastes. These gangue minerals have the capacity to buffer acid; that is, the minerals will react with and consume the hydrogen ions. Acid buffering is largely caused by the weathering of silicates, carbonates and hydroxides.

The buffering reactions occur under the same oxidizing conditions, which cause the weathering of sulfide minerals. However, unlike sulfide oxidation reactions, acid buffering reactions are independent of the oxygen concentration of the gas phase or water in which the weathering reactions take place. The individual gangue minerals dissolve at different pH values, and buffering of the solution pH by individual minerals occurs within certain pH regions (Fig. 2.3). As a consequence, depending on the type and abundance of gangue minerals within the waste (i.e. the buffering capacity of the material), not all sulfide wastes produce acidic leachates and the same environmental concerns.

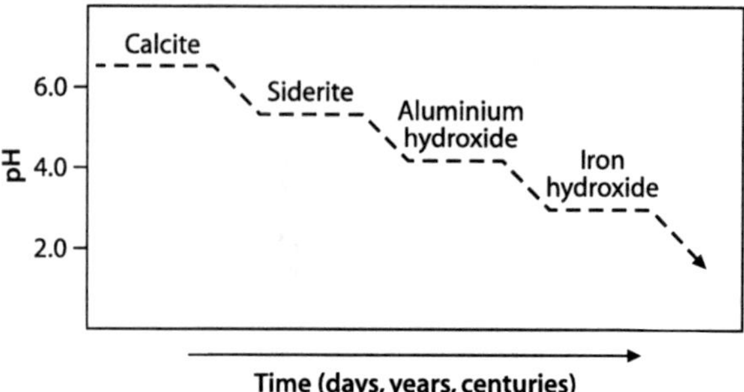

Fig. 2.3. Stepwise consumption of buffering capacity in a hypothetical sulfidic waste dump (Salomons 1995). (Reprinted from Salomons W (1995) Environmental impact of metals derived from mining activities: Processes, predictions, prevention. Journal of Geochemical Exploration 52:5–23, with permission from Elsevier Science)

2.4.1 Silicates

The major reservoir of buffering capacity in the environment are the silicate minerals which make up the majority of the minerals in the Earth's crust. Chemical weathering of silicate minerals consumes hydrogen ions and occurs via congruent or incongruent weathering. Congruent weathering involves the complete dissolution of the silicate mineral and the production of only soluble components (Reaction 2.8). Incongruent weathering is the more common form of silicate weathering whereby the silicate mineral is altered to another phase (Reaction 2.9). The chemical composition of most silicates such as olivines, pyroxenes, amphiboles, garnets, feldspars, feldspathoids, clays and micas is restricted to a range of elements. Thus, the two types of silicate weathering can be represented by the following reactions:

$$2\,MeAlSiO_{4(s)} + 2\,H^+_{(aq)} + H_2O \rightarrow Me^{x+}_{(aq)} + Al_2Si_2O_5(OH)_{4(s)} \tag{2.8}$$

$$MeAlSiO_{4(s)} + H^+_{(aq)} + 3\,H_2O \rightarrow Me^{x+}_{(aq)} + Al^{3+}_{(aq)} + H_4SiO_{4(aq)} + 3\,OH^-_{(aq)} \tag{2.9}$$

(Me = Ca, Na, K, Mg, Mn or Fe)

Chemical weathering of silicates results in the consumption of hydrogen ions, the production of dissolved cations and silicic acid, and the formation of secondary minerals (Purra and Neretnieks 2000). For example, the incongruent destruction of the sodium-rich plagioclase feldspar albite ($NaAlSi_3O_8$) may produce montmorillonite (simplified as $Al_2Si_4O_{10}(OH)_2$) or kaolinite ($Al_2Si_2O_5(OH)_4$), depending on the amount of leaching:

$$2\,NaAlSi_3O_{8(s)} + 2\,H^+_{(aq)} + 4\,H_2O_{(l)} \rightarrow 2\,Na^+_{(aq)} + Al_2Si_4O_{10}(OH)_{2(s)} + 2\,H_4SiO_{4(aq)} \tag{2.10}$$

$$2\,NaAlSi_3O_{8(s)} + 2\,H^+_{(aq)} + 9\,H_2O_{(l)} \rightarrow 2\,Na^+_{(aq)} + Al_2Si_2O_5(OH)_{4(s)} + 4\,H_4SiO_{4(aq)} \tag{2.11}$$

The incongruent destruction of other feldspars such as the calcium-rich plagioclase anorthite ($CaAl_2Si_2O_8$), and that of orthoclase, sanidine, adularia or microcline ($KAlSi_3O_8$) can be written as follows:

$$CaAl_2Si_2O_{8(s)} + 2\,H^+_{(aq)} + H_2O_{(l)} \rightarrow Ca^{2+}_{(aq)} + Al_2Si_2O_5(OH)_{4(s)} \tag{2.12}$$

$$2\,KAlSi_3O_{8(s)} + 2\,H^+_{(aq)} + 9\,H_2O_{(l)} \rightarrow 2\,K^+_{(aq)} + Al_2Si_2O_5(OH)_{4(s)} + 4\,H_4SiO_{4(aq)} \quad (2.13)$$

In most natural environments, the surface water contains dissolved carbon dioxide. The following reaction represents the incongruent weathering of K-feldspar under such conditions more accurately (Ollier and Pain 1997):

$$6\,KAlSi_3O_{8(s)} + 4\,H_2O_{(l)} + 4\,CO_{2(g)} \rightarrow 4\,K^+_{(aq)} + K_2Al_4(Si_6Al_2O_{20})(OH)_{4(s)} + 4\,HCO^-_{3(aq)} + 12\,SiO_{2(aq)} \quad (2.14)$$

In the above chemical reactions (Reactions 2.10 to 2.14), plagioclase and K-feldspar consume hydrogen ions in solution or generate bicarbonate ions. In addition, the by-products of feldspar and chlorite weathering are Na^+, K^+, Ca^{2+}, silicic acid (H_4SiO_4) and the clay minerals kaolinite ($Al_2Si_2O_5(OH)_4$), illite ($K_2Al_4(Si_6Al_2O_{20})(OH)_4$), or montmorillonite (simplified as $Al_2Si_4O_{10}(OH)_2$). The silicic acid or silica may precipitate as opaline silica or cryptocrystalline chalcedony (SiO_2). New quartz is only rarely formed, and then it usually overgrows on pre-existing quartz grains. The clay minerals may weather further and consume hydrogen ions as they dissolve. For example, the dissolution of kaolinite can be represented by the following reaction:

$$Al_2Si_2O_5(OH)_{4(s)} + 6\,H^+_{(aq)} \rightarrow 2\,Al^{3+}_{(aq)} + 2\,H_4SiO_{4(aq)} + H_2O_{(l)} \quad (2.15)$$

If the dissolved Al^{3+} is allowed to precipitate as gibbsite ($Al(OH)_3$), this neutralizing mechanism is lost because an equal amount of hydrogen will be released into solution (Deutsch 1997):

$$2\,Al^{3+}_{(aq)} + 6\,H_2O_{(l)} \leftrightarrow 2\,Al(OH)_{3(s)} + 6\,H^+_{(aq)} \quad (2.16)$$

On the other hand, if gibbsite already exists as a solid phase in the waste rocks, it provides additional neutralizing ability because it can consume dissolved hydrogen ions. Similarly, ferric hydroxide solids (Reaction 2.17) previously precipitated during pyrite oxidation can be redissolved in acidic waters, thereby consuming hydrogen ions:

$$Fe(OH)_{3(s)} + 3\,H^+_{(aq)} \leftrightarrow Fe^{3+}_{(aq)} + 3\,H_2O_{(l)} \quad (2.17)$$

Quartz (SiO_2), chalcedony (SiO_2), opal ($SiO_2 \cdot nH_2O$), and other silica minerals do not consume hydrogen when they weather to form silicic acid (Reaction 2.18). Silicic acid is a very weak acid and does not contribute significant hydrogen ions to solution. The acid is unable to donate protons to a solution unless the pH is greater than 9 (Deutsch 1997).

$$SiO_{2(s)} + 2\,H_2O_{(l)} \leftrightarrow H_4SiO_{4(aq)} \quad (2.18)$$

2.4.2 Carbonates

Carbonate minerals play an extremely important role in acid buffering reactions. Minerals such as calcite ($CaCO_3$), dolomite ($CaMg(CO_3)_2$), ankerite ($Ca(Fe,Mg)(CO_3)_2$), or magnesite ($MgCO_3$) neutralize acid generated from sulfide oxidation. Calcite is the most important neutralizing agent, because of its common occurrence in a wide range of geological environments and its rapid rate of reaction compared to dolomite. Similarly to pyrite weathering, grain size, texture and the presence of trace elements in the crystal lattice of carbonates may increase or decrease their resistance to weathering (Plumlee 1999; Strömberg and Banwart 1999). Calcite neutralizes acid by dissolving and complexing with hydrogen ion to form bicarbonate (HCO_3^-) and carbonic acid (H_2CO_3) (Stumm and Morgan 1995; Blowes and Ptacek 1994; Strömberg and Banwart 1999; Al et al. 2000). Depending on the pH of the weathering solution, acidity is consumed either by the production of bicarbonate in weakly acidic to alkaline environments (Reaction 2.19) or by the production of carbonic acid in strongly acidic environments (Reaction 2.20).

$$CaCO_{3(s)} + H^+_{(aq)} \leftrightarrow Ca^{2+}_{(aq)} + HCO^-_{3(aq)} \quad (2.19)$$

$$CaCO_{3(s)} + 2\,H^+_{(aq)} \leftrightarrow Ca^{2+}_{(aq)} + H_2CO_{3(aq)} \quad (2.20)$$

Overall, the dissolution of calcite neutralizes acidity and increases pH and alkalinity in waters. A reversal of the Reactions 2.19 and 2.20 is possible when there is a change in temperature, loss of water or loss of carbon dioxide. Reprecipitation of carbonates will occur, which in turn releases hydrogen ions, causing the pH to fall.

The presence or absence of carbon dioxide strongly influences the solubility of calcite (Sherlock et al. 1995; Stumm and Morgan 1995). Calcite dissolution can occur in an open or closed system, depending on whether carbon dioxide is available for gas exchange. If water is in contact with a gas phase, then carbon dioxide can enter the solution and calcite dissolution occurs in a so-called "*open system*" (Reaction 2.21). In the open system, there is an increased solubility of calcite (Stumm and Morgan 1995). The unsaturated zones of sulfidic waste rock piles represent such open systems. In contrast, in the water saturated zone of sulfidic waste rock piles or tailings, there is no carbon dioxide gas phase. Here, calcite dissolves in a *closed system* (Reaction 2.22):

$$CaCO_{3(s)} + CO_{2(g)} + H_2O_{(l)} \leftrightarrow Ca^{2+}_{(aq)} + 2\,HCO^-_{3(aq)} \quad (2.21)$$

$$CaCO_{3(s)} + H^+_{(aq)} \leftrightarrow Ca^{2+}_{(aq)} + HCO^-_{3(aq)} \quad (2.22)$$

Therefore, in an open mine waste environment there is increased calcite dissolution because the calcite is exposed to a carbon dioxide gas phase. More bicarbonate is generated and more hydrogen ions are consumed than it would be the case in a closed mine waste environment (Sherlock et al. 1995).

Dissolution of other carbonates such as dolomite, ankerite or magnesite will similarly result in the consumption of hydrogen ions and in the release of bicarbonate, calcium and magnesium ions and carbonic acid. However, calcite is more easily dissolved than dolomite or ankerite. Siderite ($FeCO_3$) is a common gangue mineral in coal deposits and various metal ores. The neutralizing effect of siderite depends on the redox conditions of the weathering environment. Under reducing conditions, siderite dissolves to form bicarbonate and Fe^{2+} ions. In contrast, in an open system with abundant oxygen, the dissolution of siderite has no neutralizing effect. While the generation of bicarbonate consumes hydrogen ions, any Fe^{2+} generated will undergo hydrolysis and precipitation (Reactions 2.5, 2.6). This in turn generates as much hydrogen ions as are consumed by the generation of bicarbonate (Ptacek and Blowes 1994; Blowes and Ptacek 1994; Rose and Cravotta 1999). Hence, under well oxidized conditions, the net neutralizing effect of siderite dissolution is zero (Skousen et al. 1997).

2.4.3 Exchangeable Cations

A final neutralizing source in the subsurface are the cations (Ca^{2+}, Mg^{2+}, Na^{+}, K^{+}) present on the exchange sites of micas, clays and organic matter (Deutsch 1997; Strömberg and Banwart 1999). These exchangeable cations can be replaced by cations dissolved in weathering solutions. During sulfide oxidation, dissolved hydrogen and Fe^{2+} ions are produced which will compete for the cation exchange sites. The newly generated hydrogen and Fe^{2+} ions are removed from solution and temporarily adsorbed onto the exchange sites of the solid phases. Such reactions of clays with dissolved Fe^{2+} and hydrogen ions, respectively, can be represented as (Deutsch 1997; Rose and Cravotta 1999):

$$\text{clay-}(Na^{+})_{(s)} + Fe^{2+}_{(aq)} \leftrightarrow \text{clay-}(Fe^{2+})_{(s)} + Na^{+}_{(aq)} \tag{2.23}$$

$$\text{clay-}(Ca^{2+})_{0.5(s)} + H^{+}_{(aq)} \leftrightarrow \text{clay-}(H^{+})_{(s)} + 0.5\,Ca^{2+}_{(aq)} \tag{2.24}$$

Clays may also undergo solid transformations during acid leaching whereby a potassium-bearing illite consumes hydrogen and is thereby transformed to a potassium-free smectite clay mineral (Puuru et al. 1999):

$$\text{illite}_{(s)} + H^{+}_{(aq)} \rightarrow \text{smectite}_{(s)} + K^{+}_{(aq)} \tag{2.25}$$

2.4.4 Reaction Rates

The weathering rate (i.e. weathering kinetics) of individual minerals in sulfidic wastes is influenced by: (*a*) the mineral's composition, crystal size, crystal shape, surface area, and crystal perfection; (*b*) the pH and dissolved carbon dioxide content of the weathering solution; (*c*) temperature; (*d*) redox conditions; and (*e*) access of weathering agent and removal of weathered products (Sherlock et al. 1995). For example, there is a large difference in weathering rates between fine-grained waste and larger waste rock particles (diameters >0.25 mm). Smaller particles (diameters <0.25 mm) with their larger surface areas contribute to the great majority of sulfide oxidation as well as silicate and carbonate dissolution (Strömberg and Banwart 1999).

Different minerals reacting with acidic solutions have a variable resistance to weathering (Table 2.3). Minerals such as olivine and anorthite are more reactive and less stable in the surficial environment than K-feldspar, biotite, muscovite and albite (Fig. 2.4). The rates of the different acid buffering reactions are highly variable, and the major rock-forming minerals have been classified according to their relative pH-dependent reactivity (Table 2.4). Compared with the weathering rates of even the most

Table 2.3. Mean lifetime of a 1 mm crystal at 25 °C and pH 5 (Lasaga and Berner 1998). (Reprinted from Lasaga AC, Berner RA (2000) Fundamental aspects of quantitative models for geochemical cycles. Chemical Geology 145:161–175, with permission from Elsevier Science)

Mineral	Lifetime (years)
Calcite	0.43
Wollastonite	79
Anorthite	112
Nepheline	211
Forsterite	2300
Diopside	6800
Enstatite	10100
Gibbsite	276000
Sanidine	291000
Albite	575000
Prehnite	579000
Microcline	921000
Epidote	923000
Muscovite	2600000
Kaolinite	6000000
Quartz	34000000

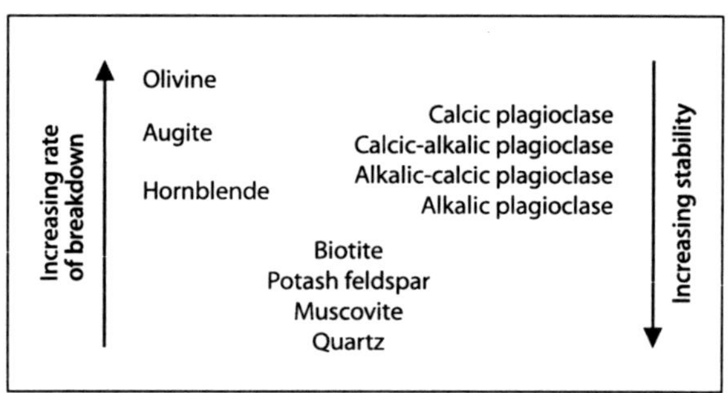

Fig. 2.4. The stability of minerals during weathering (Sherlock et al. 1995)

reactive silicate minerals, the reaction rates of carbonates are relatively rapid, particularly that of calcite (Strömberg and Banwart 1999). Carbonates can rapidly neutralize acid. In an extreme case, calcite may even be dissolved at a faster rate than pyrite. As a consequence, drainage from a calcite-bearing waste may have a neutral pH, yet the quality of the mine drainage can eventually deteriorate and turn acid as the calcite dissolves faster than the pyrite.

Silicate minerals are abundant in sulfidic wastes, and their abundance may suggest that a waste rich in silicates has a significant buffering capacity. However, silicates do not necessarily dissolve completely, and the chemical weathering rate of silicates is very slow relative to the production rate of acid by pyrite oxidation. Therefore, rock-forming silicates do not buffer acid to a significant degree, and they only contribute token amounts of additional long-term buffering capacity to sulfidic wastes (Jambor et al. 2000c). Nonetheless, silicate mineral dissolution can maintain neutral conditions if

Table 2.4. Grouping of minerals according to their relative reactivity at pH 5 (after Sverdrup 1990; Kwong 1993)

Mineral group	Relative reactivity at pH 5	Typical minerals
Dissolving	1.00	Calcite, aragonite, dolomite, magnesite, brucite, halite
Fast weathering	0.60	Anorthite, nepheline, olivine, garnet, jadeite, leucite, spodumene, diopside, wollastonite
Intermediate weathering	0.4	Ortho and ring silicates (epidote, zoisite), chain silicates (enstatite, hypersthene, hornblende, glaucophane, tremolite, actinolite, anthophyllite), sheet silicates (serpentine, chrysotile, chlorite, biotite, talc)
Slow weathering	0.02	Framework silicates (albite, oligoclase, labradorite), sheet silicates (vermiculite, montmorillonite, kaolinite), gibbsite
Very slow weathering	0.01	K-feldspar, muscovite
Inert	0.004	Quartz, rutile, zircon

the rate of acid production is quite slow and if abundant fine-grained, fast weathering silicates are present.

2.5 Coal Mine Wastes

Coal mining and processing generate the largest quantity of mine wastes (Fig. 2.5). The environmental issues related to coal wastes are attributable to the exposure of reduced earth materials (coal, sulfides, and Fe^{2+}-bearing carbonates) to oxygen (Younger 2004). The consequences of oxidation of coal and associated strata range from the release of acid waters due to pyrite oxidation to the spontaneous combustion of the wastes.

Coals were initially deposited in reduced environments such as swamps and peat bogs. This depositional environment also resulted in the presence of fine-grained sedimentary rocks enclosing the coal seams (i.e. mudstones, sandstones). Hence, coals and their associated sediments commonly contain iron sulfides including major pyrite and possible traces of marcasite, galena, chalcopyrite and sphalerite.

Coals are readily combustible sedimentary rocks, possessing significant carbon, hydrogen and sulfur contents. The total sulfur content of coals vary, ranging from a few 0.1 wt.% to extreme examples reaching 10 wt.%. Sulfur in coal occurs in three sulfur forms, pyritic sulfur, sulfate sulfur and organic sulfur. Much of the sulfur is organically bound within solid carbonaceous materials (i.e. the coal macerals), and this form of sulfur does not contribute to the acid generation of coal wastes. Sulfate sulfur is generally the result of oxidation of pyrite in the coal and is an indicator of weathering of the coal before or after mining. Thus, it is important to determine what percentage of the total sulfur is incorporated into acid-generating pyrite. Such knowledge allows an evaluation of the acid production of coal seams and associated rock types. At coal mines, AMD is commonly brought about by the oxidation of pyrite which is finely disseminated through the coals and associated sedimentary rocks.

Fig. 2.5. Open cut and spoil heaps of the Moura coal mine, Australia. Spoils of the Moura coal mine are largely non-acid generating due to the lack of pyrite.

Pyrite is not the only Fe^{2+}-bearing mineral that undergoes oxidation when coal-bearing rocks are exposed to the atmosphere (Younger 2004). Carbonate minerals such as siderite ($FeCO_3$) and ankerite ($Ca(Mg,Fe)(CO_3)_2$) are common gangue minerals of coal-bearing strata and these carbonates contain Fe^{2+}. The weathering of siderite consumes hydrogen ions as long as the released Fe^{2+} does not undergo oxidation and hydrolysis because the hydrolysis of Fe^{3+} releases hydrogen protons. Thus, siderite dissolution in an oxidizing environment has no neutralizing effect on acid waters (Sec. 2.4.2). By contrast, the dissolution of ankerite consumes more hydrogen protons than the subsequent oxidation and hydrolysis of the released iron (Younger 2004). Consequently, ankerite possesses a net neutralization potential for acid waters.

2.5.1 Spontaneous Combustion of Pyritic Wastes

Coal and certain base metal, uranium, iron and phosphate ore deposits are hosted by sedimentary sequences, some of which contain pyritic, carbonaceous shales and mudstones. The exothermic oxidation of sulfides and organic matter in these rock types can lead to a significant increase in temperature in pyritic, carbonaceous rocks. The elevated temperatures have the potential to cause premature detonation of explosives in a charged blasthole with catastrophic consequences (Briggs and Kelso 2003). This is particularly the case for ammonium nitrate-based explosive products.

The development of even higher temperatures may lead to the spontaneous ignition of coal and carbonaceous, pyritic shales and mudstones, which has been observed naturally (Mathews and Bustin 1984). It can also occur in underground workings, open pit faces, waste rock dumps, and slag heaps (Bullock and Bell 1997; Puura et al. 1999; Sidenko et al. 2001). It is visible as "smoke", comprising a variety of gases such as water steam, sulfur dioxide, carbon dioxide, carbon monoxide, and methane. In particular, colliery spoil and carbonaceous, pyritic waste rock dumps have the tendency to burn and smoke.

The rocks contain abundant, often very fine-grained, micrometer sized framboidal pyrite as well as carbon and organic matter. Spontaneous combustion of this material is initiated through its exposure to atmospheric oxygen or oxygenated ground water. This leads to the slow exothermic oxidation of pyrite, carbon and organic matter which in turn results in a gradual rise in temperature of the rock. Any fine-grained rock materials will act as heat insulators, and the heat will not be able to escape. At some stage, enough heat is generated to ignite the carbon or organic matter. The oxidation reactions are significantly accelerated as soon as significant amounts of atmospheric oxygen or oxygen dissolved in water are supplied to the carbonaceous material, and large surface areas are exposed, for example, as a result of mining. Next, rapid oxidation of this hot pyritic, carbonaceous rock is initiated, and spontaneous combustion occurs. The organic carbon and sulfur begin to burn. Smoke and steam are released resembling volcanic fumaroles. The combustion of carbon and organic matter increases the heat of the rock which in turn increases the rate of sulfide oxidation. If there is sufficient oxygen during the combustion process, the pyrite is converted to hematite and sulfur oxides:

$$2\,FeS_{2(s)} + x\,O_{2(g)} \rightarrow Fe_2O_{3(s)} + 4\,SO_{x(g)} \qquad (2.26)$$

If there is not enough oxygen for complete oxidation, hydrogen sulfide is formed. In extreme cases of oxidation, temperatures reach 1200 °C and localized melting of the rocks and wastes occurs. In such cases, the outer dump layer cracks, and surface venting of gases from sulfidic materials becomes significant. The spontaneous combustion and subsequent cooling of coal spoil and pyritic waste rock dumps produce waste materials of complex mineralogical composition, including slag-type phases, thermal metamorphic minerals, and weathering related minerals (Puura et al. 1999; Sidenko et al. 2001).

If combustion has already begun in mine waste dumps, disturbing the burning heap – by excavating or reshaping it – will only provide additional atmospheric oxygen to the waste, and the rate of combustion will increase. Various methods are used to combat combustion in mine wastes, including compaction, injection of water, and water spraying (Fig. 2.6). However, compaction may eventually lead to cracking of the seal by pressurized gases. Also, the use of excessive amounts of water may generate steam and eventually cause steam explosions.

In order to prevent premature detonation of explosives or spontaneous combustion in carbonaceous, pyritic rocks, the rocks need to be characterized for their pyrite and organic carbon contents and their temperature. Such characterization should occur before or during mining. This will ensure that any high risk material will undergo special handling pior to their finite disposal. Disposal options include dumping small

Fig. 2.6. Water spraying of spontaneous combustion at the Blair Attol coal mine, Australia (Photo courtesy of P. Crosdale)

heaps of wastes and leaving them to oxidize and cool prior to finite capping with benign wastes.

2.6 Formation and Dissolution of Secondary Minerals

The weathering of sulfides releases sulfate, metals, metalloids and other elements into solution. This water can contact more sulfide minerals and accelerate their oxidation (i.e. acid producing reactions). Alternatively, it can contact gangue minerals, some of which react to neutralize some or all of the acid (i.e. acid buffering reactions). Above all, the reactive sulfide and gangue minerals will contribute various ionic species to the weathering solution. In fact, in many sulfidic materials the acid producing, acid buffering and non-acid generating reactions release significant amounts of dissolved cations and anions into pore waters. As a result, the waters become highly saline. Some ions will remain in solution in ionic form, where they can interact with minerals and be adsorbed. Sheet silicates such as chlorite, talc, illite and smectite are especially able to adsorb metal ions from pore solutions (Dinelli and Tateo 2001). Few ions will remain in solution indefinitely and enter ground or surface waters. Other ions will interact in the weathering solution, reach saturation levels and precipitate as secondary minerals in the waste. The formation of secondary minerals is the most common form of element fixation in pore waters of sulfidic wastes. A significant fraction of the metals released by sulfide oxidation is retained in the wastes as secondary mineral precipitates (Lin 1997; Lin and Herbert 1997). Such secondary mineral formation is not exclusive to the wastes themselves; numerous salts approach saturation in ground waters, streams and leachates associated with the weathering of sulfidic wastes. Therefore, a wide range of secondary minerals are known to precipitate in oxidizing sulfidic wastes and AMD environments (Table 2.5). Also, the formation of secondary minerals is not exclusive to sulfidic wastes and AMD waters. It may occur in any saline water regardless of its pH.

2.6.1 Pre-Mining and Post-Mining Secondary Minerals

Secondary minerals are defined as those that form during weathering. Weathering of sulfides may occur before, during or after mining. Thus, a distinction has to be made between secondary minerals formed by natural processes prior to mining and those formed after the commencement of mining (Nordstrom and Alpers 1999a).

Sulfide oxidation prior to mining results in the formation of secondary minerals. For example, if a sulfide orebody has been exposed by erosion and weathered by surface waters descending through the unsaturated zone, a near-surface oxidized layer of secondary minerals forms (Williams 1990). Some of these secondary minerals are relatively insoluble in ground and surface waters. They effectively capture the metals and reduce the release of metals into the environment. Hence, leaching of completely oxidized wastes can produce non-acid mine waters. Nonetheless, an abundance of relatively soluble sulfates such as gypsum may still result in saline, sulfate-rich drainage waters.

Sulfide oxidation during and after mining results in the formation of secondary minerals. Post-mining secondary minerals form because waste and ore have been exposed to the atmosphere and subsequently weathered. Such post-mining oxidation products occur as cements and masses within the waste and as crusts at or near the waste's surface. The surface precipitates are commonly referred to as "*efflorescences*". They are particularly common in waste piles, underground workings, stream beds and seepage areas, and on pit faces (Figs. 2.7 and 2.8).

The distinction of pre- from post-mining secondary minerals can be a challenging task because some minerals, particularly the soluble sulfates, may have formed during the pre- and post-mining stage. The precipitation of post-mining secondary minerals takes place in response to one of these following processes (Nordstrom and Alpers 1999a):

- Oxidation and hydrolysis of the dissolved cation (Fe^{2+})
- Hydrolysis of the dissolved cation (e.g. Fe^{3+}, Al^{3+})
- Reaction of acid mine waters with acid buffering minerals or alkaline waters
- Mixing of acid mine waters with neutral pH waters
- Oxidation of sulfides in humid air
- Concentration of the mine water due to evaporation

Evaporation is an important mechanism in the formation of mineral salts. This process concentrates any cations and anions in mine waters until they reach mineral saturation, forming secondary minerals. Not all precipitates are crystalline, and many solids are of a poorly crystalline or even amorphous nature. The initial minerals that

Table 2.5. Examples of post-mining secondary minerals found in sulfidic mine wastes (after Alpers et al. 1994; Nordstrom and Alpers 1999a; Jambor et al. 2000a,b; Bigham and Nordstrom 2000)

1 Sulfates	
1.1 Simple hydrous metal sulfates with divalent metal cations	
Mineral	**Formula**
Poitevinite	$CuSO_4 \cdot H_2O$
Melanterite group	
Melanterite	$FeSO_4 \cdot 7H_2O$
Boothite	$CuSO_4 \cdot 7H_2O$
Bieberite	$CoSO_4 \cdot 7H_2O$
Mallardite	$MnSO_4 \cdot 7H_2O$
Epsomite group	
Epsomite	$MgSO_4 \cdot 7H_2O$
Morenosite	$NiSO_4 \cdot 7H_2O$
Goslarite	$ZnSO_4 \cdot 7H_2O$
Hexahydrite group	
Hexahydrite	$MgSO_4 \cdot 6H_2O$
Ferrohexahydrite	$FeSO_4 \cdot 6H_2O$
Chvaleticeite	$MnSO_4 \cdot 6H_2O$
Moorhouseite	$CoSO_4 \cdot 6H_2O$
Bianchite	$ZnSO_4 \cdot 6H_2O$
Chalcanthite group	
Chalcanthite	$CuSO_4 \cdot 5H_2O$
Pentahydrite	$MgSO_4 \cdot 5H_2O$
Siderotil	$FeSO_4 \cdot 5H_2O$
Rozenite group	
Rozenite	$FeSO_4 \cdot 4H_2O$
Starkeyite	$MgSO_4 \cdot 4H_2O$
Boyleite	$ZnSO_4 \cdot 4H_2O$
Kieserite group	
Kieserite	$MgSO_4 \cdot H_2O$
Szomolnokite	$FeSO_4 \cdot H_2O$
Gunningite	$ZnSO_4 \cdot H_2O$
1.2 Simple hydrous metal sulfates with trivalent metal cations	
Mineral	**Formula**
Kornelite	$Fe_2(SO_4)_3 \cdot 7H_2O$
Coquimbite	$Fe_2(SO_4)_3 \cdot 9H_2O$

1.2 Simple hydrous metal sulfates with trivalent metal cations (*continued*)	
Mineral	**Formula**
Alunogen	$Al_2(SO_4)_3 \cdot 17H_2O$
Rhomboclase	$HFe(SO_4)_2 \cdot 4H_2O$
Butlerite	$FeSO_4(OH) \cdot 2H_2O$
1.3 Mixed divalent-trivalent hydrous metal sulfates	
Mineral	**Formula**
Römerite	$Fe_3(SO_4)_4 \cdot 14H_2O$
Halotrichite group	
Halotrichite	$FeAl_2(SO_4)_4 \cdot 22H_2O$
Pickeringite	$MgAl_2(SO_4)_4 \cdot 22H_2O$
Apjohnite	$MnAl_2(SO_4)_4 \cdot 22H_2O$
Dietrichite	$(Zn,Fe,Mn)Al_2(SO_4)_4 \cdot 22H_2O$
Bilinite	$Fe_3(SO_4)_4 \cdot 22H_2O$
Copiapite group	
Copiapite	$Fe_5(SO_4)_6(OH)_2 \cdot 20H_2O$
Ferricopiapite	$Fe_5(SO_4)_6O(OH)_2 \cdot 20H_2O$
1.4 Fe and Al hydroxysulfates	
Mineral	**Formula**
Jarosite	$KFe_3(SO_4)_2(OH)_6$
Natrojarosite	$NaFe_3(SO_4)_2(OH)_6$
Hydronium jarosite	$(H_3O)Fe_3(SO_4)_2(OH)_6$
Plumbojarosite	$PbFe_6(SO_4)_4(OH)_{12}$
Alunite	$KAl_3(SO_4)_2(OH)_6$
Jurbanite	$Al(SO_4)(OH) \cdot 5H_2O$
Schwertmannite	$Fe_8O_8(SO_4)(OH)_6$
Aluminite	$Al_2(SO_4)(OH)_4 \cdot 7H_2O$
Basaluminite	$Al_4(SO_4)(OH)_{10} \cdot 4H_2O$
1.5 Other sulfates and hydroxysulfates	
Mineral	**Formula**
Anglesite	$PbSO_4$
Barite	$BaSO_4$
Strontianite	$SrSO_4$
Anhydrite	$CaSO_4$
Bassanite	$CaSO_4 \cdot 0.5H_2O$

Table 2.5. *Continued*

1.5 Other sulfates and hydroxysulfates (*continued*)	
Mineral	**Formula**
Gypsum	$CaSO_4 \cdot 2H_2O$
Thenardite	Na_2SO_4
Aphthitalite	$NaK_3(SO_4)_2$
Voltaite	$K_2Fe_8Al(SO_4)_{12} \cdot 18H_2O$
Tamarugite	$NaAl(SO_4)_2 \cdot 6H_2O$
Konyaite	$Na_2Mg(SO_4)_2 \cdot 5H_2O$
Blödite	$Na_2Mg(SO_4)_2 \cdot 4H_2O$
Löwite	$Na_{12}Mg_7(SO_4)_{13} \cdot 15H_2O$
Eugsterite	$Na_4Ca(SO_4)_3 \cdot 2H_2O$
Syngenite	$K_2Ca(SO_4)_2 \cdot H_2O$
Antlerite	$Cu_3(SO_4)(OH)_4$
Brochantite	$Cu_4(SO_4)(OH)_6 \cdot 2H_2O$
2 Oxides, hydroxides and arsenates	
Mineral	**Formula**
Fe minerals	
Goethite	α-FeOOH
Lepidocrocite	γ-FeOOH
Feroxyhyte	δ-FeOOH
Akaganéite	β-FeOOH
Ferrihydrite	$Fe_5HO_8 \cdot 4H_2O$
Al minerals	
Gibbsite	γ-AlO(OH)
Diaspore	α-AlO(OH)
Cu minerals	
Tenorite	CuO
Cuprite	Cu_2O
As minerals	
Scorodite	$FeAsO_4 \cdot 2H_2O$
Mansfeldite	$Al(AsO_4) \cdot 2H_2O$
Rauenthalite	$Ca_3(AsO_4)_2 \cdot 10H_2O$
As minerals (*continued*)	
Arsenolite	As_2O_3
Claudetite	As_2O_3
Sb minerals	
Cervantite	Sb_2O_4
Valentinite	Sb_2O_3
Senarmontite	Sb_2O_3
3 Carbonates	
Mineral	**Formula**
Calcite	$CaCO_3$
Magnesite	$MgCO_3$
Siderite	$FeCO_3$
Ankerite	$Ca(Fe,Mg)(CO_3)_2$
Dolomite	$CaMg(CO_3)_2$
Smithsonite	$ZnCO_3$
Otavite	$CdCO_3$
Cerussite	$PbCO_3$
Malachite	$Cu_2(CO_3)(OH)_2$
Azurite	$Cu_3(CO_3)_2(OH)_2$
Hydrozincite	$Zn_5(CO_3)_2(OH)_6$
Aurichalcite	$(Zn,Cu)(CO_3)_2(OH)_6$
4 Silicates	
Mineral	**Formula**
Nacrite	$Al_2Si_2O_5(OH)_4$
Kaolinite	$Al_2Si_2O_5(OH)_4$
Chrysocolla	$CuSiO_3 \cdot 2H_2O$
Plancheite	$Cu_8Si_8O_{22} \cdot 2H_2O$
Dioptase	$Cu_6Si_6O_{18} \cdot 6H_2O$
5 Native elements	
Mineral	**Formula**
Native sulfur	S
Native copper	Cu

precipitate tend to be poorly crystalline, metastable phases that may transform to more stable phases over time (Murad et al. 1994; Nordstrom and Alpers 1999a). Consequently, the collection and identification of metastable phases using conventional laboratory

Fig. 2.7. Secondary gypsum efflorescences encrusting wallrock in the Mary Kathleen open pit, Australia. Field of view 50 cm

techniques are troublesome, and materials should be collected and stored in airtight containers at temperatures resembling field conditions. By contrast, airborne and ground infrared spectrometry can be used to identify and map secondary iron minerals in the field. This approach allows the discrimination and mapping of different iron minerals in exposed outcrops, waste dumps and watersheds (Swayze et al. 2000; Dalton et al. 2000; Williams et al. 2002; Sams et al. 2003; Sams and Veloski 2003; Ackman 2003; Velasco et al. 2005). In turn, the pH value of mine drainage waters can be inferred from the colour and spectral reflectance of the precipitates because the occurrence of different iron minerals is controlled by pH as well as other parameters.

2.6.2 Solubility of Secondary Minerals

Secondary minerals can be grouped into sulfates, oxides, hydroxides and arsenates, carbonates, silicates, and native elements (Table 2.5). The type of secondary minerals formed in mine wastes is primarily controlled by the composition of the waste. For example, coal spoils commonly possess iron, aluminium, calcium, magnesium, sodium, and potassium sulfates, whereas metalliferous waste rocks tend to contain abundant iron, aluminium and heavy metal sulfate salts.

Some of the secondary minerals are susceptible to dissolution, whereby a wide range in solubility has been noted. For example, simple hydrous metal sulfates are very soluble

Fig. 2.8. Face of the Río Tinto smelting slag dump, Spain. Mineral efflorescences commonly occur as white sulfate salt precipitates (gypsum, epsomite, hexahydrite, bloedite, copiapite, roemerite) in protected overhangs and at seepage points at the base of the slag dump. The slags generate ephemeral drainage, which runs from the dump into the Río Tinto and contributes to its acidification and metal load

in water, whereas the iron and aluminium hydroxysulfates are relatively insoluble. In addition, there are a number of secondary sulfates and carbonates which are poorly soluble such as barite ($BaSO_4$), anglesite ($PbSO_4$), celestite ($SrSO_4$), and cerussite ($PbCO_3$). As a result, once these minerals are formed, they will effectively immobilize alkali earth elements as well as lead. The minerals act as sinks for sulfate, barium, strontium, and lead in oxidizing sulfidic wastes, and their precipitation controls the amount of sulfate, barium, strontium, and lead in AMD solutions.

The water soluble hydrous metal sulfates with divalent cations ($Me^{2+}SO_4 \cdot n\,H_2O$) are the most dominant secondary mineral types (Jambor et al. 2000a,b). These hydrous sulfates may redissolve in water and release their ions back into solution:

$$Me^{2+}SO_4 \cdot n\,H_2O_{(s)} \leftrightarrow Me^{2+}_{(aq)} + SO^{2-}_{4(aq)} + n\,H_2O_{(l)} \qquad (2.27)$$

(Me = Ca, Mg, Fe, Mn, Co, Ni, Cu, Zn; n = 1 to 7)

Alternatively, the hydrous sulfates may dehydrate to less hydrous or even anhydrous compositions. For example, melanterite ($FeSO_4 \cdot 7\,H_2O$) may precipitate first, which may then dehydrate to rozenite ($FeSO_4 \cdot 4H_2O$) or szomolnokite ($FeSO_4 \cdot H_2O$). Also, the hydrous Fe^{2+} sulfates may oxidize to Fe^{2+}-Fe^{3+} or Fe^{3+} sulfate salts. For instance, the Fe^{2+} mineral melanterite ($FeSO_4 \cdot 7H_2O$) may oxidize to the mixed Fe^{2+}-Fe^{3+} mineral copiapite ($Fe_5(SO_4)_6(OH)_2 \cdot 20H_2O$) (Frau 2000; Jerz and Rimstidt 2003). The newly formed secondary minerals are more stable and resistant to redissolution compared to their precursors. Thus, secondary minerals may exhibit a paragenetic sequence whereby the minerals formed in a distinct order. The general trend for the simple hydrous sulfate salts is that the Fe^{2+} minerals form first, followed by the mixed Fe^{2+}-Fe^{3+} minerals, and then the Fe^{3+} minerals (Jambor et al. 2000a,b).

Secondary minerals, be they relatively soluble or insoluble, possess large surface areas. Consequently, they adsorb or coprecipitate significant quantities of trace elements including metals and metalloids. The precipitates effectively immobilize elements in acid mine waters and hence provide an important natural attentuation and detoxification mechanism in mine waters (Lin 1997; Nordstrom and Alpers 1999a; Berger et al. 2000). However, this immobilization of metals is only temporary as many mineral efflorescences, particularly the simple hydrous metal sulfates, tend to be soluble and release their stored metals back into mine waters upon dissolution.

2.6.3
Acid Consumption and Production

The precipitation of some secondary minerals may influence the mine water pH as their formation generates or consumes hydrogen ions. Generally, the formation of Fe^{3+} or Al^{3+} hydroxides generates acid, whereas the precipitation of Fe^{2+}, Mn^{2}, Fe^{3+} and Al^{3+} sulfate salts such as jarosite ($KFe_3(SO_4)_2(OH)_6$), alunite ($KAl_3(SO_4)_2(OH)_6$), coquimbite ($Fe_2(SO_4)_3 \cdot 9H_2O$), jurbanite ($Al(SO_4)(OH) \cdot 5H_2O$), halotrichite ($FeAl_2(SO_4)_4 \cdot 22H_2O$), or melanterite ($FeSO_4 \cdot 7H_2O$) consumes acid. However, this consumption of acidity is only temporary as these minerals, particularly the simple hydrous metal sulfates, tend to be soluble and release their stored acidity upon dissolution (Cravotta 1994) (Table 2.2). A generalized reaction for this temporary acid consumption can be written as follows:

$$\text{cations}^{n+}_{(aq)} + \text{anions}^{n-}_{(aq)} + n\,H^+_{(aq)} + n\,H_2O_{(l)} \leftrightarrow \text{secondary solids-}n\,H_2O_{(s)} \quad (2.28)$$

The precipitation and redissolution of secondary minerals in sulfidic wastes may greatly influence the acidity and chemical composition of ground, surface and pore waters (Chap. 3). As a consequence, the amounts and types of secondary salts need to be determined in sulfidic mine wastes.

2.6.4
Coatings and Hardpans

The formation of secondary minerals not only influences the mine water chemistry but also impacts on potential water-rock reactions. For example, rapid precipitation of secondary minerals – during sulfide oxidation or carbonate dissolution – may coat or even encapsulate the acid producing or buffering mineral. Such coatings will make the mineral less susceptible to continued weathering and dissolution.

Prolonged precipitation of secondary minerals may occur at the surface or at a particular depth of tailings dams and waste rock piles. Such continuous precipitation results in the formation of laterally extensive or discontinuous surface or subsurface layers (Boorman and Watson 1976; McSweeney and Madison 1988; Blowes et al. 1991; Holmström et al. 1999; McGregor and Blowes 2002; Moncour et al. 2005; Alakangas and Öhlander 2006) (Fig. 2.9). Precipitated minerals include hydroxides (e.g. goethite, ferrihydrite, lepidocrocite), sulfates (e.g. jarosite, gypsum, melanterite), or sulfides (e.g. covellite), which fill the intergranular pores and cement the waste matrices.

In waste rock piles and tailings dams, secondary minerals typically precipitate below the zone of oxidation and at the interface between oxic and anoxic layers (Fig. 2.10). A distinct vertical colour change in the waste, from reddish-brown-yellow at the top to grey below, generally indicates the transition from an oxidized layer to reduced material. If the precipitation layer dries out and cements, it forms a so-called "*hardpan*". This layer acts as horizontal barrier to the vertical flow of pore waters. A hardpan may also form within the zone of oxidation at a depth where the pore water reacts with acid neutralizing carbonates. The pH of the pore water rapidly rises due to carbonate dissolution, and iron precipitates as iron hydroxides which cement the waste.

The formation of hardpans in sulfidic wastes can be induced in order to control sulfide oxidation. The addition of limestone, lime ($Ca(OH)_2$), magnesite ($MgCO_3$), brucite ($Mg(OH)_2$), or other neutralizing materials, just below the surface of sulfidic waste, will help to generate artifical hardpans or so-called "*chemical covers*" or "*chemical caps*" of gypsum, jarosite and iron hydroxides (Chermak and Runnells 1996, 1997; Ettner and Braastad 1999; Shay and Cellan 2000). Regardless whether the hardpan is naturally formed or chemically induced using neutralizing materials, a hardpan protects the underlying materials from further oxidation and limits AMD generation through various processes: (*a*) it prevents ingress of oxygenated ground and pore water into water saturated parts of the sulfidic waste; (*b*) it limits the movement of atmo-

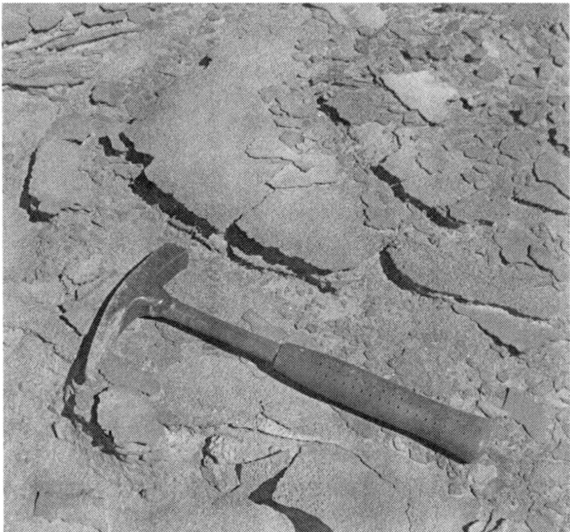

Fig. 2.9. Solid crusts of Fe-rich hardpans (hydrous ferric oxide) developed on stanniferous tailings, Jumna, Australia

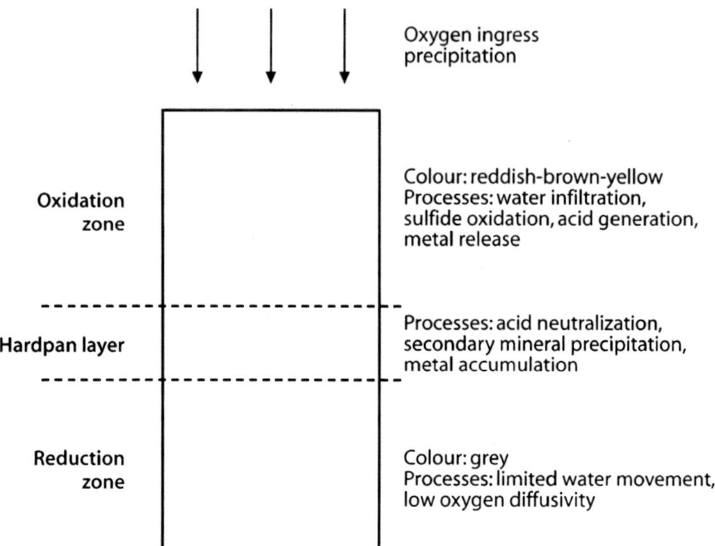

Fig. 2.10. Simplified diagram illustrating the formation of a hardpan layer in sulfidic wastes (after Jambor et al. 2000b). In this example, hardpan formation occurs at the water table between the saturated and unsaturated zone. A hardpan layer may also form within the unsaturated zone due to chemical reactions between an acidic leachate and a neutralizing layer

spheric oxygen through reactive unsaturated sulfidic wastes; (*c*) it reduces the waste's porosity; and (*d*) it accumulates heavy metals and metalloids through mineral precipitation, and adsorption and coprecipitation processes. However, elements not permanently fixed in insoluble minerals are susceptible to dissolution and mobilization back into pore waters. Such hardpans do not protect the sulfidic materials from further oxidation nor do they cause permanent sequestration of trace elements (Lottermoser and Ashley 2006).

2.7 Acid Generation Prediction

AMD generation can result in surface and ground water contamination that requires expensive water treatment and involves potential liability in perpetuity. An accurate prediction of the acid producing potential of sulfidic wastes is, therefore, essential. A prediction of acid generation requires a good understanding of the physical, geological, geochemical and mineralogical characteristics of the sulfidic wastes. Data acquisition for acid generation prediction includes the completion of

- geological modeling;
- geological, geochemical, mineralogical and petrographic descriptions;
- geochemical static and kinetic tests; and
- the use of computer models for oxygen movement and geochemical processes.

2.7.1 Geological Modeling

Geological modeling is a basic technique for assessing the acid generation potential of sulfidic wastes. It involves classification of the deposit and deduction of potential acidity problems (Table 2.6). The reasoning behind this method is that ore deposits of the same type have the same ore and gangue minerals and accordingly, the same acid producing and acid buffering materials (Kwong 1993; Plumlee 1999; Seal et al. 2000). However, the method has very limited application because it assumes that factors influencing acid generation such as pyrite surface area, abundance of sulfides or waste dump characteristics are constant for the mine sites and ore deposits being compared. The comparisons are very unreliable, yet they may provide some initial insight in the overall likelihood of acid generation. The technique may be applied to stratigraphically equivalent coal mines or ore deposits in volcano-sedimentary sequences. Thus, geological modeling and classification of an ore deposit is an initial crude step in ranking the deposit in terms of its potential to produce AMD.

2.7.2 Geological, Petrographic, Geochemical and Mineralogical Descriptions

A prediction on acid generation should begin well before sulfidic wastes are produced at mine sites. Preliminary evaluations can be performed as early as the exploration drilling and early mining of an orebody. Fundamental basic data for waste characterization and acid generation prediction include: existing lithologies; structural features;

Table 2.6. Ranking of some ore deposit types according to their AMD potential (after Kwong 1993)

Ranking	Ore deposit type	AMD potential
1	Sedimentary exhalative massive sulfide deposits; Coal	Most AMD prone
2	Volcanogenic massive sulfide deposits	
3	Epithermal gold deposits	
4	Mesothermal gold deposits	
5	Polymetallic vein deposits	
6	Calc-alkaline porphyry copper deposits	
7	Alkalic porphyry copper deposits	
8	Orthomagmatic chromium-nickel deposits	
9	Broken Hill type lead-zinc deposits	
10	Greisen tin deposits	
11	Kimberlites and lamproite diamond deposits	
12	Mississippi Valley type lead-zinc deposits	
13	Skarn deposits	
14	Carbonatite deposits	Least AMD prone

ore and gangue textures and mineralogy; particle size distribution; depth of oxidation; and whole rock geochemistry. Geological data such as pyrite content, geochemical analyses (S, C, CO_3, metals), and static test data can be used to construct a three-dimensional block model of different waste rock units prior to mining (Bennett et al. 1997).

Characterization of sulfidic waste materials involves mineralogical, mineral chemical and geochemical investigations (Lin 1997). Mineralogical observations using X-ray diffraction, optical microscopy, scanning electron microscopy, and transmission electron microscopy should note the size, shape, surface areas, degree of crystallinity, distribution, and oxidation state of sulfides, gangue minerals, and weathering products. Textural descriptions are also important as they can reveal protective encapsulation of sulfides in weathering resistant gangue minerals such as quartz.

Mineral chemical investigations using electron microprobe analyses demonstrate the abundance and siting of metals, metalloids and other elements, which may be mobilized during sulfide oxidation (Lu et al. 2005). The speciation, bioavailability and potential mobility of heavy metals in sulfidic wastes can be evaluated using partial and sequential extraction techniques (Ostergren et al. 1999; Dold 2003; Hudson-Edwards and Edwards 2005). Heavy metals may be present as cations: (*a*) on exchangeable sites; (*b*) incorporated in carbonates; (*c*) incorporated in easily reducible iron and manganese oxides and hydroxides; (*d*) incorporated in moderately reducible iron and manganese oxides and hydroxides; (*e*) incorporated in sulfides and organic matter; and (*f*) incorporated in residual silicate and oxide minerals. Thus, geological, petrographic, geochemical and mineralogical descriptions of sulfidic wastes provide important information on the nature and distribution of acid producing and acid buffering minerals, and on the mineralogical siting of metals and metalloids.

2.7.3 Sampling

The distribution of acid producing and acid consuming minerals is generally heterogeneous on micro- to macroscopic scales. Different ore lenses, coal seams and waste materials may represent acid producing or acid buffering units. Sulfidic wastes cannot be treated as a homogenous mass.

Waste samples can be obtained during exploration drilling and mining. However, representative sampling from drill cores is very difficult to achieve. The properties of vein deposits highlight the problems of sampling from drill cores for acid generation prediction (Dobos 2000). For example, a mesothermal gold vein deposit comprises of a rock mass, which is non-acid generating, and a series of acid generating veins with abundant pyrite (Fig. 2.11). Drilling and sampling of a composite over the entire drill section will yield a sample, largely comprising of the non-acid generating host rock. In contrast, blasting of this material will cause the rock to break along the veins, resulting in the exposure of a disproportionate amount of pyrite veins. If this mined material is dumped, it will generate more acid than the initially drilled and geochemically tested material. Therefore, geologically controlled sampling is most important in order to ensure that the analyzed samples are representative of the type and distribution of acid producing and acid buffering minerals (Dobos 2000). Otherwise, significant errors may occur when averages of static or kinetic test data are used: (*a*) to

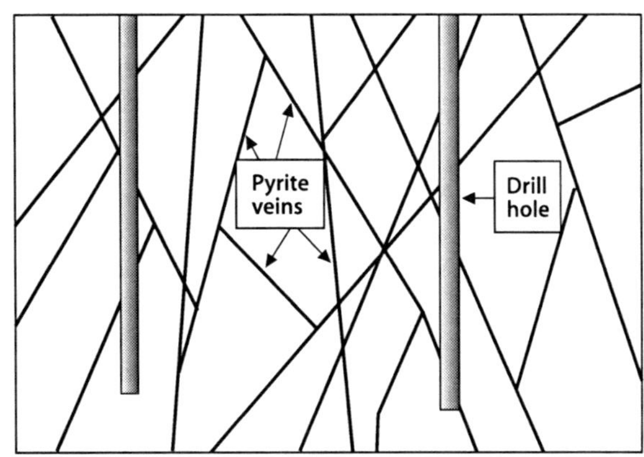

Fig. 2.11. Schematic diagram showing a macroscopic ore texture of a mesothermal gold deposit (after Dobos 2000). Acid producing pyrite veins are hosted by non-acid producing country rocks. The traces of drill holes into the veined rock are also indicated

predict the likelihood of acid generation from a particular waste pile; or (*b*) to forecast the composition of seepage waters emanating from waste dumps.

Waste rock piles and coal spoil heaps of historic mining operations commonly require characterization and acid prediction. It has been suggested that the most economic sampling strategy to adequately characterize existing waste rock piles is a homogeneous composite of 15 to 30 samples (Munroe et al. 1999; Smith et al. 2000). However, sulfidic waste piles, particularly those dumped some time ago, may have developed a vertical mineralogical and chemical zonation. Sampling restricted to dump surfaces will disregard sulfidic, partly oxidized or secondary mineral enriched wastes at depth. Hence, drilling may be required to obtain sample materials representative of the entire waste dump.

2.7.4 Geochemical Tests

Geochemical tests should not be conducted without detailed mineralogical and geochemical investigations of the material. Particularly, the acquisition of pure static and kinetic test data without a detailed knowledge of the mineralogical composition of the waste represents a waste by itself. Detailed procedures for various static and kinetic tests, and instructions on how to interpret them, are found in Morin and Hutt (1997) and Lapakko (2002). Laboratory methods for the geochemical analysis of environmental samples, including sulfidic wastes, are given by Crock et al. (1999).

2.7.4.1 *Static Tests*

Static tests are geochemical analyses of sulfidic waste which are used to predict the potential of a waste sample to produce acid. Details of these tests are documented in the literature (Sobek et al. 1978; Smith et al. 1992; Morin and Hutt 1997; White et al. 1999; Mitchell 2000; Jambor 2003). Static tests are empirical procedures, and there is a confusing array of tests to measure and to document acid production and acid neutralization. In addition, static tests and reporting conventions vary (North America: AP, NP, NNP, NPR; Australia and the Asia Pacific region: MPA, ANC, NAPP). Fortunately, static tests can be assigned to three major categories:

- *Saturated paste pH and electrical conductivity.* A representative crushed waste sample is saturated with distilled water to form a paste. The pH and electrical conductivity (EC) of the paste are determined after a period of equilibration (12 to 24 hours) (Morin and Hutt 1997). A pH value of less than 4 generally indicates that the sample is acid generating, and an EC value of greater than 20 $\mu S\ cm^{-1}$ indicates a high level of total dissolved solids in the waste's leachate. Paste pH and EC values of wastes and soils forming on waste rock dumps may change over time because sulfide minerals within the materials weather and release ions into solution and the materials are flushed by infiltration and runoff waters (Borden 2001).
- *Acid Base Accounting (ABA).* Acid Base Accounting refers to the numerical data used to predict acid generation. The three components of the ABA are: (1) determination of acid production; (2) determination of acid consumption; and (3) calculation of net acid production or consumption using the data from (1) and (2).
  1. *Determination of acid production.* The Acid Potential (AP), Acid Production Potential (APP), or Maximum Potential Acidity (MPA) tests establish the maximum amount of sulfuric acid produced from sulfidic wastes. This is measured by analyzing the sample for its sulfur content. For the MPA and APP, the weight per cent sulfur is then converted to kilograms of sulfuric acid per tonne of waste (MPA value in kg $H_2SO_4\ t^{-1}$ = wt.% S × 30.625). For the AP, the weight per cent sulfur is converted to kilograms of calcium carbonate per tonne of waste that would be required to neutralize the acidity (AP value in kg $CaCO_3\ t^{-1}$ = wt.% S × 31.25).
  2. *Determination of acid consumption.* The Neutralization Potential (NP), Acid Neutralizing Capacity (ANC) or Acid Consumption (AC) tests measure the amount of acid the sample can neutralize. This is determined by analyzing the acidity consumption of a sample in acid (HCl or H_2SO_4). Consequently, the tests establish the buffering capacity of a sample due to dissolution and weathering of gangue minerals, or in other words, the ability of a sample to neutralize acid generated from sulfide oxidation. The NP and ANC are determined by adding acid to a sample, and then back titrating with hydroxide to determine the amount of acid the sample has consumed. The ANC value is reported in the form of kilograms of sulfuric acid consumption per tonne of waste (kg $H_2SO_4\ t^{-1}$), whereas the NP value is given in the form of kilograms of calcium carbonate consumption per tonne of waste (kg $CaCO_3\ t^{-1}$).
  3. *Calculation of net acid production or consumption.* The Net Acid Production Potential (NAPP) represents the theoretical balance of a sample's capacity to generate acid. In contrast, the Net Neutralization Potential (NNP) gives the waste's capacity to neutralize any acid generated.
     3.1 *NAPP calculations* are based on the net acidity of samples (i.e. kilograms of H_2SO_4 per tonne of waste) (Environment Australia 1997). The NAPP is defined as being the difference between the Maximum Potential Acidity (MPA) and the Acid Neutralizing Capacity (ANC), whereby the MPA value is subtracted from the ANC value. A positive NAPP value indicates the sample should generate acid, whereas a negative value indicates the potential for acid neutralization.

$$NAPP = MPA - ANC$$

or

$$NAPP\ (kg\ H_2SO_4\,t^{-1}) = S\ (wt.\%) \times 30.625 - ANC\ (kg\ H_2SO_4\,t^{-1}) \qquad (2.29)$$

3.2 *NNP calculations* are based on the net neutralizing potential available in the samples (i.e. kilograms of $CaCO_3$ per tonne of waste) (Mitchell 2000; White et al. 1999; Skousen et al. 2002). The NNP or ABA is defined as being the difference between the Acid Potential (AP) and the Neutralization Potential (NP). In theory, the NNP value is the net amount of limestone required to exactly neutralize the potential acid-forming rock.

$$NNP\ or\ ABA = NP - AP$$

or

$$NNP\ or\ ABA\ (kg\ CaCO_3\,t^{-1}) = NP\ (kg\ CaCO_3\,t^{-1}) - AP\ (kg\ CaCO_3\,t^{-1}) \quad (2.30)$$

Theoretically, rocks with positive NNP values have no potential for acidification whereas rocks with negative NNP values do. In practice, a safety factor is applied and rocks with a significant positive NNP value are generally regarded as having no acidification potential (>+20 or +30 kg $CaCO_3\,t^{-1}$). Rocks with a significant negative NNP value (<−20 or −30 kg $CaCO_3\,t^{-1}$) are potentially acid generating. Materials with intermediate NNP values have uncertain acid generation potentials (−20 or −30 kg $CaCO_3\,t^{-1}$ < NNP < +20 or +30 kg $CaCO_3\,t^{-1}$).

Alternatively, the ratio NP/AP, known as the *Neutralization Potential Ratio* (*NPR*), or the ratio NP/MPA can be used as the criterion to evaluate the capacity of the material to generate AMD (Price et al. 1997; Skousen et al. 2002). Theoretically, a NP/AP ratio less than 1 generally implies that the sample will eventually lead to acidic conditions (Sherlock et al. 1995). A ratio greater than 1 is indicative that the sample will not produce acid upon weathering. In practice, a safety factor is applied and rocks with a NP/AP ratio greater than 2, 3 or 4 are non-acid generating, whereas samples with a NP/AP ratio less than 1 have a likely acidification potential (Price et al. 1997).

- *Net Acid Generation (NAG) or Net Acid Production (NAP).* The NAG test directly evaluates the generation of sulfuric acid in sulfidic wastes. It is based on the principle that a strong oxidizing agent accelerates the oxidation of sulfides. The test simply involves the addition of hydrogen peroxide to a pulverized sample and the measurement of the solution pH after 24 hours, when the oxidation reaction is thought to be complete (final NAG pH). If the NAG pH is below a critical value, then the sample has the potential to generate acid in the field (Schafer 2000). Variations of the NAG test procedure include the static, sequential and kinetic NAG test (Miller 1996, 1998a). A final NAG pH greater than or equal to 4.5 classifies the sample as non-acid forming. A final NAG pH result of less than 4.5 confirms that sulfide oxidation generates an excess of acidity and classifies the material as higher risk. The NAP test is similar to the NAG test and involves the addition of hydrogen peroxide and titration of the peroxide-sample slurry to a neutral pH using hydroxide. The amount of acidity consumed is reported in kilograms of calcium carbonate per tonne of waste (kg $CaCO_3\,t^{-1}$).

Results of NAG and NAPP tests will place a waste material into one of several categories including acid consuming (ACM), non-acid forming low sulfur (NAF-LS), non-acid forming high sulfur (NAF-HS), potentially acid forming low capacity (PAF-LC), and potentially acid forming high capacity (PAF-HC) (Table 2.7) (Miller 1996, 1998a). If a site contains PAF-HC or PAF-LC material, then kinetic test data need to be acquired, and AMD management practices have to be established (Miller 1996, 1998a). However, attention must also be given to NAF-HS and ACM material if they host soluble secondary minerals such as gypsum. Drainage from such materials may be neutral to alkaline but exceptionally saline, thereby exceeding water quality guidelines for sulfate. In addition, neutral to alkaline drainage waters may carry exceptionally high contents of metals such as zinc, molybdenum or cadmium and metalloids such as arsenic, antimony or selenium (Sec. 3.4.3).

The main advantage of these static tests is their simplicity, and most static tests can be perfomed at mine sites. However, the determination of the acid generating potential is not standardized. Also, the static tests are based on several assumptions and are, therefore, associated with many problems (Miller 1996; Morin and Hutt 1997; White et al. 1999; Paktunc 1999; Jambor 2000, 2003; Jambor et al. 2003; Weber et al. 2004):

- The tests use powdered or crushed samples for analysis which artificially increase the grain size and expose more mineral grains to reactions.
- Total sulfur analyses are not representative of the AP, APP or MPA because sulfur may also be present in non acid-producing sulfides or non-reactive or non acid-producing sulfates such as gypsum, anhydrite, barite or even organic material. It is possible to analyze for sulfidic sulfur contained in sulfides and for sulfate sulfur contained in secondary sulfate minerals (e.g. Yin and Catalan 2003). However, current bulk geochemical analytical techniques are not capable of distinguishing pyritic sulfur from sulfur present in acid-producing sulfates or in other sulfides that may or

Table 2.7. Typical classification criteria for sulfidic waste types (after Miller 1996, 1998a)

Waste type	Final NAG pH	Static NAG value [kg $H_2SO_4\,t^{-1}$]	NAPP [kg $H_2SO_4\,t^{-1}$]
Potentially acid forming high capacity (PAF-HC)	<4.5	>5	positive
Potentially acid forming low capacity (PAF-LC)	<4.5	≤5	positive
Non-acid forming (NAF)	≥4.5	0	negative
Acid consuming (AC)	≥4.5	0	<−100
Uncertain[a]	≥4.5	0	positive
Uncertain[a]	<4.5	>0	negative

[a] Samples which have conflicting NAPP and NAG results need further testing using mineralogical characterization and kinetic tests.

may not generate acid. Sulfur present in organic matter does not participate in acid generation (Casagrande et al. 1989).
- Framboidal pyrite is more reactive than euhedral forms due to the greater specific surface area (Weber et al. 2004). As a result, NAPP testing is biased by the rapid acid generating oxidation of framboidal pyrite prior to and during the ANC test.
- The possible coating of acid producing sulfides by secondary minerals is not taken into account, and it is assumed that the acid producing and acid consuming minerals will react completely.
- Organic carbon is oxidized by hydrogen peroxide during NAG testing which interferes with the acidity of the solution.
- The static tests do not allow the much slower acid buffering reactions of silicates to take place which, however, contribute only very minor amounts to the neutralization potential of sulfidic wastes (Jambor et al. 2000c).

Overall, static tests may under- or overestimate the acid production of a particular sample. As a result, numerous authors have proposed improvements and alternatives to existing static tests (e.g. Morin and Hutt 1997; Lawrence and Scheske 1997). Regardless of these modifications, static tests only predict the acid potential of individual samples and not of entire waste dumps. The tests are best used as rapid screening tools to assess the likelihood of acid generation from particular sulfidic wastes (Miller 1996, 1998a).

2.7.4.2
Kinetic Tests

Kinetic tests simulate the weathering and oxidation of sulfidic waste samples. They are generally used to follow up the findings of static testing. Kinetic tests expose the sulfidic waste over time, from several months to two to three years, to moisture and air (Smith et al. 1992; Morin and Hutt 1997; Mitchell 2000; Younger et al. 2002; Lapakko 2002; Munk et al. 2006). The experiments can be accelerated to simulate long-term weathering of waste materials in a shorter time frame. Water is thereby added to the waste more frequently than it would occur under normal field conditions.

Generally, kinetic tests involve the addition of water to a known quantity of waste. Leach columns and humidity cells are the most frequently used laboratory test techniques, whereby water is dripped or trickled onto one kilogram to one tonne of sample. The acid producing and acid buffering reactions are allowed to proceed, and the leachate is periodically collected and analyzed for its composition including pH and EC as well as sulfate, metal and metalloid concentrations. Mineralogical and geochemical characterization of the sulfidic waste has to be carried out prior to and after experimentation. In laboratory kinetic tests, relatively small samples are monitored under controlled conditions, whereas field kinetic tests monitor relatively large samples under less controlled conditions in large bins or drums (Fig. 2.12).

The main advantage of these simulated weathering techniques is that they consider the weathering rates of sulfides and gangue minerals. The tests can provide an indication of the oxidation rate, lag period for the onset of acid generation, and effectiveness of blending or layering of different wastes. The tests also provide data on the load of metals, metalloids and other elements in leachates and seepage waters from waste

Fig. 2.12. Drums used for the kinetic testing of sulfidic mine wastes at the Misima Island gold mine, Papua New Guinea. The tubes enable meteoric water and atmospheric gases to interact with the waste, and leachate can be collected from the drum base

disposal facilities. They thereby indicate the water quality in the short and long term.

Kinetic tests are not standardized, and a great number of kinetic test designs have been developed. Any interpretation of kinetic analytical results has to scrutinize experimental design, analytical techniques and local environmental conditions. Furthermore, the interpretation of kinetic data has to consider that the data can be in great contrast to the actual field data. The reason for such discrepancies may be due to the experimental designs which hardly resemble actual waste profiles where numerous variables such as oxygen diffusion, water infiltration, microbial populations, secondary mineral formation, changes in mineralogical composition, evolution of the surface state of sulfides, and other environmental conditions control AMD generation. Hence, several authors have evaluated laboratory kinetic tests for measuring rates of weathering and have proposed improvements and alternatives to existing kinetic tests (Cruz et al. 2001b; Hollings et al. 2001; Frostad et al. 2002; Kargbo and He 2004; Benzaazoua et al. 2004).

Kinetic field trials at the mine site have distinct advantages over laboratory tests. Most importantly, the tests permit accurate replication of the local climate and selection of appropriate sample material and volume (Smith et al. 1992; Morin and Hutt 1997; Bethune et al. 1997). In particular, field-based trial dumps allow the determination of acid generation parameters under actual field conditions. Small waste piles are constructed with an appropriate liner, and piezometers and lysimeters may be installed. Leachate, run-off and pore water compositions and volumes can then be investigated.

The trials reduce the inaccuracies resulting from small test samples and allow a more realistic assessment of the AMD processes and potential.

2.7.5 Modeling the Oxidation of Sulfidic Waste Dumps

Oxygen is essential for the oxidation of sulfides in waste dumps. Simple calculations can demonstrate that the availability of oxygen controls the oxidation rate of sulfidic waste (Gibson and Ritchie 1991). For example, a 50 t sulfidic waste rock pile has a sulfur concentration of 2 wt.%. The waste dump, therefore, contains 1 t of sulfur which will require, based on the stoichiometric ratio, 1.75 t of oxygen for its oxidation to sulfate. A 50 t waste pile with a porosity of 0.3 contains approximately 8×10^{-3} t of oxygen, which is only 1/200 of the 1.75 t needed for complete oxidation (Gibson and Ritchie 1991). Consequently, in order to accomplish complete oxidation of the waste, oxygen must travel into the heap from the atmosphere.

Indeed, the transport of oxygen to the oxidation sites is considered the rate limiting process in dumps and tailings deposits (Ritchie 1994a–c). The gas-phase transport of oxygen in waste dumps from the surface to the oxidation sites at depth occurs by: (*a*) diffusion (i.e. flow of oxygen induced by a gas concentration gradient); (*b*) convection (i.e. flow of air induced by wind action, barometric pressure changes, or thermal convection driven by the heat generated from the exothermic pyrite reaction); and (*c*) advection (i.e. flow of air induced by a thermal or pressure gradient) (Ritchie 1994a–c; Rose and Cravotta 1999). Minor amounts of oxygen may also be transported into the dump via liquid-phase diffusion and advection (i.e. flow of oxygen via infiltrating precipitation).

The relative contribution of diffusion, convection or advection to overall gas transport is dependent on a variety of parameters including the position of the waste within the dump, the component materials and minerals, and the way in which the dump has been constructed. Diffuse transport of oxygen through the gas-filled pore spaces is thought to dominate in unsaturated, newly built waste dumps (Ritchie 1994b; Aachib et al. 2004; Kim and Benson 2004). Uniform diffusion into such waste materials will result in oxygen profiles with horizontally flat oxygen concentration contours. Gas convection is limited to the edges of waste dumps, and since dump edges are a small fraction of the total dump volume, convection is disregarded in the modeling of the oxidation rate of pyritic waste dumps (Ritchie 1994b). However, convective gas flux has been reported from newly constructed waste dumps (Cathles 1994). In addition, localized convections have been observed in aged waste dumps, as indicated by high oxygen concentrations at depth and complex oxygen concentration profiles. The advective and convective modes of oxygen transport appear to predominate in porous waste dumps containing abundant coarse-grained rock fragments (Rose and Cravotta 1999). The diffuse mode of oxygen transport predominates in less permeable waste materials composed of small fragments.

The reactivity of a sulfidic waste pile and its oxidation behaviour in the long term can be described using the intrinsic oxidation rate (Ritchie 1994a–c). The intrinsic oxidation rate is calculated through a series of mathematical equations. These equations quantify the physical mechanisms which control the oxidation of a pyritic waste heap. For instance, the oxygen consumption rate represents the rate at which oxygen is consumed by the dump material (in units of kilograms of oxygen per cubic meter of waste per second; $kg^{-1} m^{-3} s^{-1}$ or mol $kg^{-1} s^{-1}$). The term quantifies the loss of oxygen from the pore space by oxidation reactions in the waste. A typical oxygen consumption rate value calculated for waste rock dumps is in the order of 10^{-8} to 10^{-11} $kg^{-1} m^{-3} s^{-1}$ (Bennett et al. 1994; Ritchie 1995; Hollings et al. 2001). In this model, it is assumed that oxygen is only consumed by pyrite. However, oxygen may also be consumed by the oxidation of other sulfides, native elements and organic matter. Furthermore, the sulfide oxidation rate is dependent on a large number of variables including temperature, pH, Fe^{3+} concentration, particle size distribution, mineral surface area, bacterial population, trace element substitution, degree of pyrite crystallinity and so forth. Finally, sulfide oxidation rates within a single dump appear to be variable; a dump may contain pockets of more highly oxidizing materials, particularly toward the dump edges (Linklater et al. 2005). Thus, these weathering models will need further refinement.

2.8 Monitoring Sulfidic Wastes

The recognition of sulfide oxidation does not necessarily require sophisticated equipment and measurements. In fact, some of the common indicators of sulfide oxidation can be recognized in the field:

- Abundant yellow to red staining on rocks and flocculants in seepage points, streams and ponds due to the formation of secondary iron minerals and colloids
- Sulfurous odours
- Unsuccessful colonization of waste materials by vegetation
- Abundant mineral efflorescences within and on exposed waste materials
- Increasing magnetic susceptibility due to the abundance of magnetic secondary iron oxides and carbonates
- Increasing waste temperature due to exothermic pyrite oxidation
- Decreasing oxygen concentration in pore gases due to oxygen consumption; and most importantly
- Decreasing pH, increasing EC, and increasing sulfate, metal (Cu, Zn etc) and major cation (Na, K, Ca, Mg) concentrations in drainage waters with time (Miller 1995)

The latter three indicators of sulfide oxidation are used to monitor sulfidic wastes. Sulfidic waste rock dumps, tailings dams and heap leach piles need monitoring in order to detect at the earliest point in time whether the waste material will "turn acid" and generate AMD. Also, rehabilitated waste repositories need monitoring to establish the effectiveness of the control technique used to curtail sulfide oxidation. The monitoring techniques are designed to identify the early presence of, or the changes to, any products of the acid producing reactions in sulfidic wastes. The products of the acid producing reactions are usually quantified by one or more of the following parameters (Hutchison and Ellison 1992):

- Water analyses of dissolved contaminant concentrations and loads (Sec. 3.8)
- *Temperature profiles.* Pyrite oxidation is an exothermic reaction, and the effects of heat generation can be assessed by remote or in situ sensing. Remote sensing using

thermal infrared spectrometry is best suited for identifying zones of high acid generation of exposed sulfidic materials such as open pits (Hutchison and Ellison 1992). However, remote sensing or airborne geophysical techniques are not appropriate for detecting the onset of acid generating conditions in covered or piled sulfidic wastes. In situ temperature sensing is used to detect temperatures and temperature gradients in sulfidic waste rock piles or tailings. Temperature sensitive electrical probes (thermistors) or thermocouples are placed into a series and lowered down installed PVC pipes. Rapid increases in temperature profiles are symptomatic of the exothermic oxidation reactions of pyrite.

- *Oxygen concentration within gas pores.* Sulfide oxidation reactions in the unsaturated zone of sulfidic waste piles and tailings are oxygen consuming. Hence, the depletion of oxygen within the gas phase can be indicative of sulfide oxidation, and a knowledge of pore gas compositions will allow an evaluation of sulfide oxidation reactions within the waste pile. First, pore gas sampling using appropriately constructed probe holes is performed (Lundgren 2001). Next, oxygen analyzers determine the oxygen concentration and finally, oxygen concentration profiles are acquired. Oxygen is generally supplied to the interior of a fine-grained waste rock pile by diffusion, which is induced by concentration differential from the atmosphere. The concentration profile within such a pile will show decreasing oxygen with increasing depth below the surface. Under these conditions, active oxidation zones associated with acid generation can be detected as they result in sharp oxygen partial pressure gradients.

2.9 Environmental Impacts

The visible environmental impacts of sulfidic waste dumps and spoil heaps include waste erosion and a depauperate or even absent flora. For example, the surface of coal spoil heaps with their inherent salt content, sodicity of the waste and pronounced salt crust commonly does not support any vegetation (Bullock and Bell 1997). The sparse or non-existent vegetation is also caused by a lack of soil nutrients (N, P) and organic matter, as well as the potentially high salinity and acidity and high metal content in the surface layers of metalliferous wastes (Fig. 2.13). The lack of vegetation on sulfidic wastes increases erosion rates. The erosion processes exacerbate the "moonscape appearance" of these wastes and increase the areas affected by waste particles. Reactive and unreactive waste particles are transported into soils and streams and may affect areas many kilometers downstream of the mine site (Pirrie et al. 1997; Hudson-Edwards et al. 1999; Loredo et al. 1999; Cidu and Fanfani 2002) (Case Study 2.1; Scientific Issue 2.2). The area impacted by mine wastes is then no longer confined to the immediate environments of the waste.

If no rehabilitation of waste dumps occurs, it may be several decades before slow natural revegetation of adapted local flora will eventuate, with grasses often appearing to have a pioneer function in such successions (Ashley and Lottermoser 1999a). The voluntary colonization of sulfidic wastes by native vegetation is inhibited by the harsh chemical and physical conditions in the substrate (Bordon and Black 2005). Bioaccumulation of metals and metalloids may occur in plants growing on metal and metalloid enriched substrates. Grazing animals may consume such contaminated grasses and soils (Ashley and Lottermoser 1999b; Loredo et al. 1999). This may lead to

Fig. 2.13. Partly revegetated sulfidic waste rock dump. A thin soil layer placed on the sulfidic waste has encouraged natural revegetation. In contrast, vegetation did not develop on that part of the dump without a soil cover

potential health problems for humans from the long-term exposure to contaminants in locally farmed food products. Bioaccumulation of metals and metalloids can be pronounced in certain plant species. Such plants are genetically tolerant to metal-rich substrates and have various strategies to cope with the high metal concentrations in these environments. The strategies include the preferential accumulation of heavy metals and metalloids in the plant tissue. Plants with particular capabilities to accumulate large amounts of metals in their tissue are referred to as "hyperaccumulators". They may be of possible use in the extraction of metals from low-grade ores and wastes (Scientific Issue 4.1; Sec. 4.8).

2.10 Control of Sulfide Oxidation

Uncontrolled sulfide oxidation can lead to the generation of AMD. In order to prevent sulfide oxidation and the generation of AMD, appropriate control strategies are needed. Strategies for the control of sulfide oxidation require the exclusion of one or more of the factors that cause and accompany oxidation, that is, sulfide minerals, bacteria, water, iron and oxygen. The aim of these methods is to reduce the interaction between the waste and the other reactants. Established control strategies include barriers (i.e. wet and dry covers), selective handling and isolation, co-disposal and blending with other materials, addition of organic wastes, and bacterial inhibition (Environment Australia 1997; Miller 1998b; Evangelou 1998; SMME 1998; Parker 1999; Brown et al. 2002). More technologically advanced and innovative strategies involve induced hardpan formation, grouting or mineral surface treatments (Scientific Issue 2.3, Fig. 2.14). Both established and innovative sulfide oxidation control strategies are generally designed to induce one or more of the following:

- Exclusion of water
- Exclusion of oxygen
- pH control
- Control of Fe^{3+} generation
- Control of bacterial action
- Removal and/or isolation of sulfides

No single technology is appropriate to all mine site situations, and in many cases a combination of technologies offers the best chance of success; that is, a "tool box" of technologies should be applied. Reducing oxygen availability, which is generally achieved using dry or wet covers, is the most effective control on the oxidation rate. These covers are surface barriers designed to limit the influx of oxygen and surface water to the waste body.

Alternatively, depending on the mineralogical characteristics of the tailings at a mine site, the best environmental result would be to separate different fractions of the sulfidic waste during mineral processing (Mitchell 2000). Selective concentration of pyrite or pyrrhotite during mineral processing would produce a high-sulfide concentrate. The sulfides could then be properly disposed of or used for the production of sulfuric acid.

Finally, the methods currently applied to control sulfide oxidation are not yet proven to securely prevent AMD development in the long term. Global climate change will lead to changing rainfall patterns and weathering processes at individual mine sites. In some cases, the applied control technique may only delay the onset of acid generation. Therefore, the following control techniques may only represent the first step to more sophisticated acid prevention techniques.

2.10.1 Wet Covers

Submerging sulfidic waste (i.e. tailings or waste rocks) under water is an effective counter to acid generation. The maximum concentration of dissolved oxygen in water is three orders of magnitude lower than that found in the atmosphere. The low solubility of oxygen in water and the slow transport of oxygen in water (i.e. its diffusivity) also reduces the transport of oxygen into a mass of sulfidic waste. Once the available oxygen in water is consumed, an anoxic environment is established, and the rate of sulfide oxidation is reduced because the rate of oxygen replacement is relatively slow. In addition, erosion is reduced, and the formation of reducing conditions fosters the growth of sulfate reducing bacteria which will immobilize dissolved metals as sulfides. However, oxygen can enter surface waters via vertical mixing due to the orbital motions of wind-induced waves and the turbulent mixing caused by breaking waves. Hence, the water cover has to be of substantial depth since surface waters are in constant contact and exchange with atmospheric oxygen.

Case Study 2.1. Sulfidic Mine Wastes and Their Environmental Impacts at Historical Metalliferous Mine Sites in the New England Area, Australia

Introduction

Metalliferous mine sites can represent important sources of acidity and metal pollutants in watercourses and soils, with consequent degradation of local ecosystems. This case study reports AMD caused by mining carried out during the late 19th and early 20th centuries at metalliferous mine sites in the New England area of northern New South Wales, Australia (Ashley and Lottermoser 1999a,b; Lottermoser et al. 1999). Six abandoned base metal mine sites with differing geologic, physiographic, climatic and floral regimes were investigated to provide information on the environmental behaviour of heavy metals and metalloids originating from different primary metal and acid-producing sources. Each site has the common characteristics of decomposing sulfidic waste material, acid mine drainage (AMD), limited acid buffering capacity of host rocks, and degraded soils, streams and vegetation.

Sulfidic Ore and Waste Dumps

Largely unconsolidated and unvegetated dumps of sulfidic ore and waste material occur at each mine site. They have high heavy metal contents (wt.% levels As, Cu, Fe, Pb, Zn) and are the major point sources of AMD. Ore heaps, waste dumps and tailings deposits consist of mixtures of weakly to strongly sulfidic rock and non-mineralized host rock material. Exposure of sulfide-bearing material to contemporary weathering processes has led to the dissolution of sulfide and gangue minerals, formation of post-mine oxidation minerals as efflorescences and crusts, as well as AMD with pH values as low as 2.2. Heavy metal values in the AMD leachates are largely derived from sulfide oxidation. Many post-mine oxidation minerals, particularly sulfates and arsenates, also influence the mine water chemistry. These minerals dissolve during rain events and subsequently re-form on drying.

Drainage Systems

Where AMD enters streams, there is a drop in pH, and most aquatic and bankside plant communities disappear. Green filamentous algae are common colonizing species in AMD-affected streams. Chemical mobilization from ore/waste/tailings dumps is most pronounced for arsenic, cadmium, copper and zinc, whereas lead remains largely immobile. Physical dispersion of secondary metal-bearing minerals (e.g. jarosite-bearing phases, scorodite, Fe oxides) and chemical precipitation of metals into stream sediments in the form of clays, iron and manganese oxides and organic matter are significant. Consequently, many stream sediments have metal concentrations exceeding background values by one to four orders of magnitude. Upon entering larger drainage systems the metal concentrations of waters and stream sediments may be diluted to background values, although at some locations, anomalous metal concentrations can be traced for several kilometres downstream.

Soils

Soil samples taken close to mine waste dumps and processing sites have grossly higher average heavy metal and metalloid values than background samples, commonly by two to three orders of magnitude. Highly contaminated soils may contain heavy metals incorporated into jarosite-type phases, scorodite, iron oxides, manganese oxides, clays, and organic material.

Deep-water disposal of sulfidic waste has been popular and successful in Canada for some time. Possible disposal sites are numerous and readily available there, and annual precipitation exceeds evaporation (Pedersen et al. 1998). The sulfidic wastes are thereby placed in natural or engineered water covers, including former open pits (i.e. in-pit disposal) (Mitchell 2000).

Case Study 2.1. *Continued*

Vegetation and Algae

The contaminated sites are commonly devoid of vegetation or support a depauperate assemblage, with consequent marked erosion. At mine and smelter sites, bioaccumulation and biomagnification processes of heavy metals occur in grasses and other plants but patchy recolonization of contaminated sites by metal-tolerant species is evident. Ashed plant material from contaminated sites shows heavy metal accumulation of one to three orders of magnitude compared to the same plants on background sites. Filamentous green algae flourish in AMD emanating from ore and waste dumps, shafts and adits and in drainage systems affected by acidic waters. The algae are tolerant of low pH waters with high metal, metalloid and sulfate concentrations. They have the capacity for heavy metal and metalloid accumulation, and dried material has been found to contain wt.% levels of heavy metals and metalloids.

Pollution

Environmental impacts common to all sites include: (*a*) unconstrained weathering of sulfidic mine waste with the production of AMD; (*b*) local disappearance of aquatic life and vegetation; (*c*) locally confined heavy metal pollution and acidification of streams and soils; (*d*) downstream heavy metal pollution of local drainage systems; and (*e*) bioaccumulation and biomagnification of heavy metals in aquatic algae and terrestrial plants. Bioaccumulation processes of heavy metals indicate a potential for heavy metal accumulation in the food chain of the surrounding terrestrial ecosystems, including grazing animals.

Although mining ceased many decades ago at all investigated sites, contamination of local streams by AMD remains and will continue due to the high sulfide and metal content of the exposed ore and waste dumps. In addition, the presence of post-mine oxidation minerals in topsoils, stream sediments, and ore and waste dumps creates additional contamination problems because some of the phases act as temporary storage for acidity and heavy metals and may release these upon dissolution.

The heavy metal abundances of many soils and stream sediments exceed soil quality criteria and contaminated site guidelines at certain sites by up to several orders of magnitude. Mine seepage waters contain heavy metal values well in excess of recommended concentrations in drinking waters for domestic and stock use. However, there is no significant contamination to major river systems due to massive dilution of AMD effluents.

The major concerns with subaqueous disposal are to achieve stagnant anoxic conditions and to maintain complete and continuous water saturation. Sulfidic waste should not be exposed to water containing oxygen. Such water can be moved physically to the bottom of the water layer, as a result of temperature differences or wind. In arid and semi-arid regions, the wet cover control technique is not an option because deep water bodies are rare, and there is no sufficient year-round supply of water that would ensure that the waste remains in a permanently saturated condition. Drying out of a saturated waste will lead to sulfide oxidation and AMD generation. Therefore, wet covers are unsuitable for arid and semi-arid regions. Moreover, the subaqueous deposition of partially or completely oxidized sulfidic materials is not an option. These wastes contain soluble secondary minerals which will dissolve and release sulfate, metals and metalloids when immersed in water (Li and St-Arnoud 1997).

Rapid flooding can be applied to prevent AMD from developing in underground mines and open pits. In fact, flooding has been successful for mine workings where

Scientific Issue 2.2. Trace Metal Release from Historical Smelting Slags

Characteristics of Smelting Slags

Historical copper and lead smelters, particularly those operating during the late 19th and early 20th century, produced large volumes of slag dumps. Most slag dumps are largely in the same form as it existed when smelting stopped (Fig. 2.6). The slags are known to be highly heterogeneous and mineralogically diverse materials. They may contain relict ore, gangue and flux minerals and comprise of variable amounts of glass and crystallized minerals. The composition of slag depends on the composition of the ore and the fluxes used. Thus, historical base metal smelting slags display a range in bulk chemical composition with major concentrations (wt.%) of SiO_2 and $Fe_2O_{3(total)}$, and lesser amounts of Al_2O_3, TiO_2, MnO, CaO, MgO, Na_2O, K_2O, P_2O_5 and S (e.g. Lottermoser 2002; Piatak et al. 2004). In addition, historical base metal smelting slags are commonly characterized by elevated metal and metalloid contents as a result of inefficient metal recovery technologies (wt.% levels of Cu, Pb and Zn; Parsons et al. 2001; Lottermoser 2002). The metals are thereby tied up within the silicate glass and various slag phases and minerals, including native elements, sulfides, oxides, hydroxides, chlorides, carbonates, sulfates, arsenates, and silicates (Ettler et al. 2001).

Weathering of Slags

The presence of secondary minerals on slag deposits and laboratory experiments have shown that these historical slags are not chemically inert but undergo weathering and leaching processes, thereby releasing metals to pore and seepage waters (e.g. Parsons et al. 2001; Piatak et al. 2004; Ganne et al. 2006). The slags are undergoing contemporaneous reaction with air and rainwater. The weathering results in the release of metals and metalloids from the different slag phases. The released elements either exit from the dumps into surface seepages and ground waters, or precipitate as an array of weathering related secondary minerals (Lottermoser 2005). The release of metals from slag depends on the chemical and physical properties of the pyrometallurgical wastes. Initially, the stability of the slag phases and minerals to weathering processes determines whether elements can be released. The mobility of metals from slag is also a function of the texture, for example, the degree to which reactive glass or minerals such as sulfides are exposed or encapsulated. Furthermore, coprecipitation and adsorption/desorption reactions as well as solubility equilibrium with secondary minerals will control dissolved metal concentrations in the porewaters of slag dumps (e.g. Parsons et al. 2001; Ettler et al. 2003; Piatak et al. 2004). Prevailing porewater compositions (e.g. Eh, pH of fluids present in the pore network of the slag dumps) will also affect the rate of weathering and therefore the rate of metal release. Also higher proportions of sulfides to carbonates in slag wastes favour lower pH conditions in pore and seepage waters, the more acid fluids leaching sulfides and silicates and mobilizing trace metals (Kucha et al. 1996). The abundance of sulfides and carbonate minerals in smelting slags and the texture of individual slag phases (e.g. the possible encapsulation of sulfides in other primary slag phases) will determine acid generation and the pH of leachates emanating from slag dumps. Hence, seepage waters of slag deposits are known to possess near-neutral to acid pH values.

Weathering and leaching of metalliferous smelting slags are accompanied by the mobilization of metals, metalloids, alkali earth elements and sulfate into pore and seepage waters. Evaporation of the seepage waters emanating from slag dumps may cause the precipitation of mobilized elements and compounds and leads to their temporary fixation in secondary soluble sulfate minerals (Fig. 2.8). Dissolution of these efflorescences during the next rainfall and flushing event and associated Al^{3+} and Fe^{3+} hydrolysis contributes to the acidification and metal and sulfate contamination of local waterways. Thus, historical smelting slag dumps represent long-term sources of metal pollutants to local ground and surface waters.

they are located below the water table. The submergence of underground workings and filling of open pits can eliminate atmospheric oxygen and curtail acid generation reactions.

Scientific Issue 2.3. Coating Technologies for Sulfidic Wastes

Coating Technologies

The reaction of pyrite with atmospheric oxygen is the main cause of AMD. Thus, if one can place a barrier between the two reactants, pyrite oxidation is halted. Coating sulfide grains with relatively insoluble precipitates would protect the sulfides from further oxidation and dissolution. Such natural coating has been observed during the oxidation of pyrite in carbonate buffered solutions whereby ferric hydroxides formed on pyrite surfaces (Nicholson et al. 1990). The hydroxides formed a barrier between pyrite and oxygen and limited pyrite oxidation.

Coating technologies are based on the natural encapsulation of sulfides by insoluble mineral precipitates. The techniques allow chemical reactions to take place between reagents and the sulfide minerals. Various chemical reagents such as sodium acetate, sodium metasilicate or potassium orthophosphate have been evaluated for the coating of pure pyrite and pyrrhotite samples and pyritic coal waste (Huang and Evangelou 1994; Georgopoulou et al. 1996; Zhang and Evangelou 1998; Fytas and Evangelou 1998; Evangelou 2001). The coating technology aims to produce relatively insoluble phases on sulfide surfaces. The coatings have to be insoluble and stable in order to protect the sulfides from further oxidation and acid production.

Phosphate Stabilization

Phosphate induced stabilization is an emerging technology that uses solid or liquid phosphate products to decrease the mobility and bioavailability of metals and radionuclides in contaminated soils, ground waters and sulfidic mine wastes. The principle of phosphate stabilization is based on the fact that the addition of phosphate compounds such as apatite results in the formation of new phosphate phases. These phases have low solubilities and remain stable on a geological time scale (Hodson et al. 2000). In particular, liquid phosphate stabilizers such as potassium orthophosphate are known to create stable iron phosphate coatings around pyrite and pyrrhotite grains. These coatings armour and encapsulate the sulfides and prevent them from further oxidation. Phosphate is the preferred choice of chemical because phosphate minerals of many metals (Cd, Cu, Ni, Pb, U, Zn) have exceptionally low solubilities. In addition, the minerals are stable over a wide range of Eh and pH conditions found in the natural surface environment. Therefore, the formation of phosphate minerals in metal contaminated soils, sediments, waters and wastes immobilizes metals. The metals are prevented from leaching into waters and interacting with the biosphere.

In the phosphate coating technique, sulfides are treated with a weak hydrogen peroxide solution (Georgopoulou et al. 1996; Fytas and Evangelou 1998). The role of hydrogen peroxide is to induce oxidation of a small quantity of Fe^{2+} to Fe^{3+}. Soluble phosphate ions are added in the form of orthophosphate or soluble phosphate fertilizers which results in the formation of Fe^{3+} phosphate phases on the surface of the oxidizing sulfides. These phosphates are relatively insoluble, protecting the sulfide from reactions with oxygen. The phosphate coating technology appears to be cost effective and feasible in preventing pyrite oxidation. However, these innovative techniques are still at the exploratory stage, are based on laboratory experiments (Harris and Lottermoser 2006a,b) . Field trials on polymineralic and partly oxidized sulfidic mine wastes remain to be conducted in order to prove that phosphate stabilization is a viable control strategy for sulfide oxidation.

2.10.2 Dry Covers

Capping the sulfidic wastes with a thick layer of solid material is another effective counter to acid generation. Such dry covers reduce the oxygen flux and water flow into the underlying sulfidic waste. By limiting the amount of oxygen entering the waste, the oxidation reaction can be slowed (Harries and Ritchie 1987). Likewise, by reducing the flow of water into the waste rock, the quantity of contaminated drainage can be reduced.

Fig. 2.14. Scanning electron microphotograph of phosphate coatings on pyrite and chalcopyrite. In a laboratory experiment, soluble phosphate ions were added to a polysulfidic waste. Formation of phosphate phases occurred, coating the surfaces of pyrite and chalcopyrite. These phosphates protect the sulfides from further oxidation (Photo courtesy of D. Harris)

Dry covers are constructed from low hydraulic conductivity solids. Materials used for dry covers include low-sulfide waste rock, oxide waste, clay subsoils, soils, organic wastes, and neutralizing materials (e.g. limestone, lime, dolomite, brucite, kiln dust). The solid materials are placed on the crown and sides of sulfidic waste repositories. Prior to their use, the cover materials have to be characterized for their hydraulic conductivity and evaluated for their capacity to minimize oxygen and water transfer into the waste. At their simplest, a dry cover usually consists of a layer of clay (~1 m thick), which has been compacted to give low hydraulic conductivity that allows very little infiltration. At sites, where the supply of clay limited, compaction of coarser-grained cover materials or benign mine wastes may result in the formation of a low permeability seal.

Dry covers range from simple clay barriers to complex, composite covers. The latter types have a number of layers. A possible design may have the following sequence from top to bottom:

- A soil/rock layer – which retains moisture, acts as a substrate for vegetation, and prevents erosion
- A coarse-grained layer – which provides lateral drainage for any infiltration

- A compacted clay layer (at least 30 mm thick) – which creates a low air void content, reduces the cover's permeability to water, and lowers the diffusion rate of oxygen into the waste
- A coarse-grained layer – which reduces the contact of capillary saline waste waters with the protective cover, and prevents the precipitation of secondary salts at or near the surface of the dry cover
- A compacted layer of acid buffering materials such as lime – which minimizes reaction of the waste with the overlying layers, and promotes the development of a chemical cap

In an arid region with little vegetation, the top soil layer acting as a substrate for vegetation may be replaced with either a rock cover (so-called "riprap") or a layer of coarse-grained material that will reduce erosion.

Non-conventional dry capping solutions include the use of epoxy resins, chemical caps (i.e. chemically induced surficial hardpan layers), wood chips, bark, municipal solid waste compost, sewage sludge, peat, pulp and paper mill residues, grouts, fly ash mixtures, and non-acid generating or low-sulfides tailings, some of which have been applied with variable success (Elliott et al. 1997; SMME 1998; Xenidis et al. 2002; Bussière et al. 2004: Forsberg and Ledin 2003, 2006; Pond et al. 2005; Hulshof et al. 2006). In addition, permafrost has been used in cold climates as a sulfide oxidation control strategy (Scientific Issue 2.1).

The construction of effective dry covers has to consider the climatic conditions at the mine site. Depending on the prevailing climate, dry covers are either designed: (*a*) to maximize run-off using *unsaturated covers*; (*b*) to store relatively large volumes of infiltrating water for long periods of time using *saturated covers*; or (*c*) to store relatively large volumes of infiltrating water for short periods of time using *sponge covers*.

2.10.2.1
Unsaturated Covers

In areas where evaporation exceeds rainfall (semi-arid to arid), only unsaturated dry covers can be used. Unsaturated covers comprise a variety of geological materials (e.g. alluvium, topsoil, oxide waste). They contain a compacted fine-grained layer or low permeability clay seal, and they may have a capillary break of coarse-grained material and a layer of acid buffering materials (Fig. 2.15a). The covers are designed to maximize rainfall run-off and to minimize water infiltration and oxygen diffusion into the waste. The cover is topped with a loose soil or benign waste layer, needed to promote the establishment of vegetation. However, a relatively thin top layer means that trees need to be removed regularly to prevent roots penetrating and damaging the layer design and allowing access of oxygen to the sulfidic waste.

2.10.2.2
Saturated Covers

For mine sites with a wet climate, water saturated covers prevent infiltration of oxygen to potentially acid generating materials. The capping consists of carefully designed

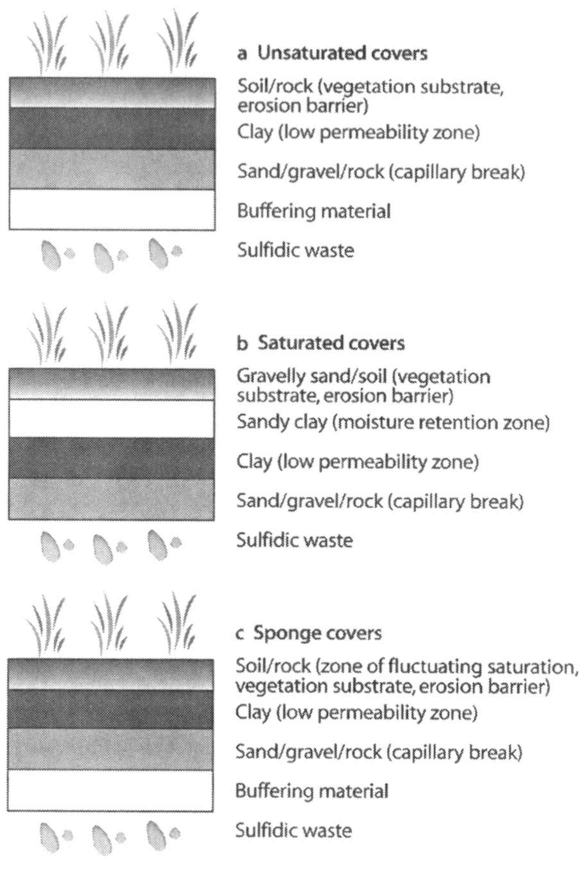

Fig. 2.15. Schematic cross-sections illustrating complex dry cover designs; **a** unsaturated covers; **b** saturated covers; **c** sponge covers

layers of soil and clay, which maintain a saturated layer throughout the year, with the water being provided by natural rainfall. The basic design involves a medium-grained material such as sandy clay with medium hydraulic conductivity underlain by fine-grained materials such as clay with low hydraulic conductivity (Fig. 2.15b). The layer of sandy clay is designed to hold water from infiltrating rainfall and to act as a water reservoir keeping the pores close to saturation; that is, the layer acts as a moisture retention layer. The clay layer may be compacted or uncompacted. Capillary suction forces prevent drainage of this layer with low hydraulic conductivity. A coarse-grained layer of rock, below the clay and at the base of the cover, drains first, and provides a capillary break to the movement of any AMD waters rising from the sulfidic material below. An additional coarse-grained layer may also be installed above the clay layer in order to reduce evaporation of the clay layer. At the surface, a layer of gravelly sand/soil is placed above the sandy clay zone. The soil is not only substrate for the vegetation but also protects the underlying cover from erosion.

A good saturated cover promotes run-off and maintains a high degree of water saturation within the sandy clay layer (Taylor et al. 1998). Transport of oxygen in the pores of this saturated layer is then governed by the low solubility and slow transport of oxygen in water rather than air. This in turn limits the movement of oxygen into the sulfidic waste.

Saturated covers are used in humid, wet climates where the cover remains saturated due to the high rainfall. The Rum Jungle uranium mine is an example where the application of a saturated cover system on sulfidic waste rocks has had limited success (Harries and Ritchie 1988; Bennett et al. 1999). The cover system was particularly effective in reducing the oxygen flux during the wet season; however, during the dry season the clay seal cracks resulting in AMD release and environmental impacts downstream (Figs. 2.16, 2.17). Thus, clay as part of a cover design may work well in wet climates but not necessarily in dry climates or seasonal climates due to drying, shrinking and cracking of the clay seal. Also, in many mining districts, the soil profiles can be notably deficient in clays. Consequently, the lack of suitable cover materials such as clays makes the construction of saturated covers as oxygen diffusion barriers impossible.

2.10.2.3
Sponge Covers

Sponge covers or so-called "store-release" covers are suitable for climates with distinctly seasonal rainfall (Williams et al. 1997; Currey et al. 1999). The covers are designed to store water in an upper cover layer (Fig. 2.15c). An irregular topography prevents surface run-off, and much of the drainage flows into the waste. The porous, loose top layer becomes saturated with water during a precipitation event. It then functions as an oxygen ingress barrier for the underlying sulfidic waste. The barrier uses the low solubility and slow transport of oxygen in water, reducing oxygen ingress in the same manner as a water cover does. Percolation of water into the waste is limited because the majority of the stored water is removed through evapotranspiration. In fact, vegetation plays a significant role in using and pumping water from these covers (Williams et al. 1997; Currey et al. 1999). The pumping action of plants prevents the stored water from infiltrating the underlying sulfidic waste. Nonetheless, cover failures may still be possible. Prolonged droughts or bushfires may cause significant die-off to plants, and subsequent infiltration would lead to a significant flow through and out of the waste materials (Dobos 2000).

2.10.3
Encapsulation, In-Pit Disposal and Mixing

Mining of sulfidic ores generally produces wastes with different acid generation potential. Selective handling of these different waste types allows the construction of waste rock dumps according to their acid generation potential (Cravotta et al. 1994; Environment Australia 1997). The disposal practice may utilize the buffering capacity of any benign waste to control acid production. Potentially acid generating material

Fig. 2.16. White's waste heap, Rum Jungle uranium mine, Australia. The sulfidic waste rock dump has a saturated cover design which comprises compacted clay, sandy clay loam and gravelly sand on top. The crown of the waste cover is covered with grass while the sides of the waste pile did not develop a complete vegetation cover

Fig. 2.17. Scalded seepage area at the base of White's waste heap (Fig. 2.16). The clay seal has cracked, and water infiltration into the dump has increased since the installation of the cover in 1984. Saline, acid seepages emanate from the base of the dump and precipitate abundant sulfate and iron oxyhydroxide efflorescences

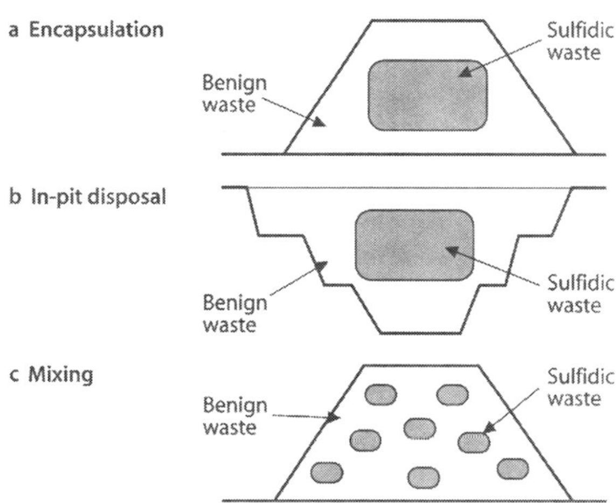

Fig. 2.18. Control of sulfide oxidation in mine waste using **a** encapsulation, **b** in-pit disposal, and **c** mixing techniques (after Environment Australia 1997)

is thereby enclosed in non-reactive benign material such as oxide waste or even neutralizing material (Fig. 2.18a) (i.e. encapsulation method). Alternatively, the waste is backfilled into open pits (Fig. 2.18b) (i.e. in-pit disposal method) (cf. Sec. 4.5). The sulfidic waste needs to be placed below the post-mining ground water table (cf. Fig. 4.6). Mixing highly sulfidic and benign wastes represents an additional disposal option (Fig. 2.18c). Encapsulation and mixing practices do not usually prevent sulfide oxidation and acid generation unless additional control measures are applied. However, the techniques may reduce AMD intensity substantially.

2.10.4 Co-Disposal and Blending

Sulfidic waste may also be blended and/or co-disposed with benign or alkaline material. Co-disposal refers to the mixing of coarse-grained waste rock with fine-grained tailings or coal washery wastes (Williams 1997; Wilson et al. 2000; Rensburg et al. 2004). Such a disposal practice has distinct advantages. It allows filling of the large pores of waste rock with fine tailings. As a result, the hydraulic properties of the wastes are altered, the water retention and saturation is increased, and the oxygen transfer into the waste rock is decreased. Sulfide oxidation can be curtailed.

Blending is generally conducted in conjunction with other control measures such as dry covers. Blending refers to the addition of alkaline material, which is used to raise the neutralization potential of the mine waste. The objectives are: (*a*) to balance the acid neutralization and acid generation potentials; (*b*) to minimize the risk of AMD; and (*c*) to immobilize any soluble or potentially soluble metals and metalloids as insoluble or sparingly soluble sulfates, carbonates and hydroxides. Neutralizing materials are mixed with acid producing waste. The neutralizing materials may be limestone ($CaCO_3$) or lime (CaO) as well as any acid buffering waste produced at the mine site. The alkaline materials are applied to ensure that the metals are immobilized permanently by converting them into sparingly soluble minerals such as sulfates, carbonates and hydroxides:

$$CaCO_{3(s)} + H^+_{(aq)} + SO^{2-}_{4(aq)} + Pb^{2+}_{(aq)} \rightarrow PbSO_{4(s)} + HCO^-_{3(aq)} \quad (2.31)$$

$$CaCO_{3(s)} + Pb^{2+}_{(aq)} \rightarrow PbCO_{3(s)} + Ca^{2+}_{(aq)} \quad (2.32)$$

$$CaO_{(s)} + Zn^{2+}_{(aq)} + H_2O_{(l)} \rightarrow Zn(OH)_{2(s)} + Ca^{2+}_{(aq)} \quad (2.33)$$

Surface applications of some alkaline materials or applications under thin soil cover have not been successful (Smith and Brady 1999). Possible explanations may include: (*a*) the dissolution of calcite at surface conditions is limited by the partial pressure of carbon dioxide; and (*b*) preferential flow of water occurs through the waste dump, bypassing much of the near surface alkaline material. Thus, uniform mixing of acid buffering waste or neutralizing materials with sulfidic waste is of paramount importance in order to achieve consumption of acidity (Smith and Brady 1999). In an already established sulfidic waste dump, the use of neutralizing agents is limited by the difficulty of blending them through waste layers. Thorough mixing is usually difficult, inefficient and expensive. In addition, a treatment relying on deep disturbance of the waste has the risk of exposure of additional unoxidized sulfidic mine waste. Therefore, blending has to occur while the waste materials are being dumped at their disposal sites.

The major disadvantage of blending is that it does not prevent sulfide oxidation. While blending helps to immobilize dissolved metals such iron, aluminium, copper and lead as insoluble minerals in sulfidic wastes, it may not prevent the release of sulfate as well as other metals and metalloids (e.g. As, Cd, Mo, Zn) into pore waters. These elements are potentially mobile under neutral to alkaline pore water conditions.

2.10.5 Addition of Organic Wastes

The addition of organic wastes may also prevent sulfide oxidation. The wastes can be used as a compacted subsurface layer in dry cover designs or as amendments to create reactive, low permeability biomass surfaces. Trialed organic materials include sewage sludge, wood chips, sawdust, manure, peat, pulp and paper mill residues, and municipal solid waste compost (Cabral et al. 1997; Elliott et al. 1997; SMME 1998; Hulshof et al. 2006). The wastes inhibit pyrite oxidation via various mechanisms. Firstly, organic wastes provide a pH buffer and create reducing conditions which inhibit sulfide oxidizing bacteria, reduce sulfate to sulfide, and immobilize metals as sulfides. Secondly, dissolved organic compounds form stable iron-organic complexes or combine with iron to form stable precipitates. Thirdly, the organic compounds are adsorbed on pyrite surfaces, preventing oxidation (Evangelou 1995, 1998). However, if organic waste such as sewage sludge is part of a dry cover design, organic acids (simplified stoichiometrically as the molecules CH_2O and H^+) may dissolve iron hydroxide phases (Blowes et al. 1994):

$$4\,Fe(OH)_{3(s)} + CH_2O_{(aq)} + 7\,H^+_{(aq)} \rightarrow 4\,Fe^{2+}_{(aq)} + HCO^-_{3(aq)} + 10\,H_2O_{(l)} \quad (2.34)$$

Adsorbed and coprecipitated metals, originally present in the iron precipitates, are also released into the aqueous phase. The iron and other metals can form stable organic complexes which may migrate into ground and surface waters. Therefore, it is possible that the use of organic material enhances the migration of contaminants from sulfidic wastes (Mitchell 2000).

Organic waste placed at the surface of waste repositories is in constant contact with the atmosphere and will decompose over time. Such oxidation ensures that the added organic material is eventually used up. Consequently, the addition of organic waste, unless regularly done, is a short-term fix to a long-term waste problem.

2.10.6 Bactericides

Certain bacteria are known to increase the rate of pyrite oxidation. Hence, anti-bacterial agents, so-called "*bactericides*", have been used to inhibit the growth of these microorganisms (Ledin and Pedersen 1996; Kleinmann 1997, 1999). Many compounds including anionic surfactants, cleaning detergents, organic acids, and food preservatives have been screened as selective bactericides. The anionic surfactants sodium lauryl sulfate and alkyl benzene sulfonate are considered the most reliable and cost effective inhibitors (Kleinmann 1997, 1999). In the presence of such compounds, hydrogen ions in the acidic environment move freely into or through bacteria cell membranes, causing their deterioration.

While there are clear advantages in the use of bactericides including decreased pyrite oxidation and metal mobility, there are also disadvantages and potential risks. The applied compound may cause toxicity to other organisms; there is the possibility of resistance development; and it is difficult to reach all zones of the sulfidic waste (Kleinmann 1999). Bactericides are generally water soluble and leach from the waste, and they may be adsorbed on the surfaces of other minerals. As a result, repetitive treatments are necessary to prevent repopulation of the waste by bacteria when the bactericide is depleted (Kleinmann 1999). Alternatively, slow-release pellets may help to provide long-term bacterial inhibition (Kleinmann 1999). Thus, bactericides are another short-term solution to a long-term waste problem, and if applied, they should be part of other control measures (Environment Australia 1997).

2.11 Summary

Sulfidic mine wastes, especially those which contain high concentrations of pyrite, are the major sources of AMD. Pyrite oxidation may occur via biotic or abiotic and direct or indirect oxidation processes. Biotic indirect oxidation of pyrite is an important acid generating process whereby pyrite is oxidized by oxygen and Fe^{3+} in the presence of microorganisms. Pyrite oxidation is a complex process because it not only involves chemical, electrochemical and biochemical reactions, but it also varies with environmental conditions. The following factors all work to enhance the speed of pyrite oxidation: large surface area; small particle size; high porosity; poor crystallinity; significant trace element substitutions of and solid inclusions within the pyrite; acidic pH values of the solution in contact with pyrite; no direct physical contact with other sulfides; high oxygen and high Fe^{3+} concentrations in the oxidizing medium; high temperature; abundant microbial activity; and alternate wetting and drying of the sulfide grain.

Sulfides other than pyrite have different acid production potentials, stabilities and rates of reaction. The metal/sulfur ratio in sulfides influences how much sulfuric acid is liberated by oxidation. Also, the amount of iron sulfides present strongly influences whether and how much acid is generated during weathering. Iron sulfides (e.g. pyrite, marcasite, pyrrhotite) or sulfides having iron as a major constituent (e.g. chalcopyrite, iron-rich sphalerite) generate the most acidity. In contrast, sulfide minerals which do not contain iron in their crystal lattice (e.g. covellite, galena or iron-poor sphalerite) do not have the capacity to generate significant amounts of acid. This is because Fe^{3+} is not available as the important oxidant, and iron hydrolysis cannot occur which would generate additional hydrogen ions. The production of acid also occurs through the dissolution of secondary soluble Fe^{2+}, Mn^{2+}, Fe^{3+} and Al^{3+} sulfate salts, and the precipitation of secondary Fe^{3+} and Al^{3+} hydroxides.

Any acid generated can be consumed through the reaction of the hydrogen ions with gangue minerals. The weathering of silicates results in the consumption of hydrogen ions, formation of secondary minerals, and release of dissolved cations and silicic acid. The dissolution of hydroxides and carbonates as well as cation exchange processes on clay minerals also consume acid. The gangue minerals reacting with the acidic solutions have a variable resistance to weathering and therefore, exhibit different reaction rates. Carbonate minerals show the highest reactivity and highest acid consumption, whereas silicates have significantly slower reaction rates and provide only token amounts of additional long-term buffering capacity to sulfidic wastes.

Weathering of sulfidic wastes produces mine waters laden with dissolved salts. The dissolved salts may approach saturation in pore waters, ground waters, streams and leachates associated with the oxidation and leaching of sulfidic wastes. In fact, numerous secondary minerals are known to precipitate in AMD environments. Some secondary minerals may redissolve in AMD waters thereby influencing the mine water chemistry; others may precipitate and coat acid buffering or acid producing minerals. Massive precipitation of secondary minerals in wastes results in the formation of laterally extensive or discontinuous subsurface or surface layers which act as horizontal barriers to vertical water movements. Thus, the secondary mineralogy of sulfidic wastes plays an important role in controlling how readily and how much acid, metals and sulfate are liberated to drainage waters.

The prediction of AMD generation from sulfidic wastes is possible using geological and petrographic descriptions, geological modeling, static and kinetic tests, and mathematical models. These tools may be used to estimate potential sulfide oxidation and dissolved metal mobility. Sulfidic waste dumps are major sources of AMD. Oxygen advection, convection and diffusion occur in such wastes, which can result in the production of acid. Fine-grained wastes have much greater surface areas and hence, a greater acid generation potential than coarse-grained wastes, yet the fine grain size limits oxygen diffusion, water ingress and acid generation.

Monitoring techniques of sulfidic wastes are designed to identify the early presence of or the changes to any products of the acid producing reactions. The products can be identified by obtaining waste temperature measurements, oxygen pore gas concentration profiles, and leachate analyses for dissolved contaminant concentrations and loads. Rapid increases in temperature profiles of waste dumps indicate the exothermic oxidation of sulfides, whereas the depletion of oxygen concentration within gas pores is also indicative of sulfide oxidation.

Control techniques of sulfide oxidation are based on the exclusion of one or more of the factors that cause and accompany oxidation; that is, sulfide minerals, bacteria, water, iron and oxygen. Controlling sulfide oxidation may reduce or even eliminate the possibility of AMD generation. The destructive sulfide oxidation processes are driven by the ready exchange of oxygen with the atmosphere. Hence, reducing oxygen availability is the most effective control on the oxidation rate. This is generally achieved using dry or wet covers. The advantage of wet covers is that oxygen diffuses very slowly and has limited solubility in water. In contrast, a dry cover with a low oxygen permeability restricts water and oxygen movement into and through the waste. The dry cover reduces both the oxidation rate of sulfides and the transport of leachates from the waste. Other established and experimental methods for the prevention of sulfide oxidation and AMD development include selective handling and isolation, co-disposal and blending, mineral surface treatments, addition of organic wastes, and bacterial inhibition.

Further information on sulfidic wastes can be obtained from web sites shown in Table 2.8.

Table 2.8. Web sites covering aspects of sulfidic wastes

Organization	Web address and description
The International Network for Acid Prevention (INAP)	*http://www.inap.com.au* Industry based initiative to coordinate research and development into the management of sulfidic wastes
US Environmental Protection Agency (EPA)	*http://www.epa.gov/epaoswer/other/mining/techdocs/amd.pdf* Technical report on AMD prediction
US Geological Survey (USGS)	*http://crustal.usgs.gov/projects/minewaste/index.html* Mine waste characterization
Mine Environment Neutral Drainage Program (MEND), Canada	*http://www.nrcan.gc.ca/mms/canmet-mtb/mmsl-lmsm/mend* Extensive technical reports on sulfide oxidation prevention techniques
Department of Environmental Protection, Pennsylvania	*http://www.dep.state.pa.us/dep/deputate/minres/Districts/homepage/Default.htm* AMD and coal mine drainage prediction and pollution prevention
Mining Information Service EnviroMine	*http://technology.infomine.com/enviromine* News and information on sulfidic wastes including static and kinetic tests

Chapter 3

Mine Water

3.1 Introduction

Water is needed at mine sites for dust suppression, mineral processing, coal washing, and hydrometallurgical extraction. For these applications, water is mined from surface water bodies and ground water aquifers, or it is a by-product of the mine dewatering process. Open pits and underground mining operations commonly extend below the regional water table and require dewatering during mining. In particular, mines intersecting significant ground water aquifers, or those located in wet climates, may have to pump more than 100 000 liters per minute to prevent underground workings from flooding. At some stage of the mining operation, water is unwanted and has no value to the operation. In fact, unwanted or used water needs to be disposed of constantly during mining, mineral processing, and metallurgical extraction.

At modern mine sites, water is collected and discharged to settling ponds and tailings dams. In contrast, at historic mine sites, uncontrolled discharge of mine water commonly occurs from adits and shafts into the environment. Generally, the volume of mine water produced, used and disposed of at mine sites is much larger than the volume of solid waste generated. At mine sites, water comes in contact with minerals and dissolves them. Hence, water at mine sites often carries dissolved and particulate matter. When such laden waters reach receiving water bodies, lakes, streams or aquifers, the waters can cause undesirable turbidity and sedimentation, they may alter temperatures, or their chemical composition may have toxic effects on plants and animals. For example, in the United States, it has been estimated that 19 300 km of streams and 72 000 ha of lakes and reservoirs have been seriously damaged by mine effluents from abandoned coal and metal mines (Kleinmann 1989).

The worst example of poor mine water quality and associated environmental impacts is acid mine drainage (AMD) water, which originates from the oxidation of sulfide minerals (Sec. 2.3). Sulfide oxidation is an autocatalytic reaction and therefore, once AMD generation has started, it can be very difficult to halt. AMD is the most severe in the first few decades after sulfide oxidation begins, and the systems then produce lower levels of contaminants (Lambert et al. 2004; Demchak et al. 2004). In extreme cases, however, AMD may continue for thousands of years (Case Study 3.1, Fig. 3.1).

This chapter summarizes information on AMD waters and gives the principles of AMD characterization, monitoring, prediction, environmental impact, and treatment. Such aspects of AMD waters are important issues for any mining operation, regardless whether the mine waters are acid or not.

Case Study 3.1. Acid Mine Drainage at the Rio Tinto Mines, Spain

The Iberian Pyrite Belt

The Iberian Pyrite Belt corresponds to an area of volcanic and sedimentary rocks containing massive sulfide deposits. This area forms an arcuate belt, about 250 km long and up to 60 km wide, from Seville in southern Spain to the Atlantic coast of Portugal. The belt contains about 90 known massive sulfide deposits which have been mined for pyrite, base metals (Cu, Pb, Zn), and various trace elements (e.g. Ag, As, Au, Hg, Sn). The orebodies range from small lenses with thousands of tonnes to giant bodies with hundreds of million tonnes. Prior to erosion, these deposits contained more than 1700 Mt of massive sulfides. About 20% of this amount has been mined, and 10 to 15% has been lost to erosion (Barriga and Carbalho 1997). The Iberian Pyrite Belt represents the largest concentration of metals and sulfides in the world.

The Rio Tinto Mining District

The Rio Tinto district (i.e. Nerva–Minas de Riotinto) has been the most important mining district of the Iberian Pyrite Belt. Various civilizations have been living in the Rio Tinto region, exploiting its natural resources and establishing one of the oldest mines in the world. Mining dates back to the Copper Age 5000 years ago and continued during Tartessian and Phoenician times. The mineral wealth of the region also drew Greek, Carthaginian and Roman invasions – the greatest mining activity took place during the Roman period. The mines were abandoned after the Roman era, and little mining was conducted after this time. In 1873, British mining companies began to exploit the ores on a large scale, and a significant industrial complex was established by the late 19th century. Exploitation of the mines continued into modern times.

These successive mining activities have resulted in the creation of a unique "mining landscape". The region has numerous abandoned mine workings and is littered with derelict buildings and disused mining and processing equipment. There are also uncountable waste rock heaps, ore stockpiles, tailings dumps, slag deposits, and settling ponds, most of which do not support any vegetation. The lack of vegetation on sulfidic wastes and local soils increases erosion rates into the headwaters of the Rio Tinto. Pyritic waste particulates are transported into the river tens of kilometers downstream of the mine and processing sites (Hudson-Edwards et al. 1999). The erosion processes exacerbate the "moonscape appearance" of the area affected by mining, mineral processing and smelting activities. Most importantly, the Rio Tinto mining district is characterized by uncontrolled pyrite oxidation in exposed mine workings and waste materials (Romero et al. 2006). The oxidation of pyrite causes formation of sulfuric acid and the dissolution of many metals. Mine waters are quite acid (pH 2 or less) due to the oxidation of abundant pyrite.

The Tinto River

The Rio Tinto mining district is drained by the Odiel and Tinto rivers. The Tinto river is 90 km long and remains strongly acid ($pH < 3$) for its entire length (Leblanc et al. 2000; Braungardt et al. 2003). Tinto is Spanish for "red wine" and clearly refers to its turbid red, acid water. The stream's distinct turbidity is the result of abundant iron-rich suspended colloids and gelatinous flocculants. The river also carries very high dissolved sulfate, metal, and metalloid loads (As, Fe, Cu, Cd, Ni, Pb, Zn) from the headwaters to its estuary. Part of this dissolved metal load is precipitated into fluvial and estuarine sediments. Another part of the metal load enters continental shelf sediments and waters of the Gulf of Cadiz and contributes to the metal content of the Mediterranean Sea through the Strait of Gibraltar (Nelson and Lamothe 1993; Elbaz-Poulichet and Leblanc 1996). The waters and sediments of the Rio Tinto are strongly polluted with metals and metalloids. In fact, the Tinto river is one of the most polluted streams in the world. Such pollution began with the exploitation of the ores 5000 years ago (Davis et al. 2000). Despite its pollution, the Rio Tinto acts as an ecological niche for at least 1300 different microorganisms including algae, fungi, bacteria, yeast and protists (Lopez-Archilla et al. 1993; Ariza 1998). Some of these microorganisms actively participate in the oxidation of detrital pyrite particles present in the stream bed. The Rio Tinto is unique in the sense that the pH value of the river water does not increase downstream, nor does it change significantly during rainfall events. In general, the pH of AMD affected streams generally increases downstream due to the inflow of

Case Study 3.1. *Continued*

unpolluted surface or ground waters. The unique conditions of the Rio Tinto are likely due to a number of factors including: (*a*) a constant input of low pH waters from the mine workings and waste dumps into the headwaters of the river; (*b*) the presence of abundant detrital pyrite grains in Rio Tinto's alluvium; (*c*) an pleothora of microorganisms which help to oxidize the pyrite grains to acid; (*d*) the presence of acid producing secondary minerals such as copiapite and coquimbite in the fluvial sediments; and (*e*) the hydrolysis of iron, precipitation of secondary iron minerals and associated production of hydrogen ions in the Rio Tinto (Lopez-Archilla et al. 1993; Hudson-Edwards et al. 1999; Lottermoser 2005; Buckby et al. 2003; España et al. 2006).

Mining may not be entirely responsible for the generation of AMD and its impact on the Rio Tinto. Historical records refer to the river's long-standing acidity. The Romans called the Rio Tinto "urbero", Phoenician for "river of fire", and the Arab name for it was "river of sulfuric acid" (Ariza 1998). There is also geological evidence that the sulfide orebodies experienced long-term weathering and erosion at some stage in their geological history. The presence of thick jarosite-rich gossans capping the pyritic ores indicates that acid weathering of outcropping sulfide ores could have produced natural ARD prior to mining. The unique red colour of the river may have attracted the very first miners to the region (Ariza 1998). Consequently, the water's conditions today could be a combination of natural ARD and mining induced AMD.

Fig. 3.1. Slag heap, sulfidic waste dumps, and abandoned railway carriages at Rio Tinto, Spain. Mankind has exploited the Rio Tinto ores since the Copper Age 5 000 years ago. The mining activities have left uncountable waste rock heaps, ore stockpiles, tailings dumps, slag deposits, and settling ponds, most of which do not support any vegetation. The exploitation of sulfidic ores has created a unique "mining landscape" and caused massive AMD flowing into the Rio Tinto

3.2 Sources of AMD

Mining of metallic ore deposits (e.g. Cu, Pb, Zn, Au, Ni, U, Fe), phosphate ores, coal seams, oil shales, and mineral sands has the potential to expose sulfide minerals to oxidation and generate AMD water. Coal and ore stockpiles, tailings storage facilities, as well as waste rock and heap leach piles are all potential sources for acid generation as are underground workings, mine adits, shafts, pit walls, and pit floors (Fig. 3.2). At these sites, mine waters can become acidic through reactions of meteoric water or ground water with exposed sulfides. Consequently, AMD water can form as the result of numerous processes such as:

- ground water enters underground workings located above the water table and exits via surface openings or is pumped to the surface (i.e. mining water);
- ground water enters pits and surface excavations;
- meteoric precipitation comes in contact with pit faces;
- meteoric precipitation infiltrates coal and ore stockpiles, heap leach piles, coil spoil heaps, and waste rock dumps;
- meteoric precipitation and flood inflow enter tailings disposal facilities;
- run-off from rainfall interacts with mining, mineral processing, and metallurgical operations;
- surface water and pore fluids of tailings, heap leach piles, ore stockpiles, coal spoil heaps, and waste rock dumps may surface as seepage waters or migrate into ground water aquifers; and
- uncontrolled or controlled discharge of spent process waters occurs from tailings dams, stacks, ponds, and heap leach piles.

AMD waters can form rapidly, with evidence such as iron staining or low pH run-off often appearing within months or even weeks. AMD generation is thereby independent of climate and is encountered at mine sites in arid to tropical climates

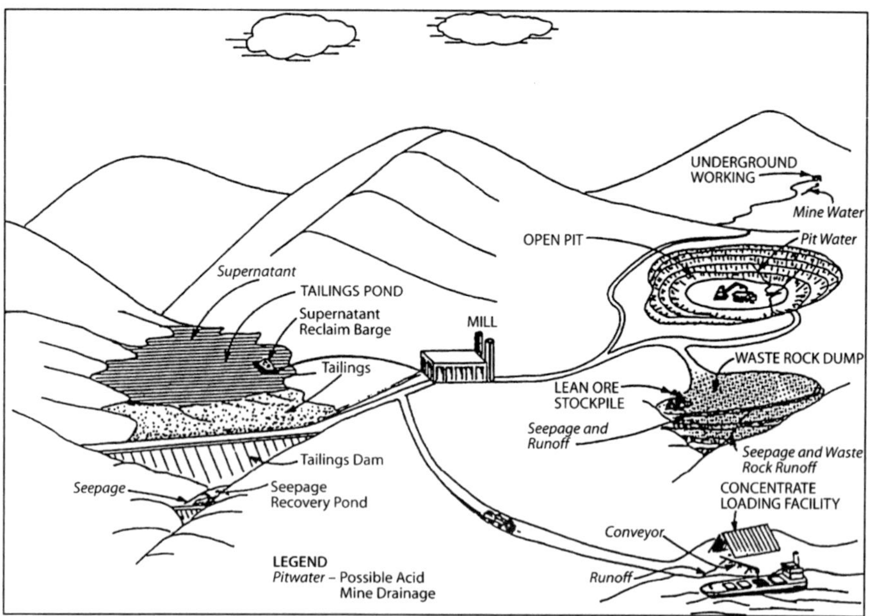

Fig. 3.2. Sources of AMD at a metal mine (Ferguson and Erickson 1988)

from the Arctic Circle to the equator (Scientific Issue 2.1). However, not all mining operations that expose sulfide-bearing rock will cause AMD. In addition, contaminant generation and release are not exclusive to AMD environments. They also occur in neutral and alkaline drainage environments as shown in the following sections.

3.3 Characterization

Constituents dissolved in mine waters are numerous, and mine waters are highly variable in their composition. Some waters contain nitrogen compounds (nitrite, NO_2^-; nitrate, NO_3^-; ammonia, NH_3) from explosives used in blasting operations and from cyanide heap leach solutions used for the extraction of gold (Sec. 5.4). Other mine waters possess chemical additives from mineral processing and hydrometallurgical operations (Sec. 4.2.1). For instance, metallurgical processing of many uranium ores is based on leaching the ore with sulfuric acid (Sec. 6.5.1). Spent process waters are commonly released to tailings repositories, so the liquids of uranium tailings dams are acid and sulfate rich. Also, coal mining may result in the disturbance of the local aquifers and the dissolution of chloride and sulfate salts that are contained in the marine sedimentary rocks present between the coal seams. As a result, coal mine waters can be exceptionally saline.

Therefore, depending on the mined ore and the chemical additives used in mineral processing and hydrometallurgical extraction, different elements and compounds may need to be determined in waters of individual mine sites. Regardless of the commodity extracted and the mineral processing and hydrometallurgical techniques applied, major cations (i.e Al^{3+}, Si^{4+}, Ca^{2+}, Mg^{2+}, Na^+ and K^+) and anions (i.e Cl^-, SO_4^{2-}, CO_3^{2-}, HCO_3^-) are important constituents of any mine water. Other constituents such as nitrogen or cyanide compounds, or dissolved and total organic carbon concentrations, should be determined depending on site specific conditions. Additional parameters analyzed and used for the study of mine waters are given in Table 3.1.

3.3.1 Sampling and Analysis

Detailed procedures for water sampling, preparation and analysis are found in manuals and publications (e.g. Ficklin and Mosier 1999; Appelo and Postma 1999). Laboratory methods for the geochemical analysis of environmental samples including mine waters are given by Crock et al. (1999). Quality assurance/quality control of the analytical results must be ensured using established procedures. The submission of duplicates or even triplicates of the same sample will allow an evaluation of the analytical precision (i.e. repeatability). Blanks of deionized water should be included in order to check for unclean sample processing or inaccurate chemical analysis. The low pH of AMD waters will aid in preservation of dissolved metals; otherwise, neutral or alkaline waters need to be acidified to keep metals in solution. Degassing of CO_2-rich samples is possible after sampling, so containers should be completely filled and tightly closed.

The longer the period of time between collection and analysis, the more likely it is that unreliable analytical results will be measured. Exposure to light and elevated tem-

Table 3.1. Selected parameters important to mine waters (after Brownlow 1996; Drever 1997; Appelo and Postma 1999; Ficklin and Mosier 1999)

Parameter	Explanation
Eh	Reduction-oxidation potential of a solution. The redox state of mine waters is best analyzed by determining the concentrations of multiple redox species such as Fe^{2+} and Fe^{3+} in the sample. Expressed as volts (V).
pH	Negative logarithm of hydrogen activity; $-\log[H^+]$.
Electrical conductivity (EC)	Ability to conduct electrical current, depending on the amount of charged ions in solution. Expressed as $\mu S\ cm^{-1}$. For freshwater, the conductivity is approximately related to the quantity of total dissolved solids by the expression: $EC\ 0.65^{-1} = mg\ l^{-1}$ TDS
Hardness	Sum of the ions which can precipitate as "hard particles" from water. It is expressed as mg $CaCO_3\ l^{-1}$ and related to the sum of calcium and magnesium ions by the equation: mg $CaCO_3\ l^{-1} = 2.5$ (mg Ca l^{-1}) + 4.1 (mg Mg l^{-1})
Alkalinity	Capacity of a solution to neutralize acid. Determined by titrating with acid down to a specific pH, e.g. pH 4.5. In most natural waters, alkalinity is equal to the molality concentrations of bicarbonate and carbonate ions. Other ions such as ammonia, borate, silicic acid, bisulfides, organic anions, and hydroxide ions can combine with hydrogen and contribute to alkalinity. Expressed as mg $CaCO_3$ or $HCO_3^-\ l^{-1}$.
Acidity	Capacity of a solution to donate protons. Determined by titrating with a base up to a specific pH value, e.g. pH 8.3. A number of dissolved ions (H^+, Fe^{2+}, HSO_4^-), gases (CO_2, H_2S), humic and fulvic acids, and suspended matter (metal hydroxides, clays) may contribute to an acidity value. Mine waters generally have very low organic acid contents. Expressed as mg $CaCO_3$ or $HCO_3^-\ l^{-1}$.
Total dissolved solids (TDS)	Amount of dissolved solids. Determined by evaporating and weighing the dry residue. Expressed as mg l^{-1}.
Dissolved oxygen (DO)	Amount of dissolved oxygen. Expressed as mg l^{-1}.
Turbidity	Macroscopic, visual evidence of suspended matter in water. Expressed by using an empirical number scale.
Temperature	°C
Salinity	Amount of total dissolved solids. Freshwater has <1 000 or 1 500 mg l^{-1} TDS, brackish waters have 1 000 to 10 000 mg l^{-1} TDS, saline waters have 10 000 to 100 000 mg l^{-1} TDS, whereas brines have even higher salinities.

peratures will cause precipitation of salts, or dissolution of transitional and solid species. Consequently, it is of paramount importance to preserve water samples on ice in a closed container and to submit collected samples as soon as possible to the laboratory. Upon receipt of the analytical results, analytical values of duplicates/triplicates and blanks should be evaluated, and the charge balance of anions and cations should be confirmed (Appelo and Postma 1999).

The concentrations of dissolved substances in water samples are presented in different units. The most commonly used units are mg l^{-1} and ppm or ppb. The units mg l^{-1} and ppm are numerically equal, assuming that 1 l of water weighs 1 kg. Such a conversion is only valid for dilute freshwaters, yet many mine waters are saline. Thus, any conversion has to consider the increased density (Appelo and Postma 1999). The density of waters needs to be determined if it is desired to convert analytical values from mg l^{-1} to ppm.

With the advent of modern field equipment, many mine water parameters (i.e. pH, dissolved oxygen, temperature, electrical conductivity, turbidity) should be determined in the field since these values can quickly change during sample storage (Ficklin and Mosier 1999). If possible, an elemental analysis should be accompanied by the measurement of the reduction-oxidation (redox) potential (i.e. Eh), or of a redox pair such as Fe^{2+}/Fe^{3+}. Such an analysis is sufficient to define the redox state of the AMD water and allows the simulation of redox conditions during geochemical modeling.

3.4 Classification

There is no typical composition of mine waters and as a result, the classification of mine waters based on their constitutents is difficult to achieve. A number of classification schemes of mine waters have been proposed using one or several water parameters:

- *Major cations and anions.* This is a standard technique to characterize ground and surface waters (e.g. Brownlow 1996; Drever 1997; Appelo and Postma 1999). It involves plotting the major cation (Ca^{2+}, Mg^{2+}, Na^{+}, K^{+}) and anion (Cl^{-}, SO_4^{2-}, CO_3^{2-}, HCO_3^{-}) chemistry on a so-called "Piper" or trilinear diagram. The plotted waters are then classified according to their cation and anion abundances.
- *pH.* A basic scheme labels mine waters according to their pH as acidic, alkaline, near-neutral, and others (Morin and Hutt 1997).
- *pH and Fe^{2+} and Fe^{3+} concentration.* This classification technique requires a knowledge of the pH and of the amount of Fe^{2+} and Fe^{3+} present (Glover 1975; cited by Younger 1995).
- *pH vs. combined metals.* Mine waters can also be classified according to pH and the content of total dissolved metals (Ficklin et al. 1992; Plumlee et al. 1999).
- *Alkalinity vs. acidity.* This scheme has been devised to allow classification of mine waters according to their treatability using passive treatment methods (Hedin et al. 1994a). It requires a knowledge of the alkalinity and acidity of the waters as determined by titration (Kirby and Cravotta 2005a,b). The categorization is useful for the selection of aerobic or anaerobic treatment methods as net acid waters require anaerobic treatment and net alkaline waters require aerobic remediation.
- *Alkalinity vs. acidity and sulfate concentration.* This classification considers both the alkalinity and acidity as well as the sulfate content of mine waters (Younger 1995).

The above classifications have one or several short-comings: (*a*) the classifications do not include waters with neutral pH values and extraordinary salinities; (*b*) the schemes do not consider mine waters with elevated concentrations of arsenic, antimony, mercury, cyanide compounds, and other process chemicals; (*c*) the categorizations do not consider iron, manganese and aluminium which are present in major concentrations in AMD waters; and (*d*) routine water analyses do not include determinations of the Fe^{2+} and Fe^{3+} concentrations. Therefore, the categorizations are not inclusive of all mine water types. In this work, the simple classification scheme of Morin and Hutt (1997) has been modified (Table 3.2), and the following presentation of mine waters is given according to their pH.

Table 3.2. Classification of mine waters based on pH (after Morin and Hutt 1997)

Class	Characteristics
Extremely acid	pH < 1. Extreme examples of acidities generated through sulfide oxidation and hydrolysis reactions. The rocks are distinctly enriched in pyrite and depleted in acid buffering materials.
Acid	pH < 5.5. Acidity generated through oxidation of Fe-rich sulfides. Commonly found at base metal, gold and coal mines.
Neutral to alkaline	pH 6–10. Acid producing and acid buffering reactions keep a pH balance, or abundant Fe-rich sulfides are absent. The drainage may become acid or alkaline with time upon exhaustion of acid producing or acid buffering minerals. High levels of alkalinity generated through dissolution of carbonates, alkali oxides, hydroxides, and silicates. Commonly found at diamond, base metal, gold, uranium, iron, coal and mineral sand mines.
Saline	pH highly variable which influences the concentrations of aqueous ions. Associated with the mining of coal and industrial minerals, including evaporites such as potash, halite and borate.

3.4.1 Acid Waters

Oxidation of pyrite and other sulfides is the major contributor of hydrogen ions in mine waters, but a low pH is only one of the characteristics of AMD waters (Fig. 3.3). The oxidation of sulfide minerals does not only create acid, but it also liberates metals and sulfate into waters and accelerates the leaching of other elements from gangue minerals. As a consequence, AMD is associated with the release of sulfate, heavy metals (Fe, Cu, Pb, Zn, Cd, Co, Cr, Ni, Hg), metalloids (As, Sb), and other elements (Al, Mn, Si, Ca, Na, K, Mg, Ba). In general, AMD waters from coal mines typically contain much lower concentrations of heavy metals and metalloids than waters from base metal or gold deposits (Geldenhuis and Bell 1998).

AMD waters are particularly characterized by exceptionally high sulfate (>1 000 mg l^{-1}), high iron and aluminium (>100 mg l^{-1}), and elevated copper, chromium, nickel, lead and zinc (>10 mg l^{-1}) concentrations. Dissolved iron and aluminium typically occur in significantly higher concentrations than the other elements. Elements such as calcium, magnesium, sodium, and potassium may also occur in strongly elevated concentrations. These latter elements are not of environmental concern themselves. However, they may limit the use of these waters because of their sodium content or their hardness. High sodium levels prevent the use of these waters for irrigation of soils, and the hardness influences the toxicity of heavy metals such as zinc.

Sulfide oxidation and the AMD process also form the basis for modern heap leach operations used to recover copper and uranium from geological ores. In these hydrometallurgical processes, copper and uranium ores are piled into heaps and sprinkled with acid leach solutions. Sulfuric acid is applied to dissolve the ore minerals (e.g. malachite, azurite, uraninite). Once the recovery of metals is complete, the heap leach piles are rinsed to reduce any contaminant loads (Li et al. 1996; Shum and Lavkulich 1999; Ford 2000). Despite rinsing, drainage waters emanating from spent heap leach piles can have high acidity, sulfate, metal, metalloid, and aluminium con-

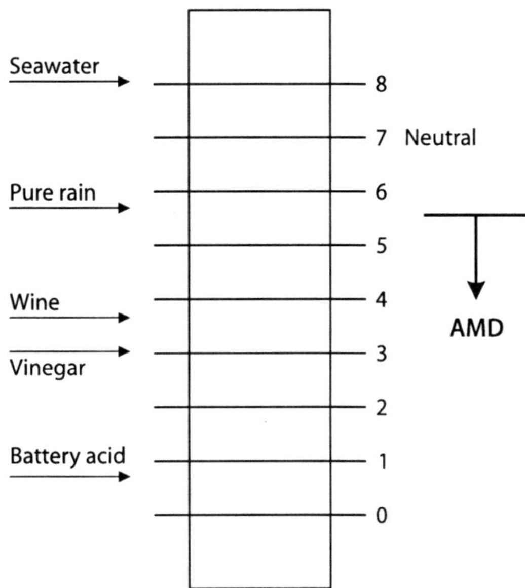

Fig. 3.3. pH scale and comparison of AMD water with other familiar fluids (after Jambor et al. 2000b)

centrations. In addition, sulfuric acid is used for the extraction of nickel from nickel laterite deposits and the production of synthetic rutile from placer deposits. Both processes result in the formation of acidic tailings. Finally, the presence of acid conditions in surface waters should not always be attributed to anthropogenic processes. Acidity of streams may also be caused by naturally occurring organic acids that are flushed from soils into surface waters. Therefore, acidic drainage waters are not exclusive to sulfidic wastes. In most cases, the acidity of mine waters is the result of sulfide oxidation.

3.4.2 Extremely Acid Waters

The pH of most drainages is buffered by acid neutralizing minerals. The buffering reactions ensure that AMD waters have pH values of greater than 1. There are, however, rare examples with drainage acidities of below pH 1, in extreme cases even with negative pH values (Nordstrom and Alpers 1999b; Williams and Smith 2000; Nordstrom et al. 2000). These waters not only contain exceptionally low pH values – in rare cases as low as minus 3 – they also exhibit extraordinarily high concentrations of iron, aluminium, sulfate, metals, and metalloids. The concentrations are so high that the waters are significantly over-saturated with mineral salts. Theoretically, precipitation of secondary minerals should occur. Precipitation of mineral salts from these waters is very slow, and the total ionic strengths of the waters exceed their theoretical maximum. Such conditions are referred to as "*super-saturation*". Super-saturated AMD waters are generated from rocks distinctly enriched in pyrite and depleted in acid buffering carbonates. The acid buffering capacity of such rocks is minimal, and the formation of extremely acid mine waters is favoured by unhindered sulfide oxidation and hydrolysis reactions.

3.4.3 Neutral to Alkaline Waters

A low pH is not a universal characteristic of waters influenced by mining. The pH of mine waters extends to alkaline conditions, and the aqueous concentrations of anions and cations range from less than 1 mg l^{-1} to several 100 000 mg l^{-1}. In acid waters, sulfate is the principal anion, and iron, manganese and aluminium are major cations. In alkaline waters, sulfate and bicarbonate are the principal anions, and concentrations of calcium, magnesium and sodium are generally elevated relative to iron and aluminium (Rose and Cravotta 1999). Substantial concentrations of sulfate, metals (Cd, Hg, Mn Mo, Ni, U, Zn), and metalloids (As, Sb, Se) have been documented in oxidized, neutral to alkaline mine waters (Carroll et al. 1998; Lottermoser et al. 1997b, 1999; Pettit et al. 1999; Plumlee 1999; Plumlee et al. 1999; Younger 2000; Schmiermund 2000; Scharer et al. 2000; Ashley et al. 2003; Craw et al. 2004; Wilson et al. 2004). Such waters are of environmental concern as they may adversely impact on the quality of receiving water bodies. Neutral to alkaline mine waters with high metal, metalloid, and sulfate contents can be caused by:

- drainage from tailings repositories containing residues of alkaline leach processes or neutralized acidic tailings;
- drainage from non-sulfidic ores and wastes;
- drainage from sulfidic ores or wastes that have been completely oxidized during pre-mining weathering;
- drainage from pyrite-rich ores and wastes with abundant acid neutralizing minerals such as carbonate; and
- drainage from sulfide ores or wastes depleted in acid producing sulfides (e.g. pyrite, pyrrhotite) and enriched in non-acid producing sulfides (e.g. galena, sphalerite, arsenopyrite, chalcocite, covellite, stibnite).

3.4.4 Coal Mine Waters

AMD waters of coal mines are characterized by low pH as well as high electrical conductivity, total dissolved solids, sulfate, iron and aluminium values. In addition, individual mine sites may have waters with elevated manganese and trace metal and metalloid values (Cravotta and Bilger 2001; Larsen and Mann 2005). Coal contains a range of trace elements and leaching of these trace metals (e.g. Cu, Pb, Zn, Ni, Co) and metalloids (e.g. As, Se) may impact on the receiving environment (e.g. Lussier et al. 2003).

Mine waters of coal mines are not necessarily acid. Many mine waters of coal mines have near neutral pH values. However, such waters typically contain elevated total dissolved solids and exhibit high electrical conductivities (Foos 1997; Szczepanska and Twardowska 1999). Salt levels, particularly chloride concentrations, can be extreme. These saline waters originate from saline aquifers as dewatering of the mine may intersect deep saline formation waters. Also, atmospheric exposure of saline coals and marine sediments within the stratigraphic sequence, containing abundant salt crystals, will lead to the generation of saline mine waters. Such waters need to be contained on site.

Discharge off-site should occur when suitable flow conditions in the receiving streams are achieved, and dilution of saline waters is possible.

In rare cases coals have significant concentrations of uranium, thorium, and radioactive daughter products of the uranium and thorium decay series. Mine waters of such coals possess elevated radium-226 (Ra-226) levels. The dissolution of Ra-226 is possible if the waters contain low sulfate concentrations. This allows the dissolution of barium and radium (Ra-226) ions and causes elevated radiation levels (Pluta 2001; Schmid and Wiegand 2003).

3.5 Processes

There are several geochemical and biogeochemical processes which are important to mine waters, particularly to AMD waters. These processes, directly or indirectly, influence the chemistry of AMD waters. The processes are not exclusive to surface AMD environments and also operate below the surface in acid ground waters (e.g. Paschke et al. 2001).

3.5.1 Microbiological Activity

AMD waters are generally thought to be biologically sterile; however, they are hardly lifeless. Microorganisms such as bacteria, fungi, yeasts, algae, archaea, and protozoa are common and abundant in AMD waters (Johnson 1998a,b). For example, there are over 1300 different forms of microorganisms identified in the infamous acid waters of the Rio Tinto, Spain (Ariza 1998) (Case Study 3.1).

Bacteria isolated from AMD environments are numerous and include *Acidithiobacillus thiooxidans*, *Acidithiobacillus ferrooxidans*, *Leptospirillum ferrooxidans*, and *Thiobacillus thioparus* (Gould et al. 1994; Ledin and Pedersen 1996; Blowes et al. 1998; Johnson 1998a,b; Nordstrom and Alpers 1999a; Gould and Kapoor 2003). These bacteria function best in an acid, aerobic environment (pH < 4). The bacteria need minor nitrogen and phosphor for their metabolism, and they depend on the oxidation of Fe^{2+}, hydrogen sulfide, thiosulfate, sulfur, and metal sulfides for energy. They also transform inorganic carbon into cell building material (Ledin and Pedersen 1996). The inorganic carbon may originate from the atmosphere or from the dissolution of carbonates. The bacterial activity produces metabolic waste (i.e. sulfuric acid, Fe^{3+}) that accelerates the oxidation of sulfides (Sec. 2.3.1).

Algae are common organisms in AMD waters (Fig. 3.4). Such algae are not only capable of thriving in hostile AMD waters, they also remove metals and metalloids from solution. In addition, algae such as the protozoa *Euglena mutabilis* photosynthesize oxygen and contribute to dissolved oxygen in mine waters. This facilitates inorganic precipitation of iron and hence, the algae indirectly remove iron from AMD waters (Brake et al. 2001a,b). There are other life forms apart from bacteria and algae identified in AMD environments. For instance, a species of Archaea, *Ferroplasma acidarmanus*, has been found to thrive in exceptionally acid (pH 0), metal-rich waters (Edwards et al. 2000).

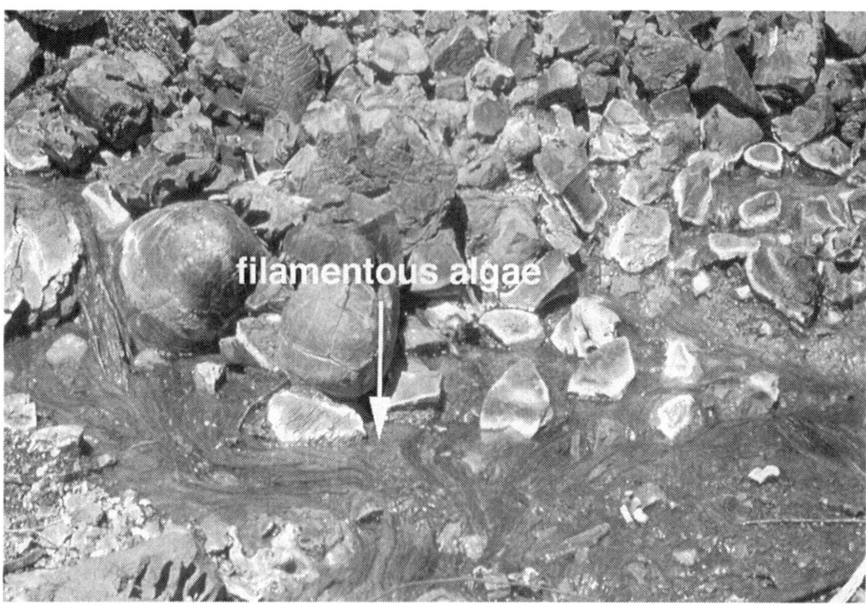

Fig. 3.4. Streamers of filamentous algae (*Klebsormidium* sp.) growing in AMD waters (pH 4.2, 7.4 mg l^{-1} Cu in solution), Gulf Creek, Australia (Lottermoser et al. 1999). The algae contain up to 0.25 wt.% copper. Largest cobble is 20 cm long

Certain microorganisms survive or even thrive in AMD environments because: (*a*) they tolerate elevated concentrations of dissolved metals and metalloids; and (*b*) they use the energy from the chemical oxidation reactions for their own growth. Furthermore, the microbes are capable of removing elements from AMD waters through adsorption and precipitation processes. The microbes thereby participate, actively or passively, in the removal of metals and metalloids from mine waters (Ferris et al. 1989; Leblanc et al. 1996; Johnson 1998a,b). For example, the bacterium *Acidithiobacillus ferrooxidans* oxidizes Fe^{2+} and promotes precipitation of iron as iron oxides and hydroxides (Ferris et al. 1989). Other microbes produce oxygen, reduce sulfate to sulfide, actively precipitate metals outside their cells, or incorporate metals into their cell structure. Moreover, some microorganisms are capable of inducing the formation of "*microbial*" minerals such as ferrihyrite, schwertmannite and hydrozincite in AMD affected waters (Kim et al. 2002; Zuddas and Podda 2005). In extreme cases, the metals and metalloids accumulated by living microorganisms, or the dead biomass, may amount to up to several weight percent of the cell dry weight. In addition, organic matter and dead cells indirectly participate in the immobilization of metals. If any dead biomass accumulates at the bottom of an AMD stream or pond, its degradation will lead to anaerobic and reducing conditions. Under such conditions, most metals may precipitate as sulfides and become both insoluble and unavailable for mobilization processes.

In summary, all three major life groups (Archaea, Eukaraya, Bacteria) on Earth are present as microorganisms in AMD environments. Some of these microorganisms accelerate the oxidation of sulfides whereas others adsorb and precipitate metals and metalloids from mine waters. Hence, microbes play an important role in the solubilization as well as immobilization of metals and metalloids in AMD waters.

3.5.2 Precipitation and Dissolution of Secondary Minerals

The precipitation of secondary minerals and of poorly crystalline and amorphous substances is common to AMD environments (Fig. 3.5) (McCarty et al. 1998) (Sec. 2.6). The precipitation of solids is accompanied by a decrease of individual elements and compounds, resulting in lower total dissolved solids (TDS) in the mine waters. The precipitated salts can also be redissolved. In particular, the exposure of soluble mineral salts to water, through ground water flow changes or rainfall events, will cause their dissolution.

Secondary salts can be classified as readily soluble, less soluble, and insoluble. Examples of readily soluble secondary salts are listed in Table 3.3. Soluble salts can be further classified as acid producing, non-acid producing, and acid buffering phases. Above all, the formation of soluble Fe^{3+} and Al^{3+} salts as well as of Fe^{2+}, Fe^{3+} and Mn^{2+} sulfate salts influences the solution pH since their formation can consume or generate hydrogen ions (Sec. 2.6.3). However, such a classification scheme is too simplistic and does not consider the physical, chemical and biological environments in which the minerals dissolve. The solubility of secondary minerals is highly variable and primarily pH, Eh and solution chemistry dependent.

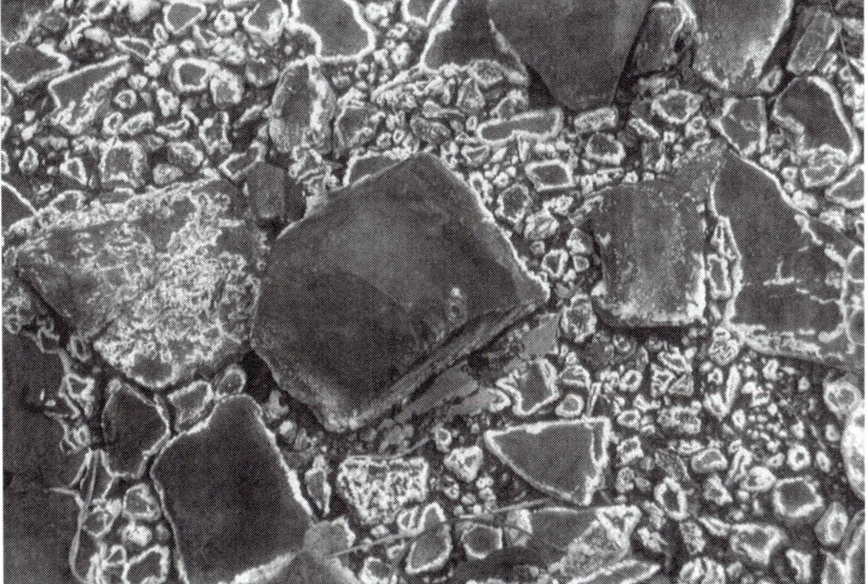

Fig. 3.5. Secondary minerals (iron oxyhydroxides, alumnium hydroxides, gypsum, jarosite) encrusting stream sediments of the acid Dee River (pH 3), downstream of the historic Mt. Morgan copper mine, Australia. Field of view 70 cm

Table 3.3. Examples of soluble secondary minerals classified according to their ability to generate or buffer any acid upon dissolution (after Alpers et al. 1994; Keith et al. 1999)

Acid producing soluble mineral		Formula
Fe^{2+} and Mn^{2+} sulfates	Melanterite	$FeSO_4 \cdot 7H_2O$
	Ferrohexahydrite	$FeSO_4 \cdot 6H_2O$
	Siderotil	$FeSO_4 \cdot 5H_2O$
	Rozenite	$FeSO_4 \cdot 4H_2O$
	Szomolnokite	$FeSO_4 \cdot H_2O$
	Apjohnite	$MnAl_2(SO_4)_4 \cdot 22H_2O$
	Halotrichite	$FeAl_2(SO_4)_4 \cdot 22H_2O$
Mixed Fe^{2+}-Fe^{3+} sulfates and hydroxysulfates	Copiapite	$Fe_5(SO_4)_6(OH)_2 \cdot 20H_2O$
	Ferricopiapite	$Fe_5(SO_4)_6O(OH)_2 \cdot 20H_2O$
	Bilinite	$Fe_3(SO_4)_4 \cdot 22H_2O$
	Römerite	$Fe_3(SO_4)_4 \cdot 14H_2O$
	Voltaite	$K_2Fe_8Al(SO_4)_{12} \cdot 18H_2O$
Fe^{3+} sulfates	Rhomboclase	$HFe(SO_4)_2 \cdot 4H_2O$
	Kornelite	$Fe_2(SO_4)_3 \cdot 7H_2O$
	Coquimbite	$Fe_2(SO_4)_3 \cdot 9H_2O$
Al^{3+} hydroxysulfate	Jurbanite	$Al(SO_4)(OH) \cdot 5H_2O$
Al sulfates	Kalinite	$K_2Al_2(SO_4)_4 \cdot 22H_2O$
	Pickeringite	$MgAl_2(SO_4)_4 \cdot 22H_2O$
	Alunogen	$Al_2(SO_4)_3 \cdot 17H_2O$
Non acid producing soluble mineral		**Formula**
Mg sulfates	Epsomite	$MgSO_4 \cdot 7H_2O$
	Hexahydrite	$MgSO_4 \cdot 6H_2O$
Zn sulfates	Goslarite	$ZnSO_4 \cdot 7H_2O$
	Bianchite	$ZnSO_4 \cdot 6H_2O$
	Gunningite	$ZnSO_4 \cdot H_2O$
Ca sulfates	Gypsum	$CaSO_4 \cdot 2H_2O$
	Bassanite	$CaSO_4 \cdot 0.5H_2O$
	Anhydrite	$CaSO_4$
Cu sulfates	Chalcanthite	$CuSO_4 \cdot 5H_2O$
As minerals	Arsenolite	As_2O_3
Acid buffering soluble mineral		**Formula**
Metal carbonates	Smithsonite	$ZnCO_3$
	Otavite	$CdCO_3$
	Malachite	$Cu_2(CO_3)(OH)_2$
	Azurite	$Cu_3(CO_3)_2(OH)_2$
	Hydrozincite	$Zn_5(CO_3)_2(OH)_6$

Jarosite-type phases can be viewed as less soluble phases as their dissolution is strongly influenced by the solution's pH (Smith et al. 2006). Their dissolution can be a two-step process. For example, alunite ($KAl_3(SO_4)_2(OH)_6$) and jarosite ($KFe_3(SO_4)_2(OH)_6$) dissolution initially consumes acid (Reaction 3.1). This may be followed by the precipitation of gibbsite ($Al(OH)_3$), which generates acid (Reaction 3.2). The overall combined Reaction 3.3 illustrates that the dissolution of alunite and jarosite produces acid:

$$KAl_3(SO_4)_2(OH)_{6(s)} + 6\,H^+_{(aq)} \leftrightarrow K^+_{(aq)} + 3\,Al^{3+}_{(aq)} + 6\,H_2O_{(l)} + 2\,SO^{2-}_{4(aq)} \tag{3.1}$$

$$3\,Al^{3+}_{(aq)} + 9\,H_2O_{(l)} \leftrightarrow 3\,Al(OH)_{3(s)} + 9\,H^+_{(aq)} \tag{3.2}$$

(Reaction 3.1 + Reaction 3.2 = Reaction 3.3)

$$KAl_3(SO_4)_2(OH)_{6(s)} + 3\,H_2O_{(l)} \leftrightarrow K^+_{(aq)} + 3\,Al(OH)_{3(s)} + 2\,SO^{2-}_{4(aq)} + 3\,H^+_{(aq)} \tag{3.3}$$

Sulfate salts are particularly common in AMD environments and soluble under oxidizing conditions, especially the Ca, Mg, Fe^{2+}, Fe^{3+} and Mn^{2+} sulfate salts (Cravotta 1994; Jambor et al. 2000a,b). A decrease in pH is principally caused by the dissolution of Fe^{2+} sulfate salts, which are capable of producing acidity due to the hydrolysis of Fe^{3+}. For instance, melanterite ($FeSO_4 \cdot 7\,H_2O$) can control the acidity of mine waters (Frau 2000). Melanterite dissolution releases hydrogen ions as shown by the following equations (White et al. 1999):

$$FeSO_4 \cdot 7\,H_2O_{(s)} \leftrightarrow Fe^{2+}_{(aq)} + SO^{2-}_{4(aq)} + 7\,H_2O_{(l)} \tag{3.4}$$

$$4\,Fe^{2+}_{(aq)} + 4\,H^+_{(aq)} + O_{2(g)} \rightarrow 4\,Fe^{3+}_{(aq)} + 2\,H_2O_{(l)} \tag{3.5}$$

$$Fe^{3+}_{(aq)} + 3\,H_2O_{(l)} \leftrightarrow Fe(OH)_{3(s)} + 3\,H^+_{(aq)} \tag{3.6}$$

The release of Fe^{2+} into water does not necessarily result in only the precipitation of iron hydroxides but can also trigger more sulfide oxidation (Keith et al. 1999; Alpers and Nordstrom 1999; Plumlee 1999). The dissolution of melanterite releases Fe^{2+} which can be oxidized to Fe^{3+}. Any pyrite may subsequently be oxidized by Fe^{3+} as shown by the following equation:

$$FeS_{2(s)} + 14\,Fe^{3+}_{(aq)} + 8\,H_2O_{(l)} \rightarrow 15\,Fe^{2+}_{(aq)} + 16\,H^+_{(aq)} + 2\,SO^{2-}_{4(aq)} \tag{3.7}$$

Similarly, the dissolution of römerite ($Fe_3(SO_4)_4 \cdot 14\,H_2O$), halotrichite ($FeAl_2(SO_4)_4 \cdot 22\,H_2O$), and coquimbite ($Fe_2(SO_4)_3 \cdot 9\,H_2O$) generates acid (Cravotta 1994; Rose and Cravotta 1999):

$$Fe_3(SO_4)_4 \cdot 14\,H_2O_{(s)} \leftrightarrow 2\,Fe(OH)_{3(s)} + Fe^{2+}_{(aq)} + 4\,SO^{2-}_{4(aq)} + 6\,H^+_{(aq)} + 8\,H_2O_{(l)} \tag{3.8}$$

$$\begin{aligned}4\,FeAl_2(SO_4)_4 \cdot 22\,H_2O_{(s)} + O_{2(aq)} \leftrightarrow\ & 4\,Fe(OH)_{3(s)} + 8\,Al(OH)_{3(s)} + 54\,H_2O_{(l)} \\ & + 16\,SO^{2-}_{4(aq)} + 32\,H^+_{(aq)}\end{aligned} \tag{3.9}$$

$$Fe_2(SO_4)_3 \cdot 9\,H_2O_{(s)} \leftrightarrow 2\,Fe(OH)_{3(s)} + 3\,SO^{2-}_{4(aq)} + 6\,H^+_{(aq)} + 3\,H_2O_{(l)} \tag{3.10}$$

Generalized reactions for the dissolution of Fe^{3+} and Al^{3+} salts and of Fe^{2+}, Fe^{3+}, and Mn^{2+} sulfate salts can be written as follows:

$$(Fe^{3+} \text{ and } Al^{3+} \text{ salts; } Fe^{2+}, Fe^{3+}, \text{ and } Mn^{2+} \text{ sulfate salts})_{(s)} + n\,H^+_{(aq)} \leftrightarrow (Fe^{2+}, Fe^{3+}, Al^{3+}, Mn^{2+})^{n+}_{(aq)} + \text{anions}^{n-}_{(aq)} + n\,H^+_{(aq)} + n\,H_2O_{(l)} \quad (3.11)$$

$$(Fe^{2+}, Fe^{3+}, Al^{3+}, Mn^{2+})^{n+}_{(aq)} + n\,H_2O_{(l)} \leftrightarrow \text{salts-}n\,H_2O_{(s)} + n\,H^+_{(aq)} \quad (3.12)$$

(Reaction 3.11 + Reaction 3.12 = Reaction 3.13)

$$(Fe^{3+} \text{ and } Al^{3+} \text{ salts; } Fe^{2+}, Fe^{3+}, \text{ and } Mn^{2+} \text{ sulfate salts})_{(s)} + n\,H_2O_{(l)} \leftrightarrow \text{cations}^{n+}_{(aq)} + \text{anions}^{n-}_{(aq)} + n\,H^+_{(aq)} + \text{salts-}n\,H_2O_{(s)} \quad (3.13)$$

Iron sulfate minerals can be significant sources of acidity and sulfate when later dissolved. Release of Fe^{2+} from these salts can also trigger more sulfide oxidation. Furthermore, other forms of sulfur such as native sulfur (S^0) and thiosulfate ($S_2O_3^{2-}$) can be intermediate products that tend to be oxidized to sulfate under oxidizing conditions. Moreover, many of the secondary minerals allow substitution of iron and aluminium by numerous other metals (e.g. substitution of Fe by Cu and Zn in melanterite). As a result, dissolution of secondary minerals will lead to the release of major and minor metals, and metalloids (Lin 1997). In contrast, the dissolution of soluble aluminium (e.g. alunogen: $Al_2(SO_4)_3 \cdot 17\,H_2O$), magnesium (e.g. epsomite: $MgSO_4 \cdot 7\,H_2O$), or calcium sulfate minerals (e.g. gypsum: $CaSO_4 \cdot 2\,H_2O$) does not generate any acid. Their dissolution does not influence mine water pH (Keith et al. 1999). Other soluble secondary minerals are acid buffering, and a variety of metal carbonates such as smithsonite ($ZnCO_3$), malachite ($Cu_2(CO_3)(OH)_2$), and azurite ($Cu_3(CO_3)_2(OH)_2$) are effective acid consumers (Table 3.3).

The presence of soluble salts in unsaturated ground water zones of waste rock dumps, tailings dams, and other waste repositories is important because their dissolution will lead to a change in the chemistry of drainage waters. Evaporation, especially in arid and seasonally dry regions, causes the precipitation of secondary minerals which can store metal, metalloids, sulfate, and hydrogen ions. The formation of soluble secondary, sulfate-, metal- and metalloid-bearing minerals slows down sulfate, metal and metalloid mobility but only temporarily until the next rainfall (Bayless and Olyphant 1993; Keith et al. 1999). Rapid dissolution of soluble salts and hydrolysis of dissolved Fe^{3+} may occur during the onset of the wet season or the beginning of spring. This in turn can result in exceptionally high sulfate, metal and metalloid concentrations as well as strong acidity of waters during the initial flushing event (Kwong et al. 1997; Keith et al. 2001). In particular, the dissolution of the iron sulfates releases incorporated sulfate, metals, metalloids, and acidity to ground and surface waters. The pH of drainage waters may eventually change to more neutral conditions due to increased dilution. Such neutral pH values will limit heavy metal mobility. Upon changes to drier conditions, evaporation will again cause the precipitation of secondary minerals. This type of wetting and drying cycle can result in dramatic seasonal variations in acidity, and metal and metalloid loads of seepages and local streams (Bayless and Olyphant 1993; Keith et al. 1999). The production of contaminant pulses at the onset of rainfall is common to mine sites in seasonally dry climates. In these environments, seasonal variations in the chemistry of drainage waters from sulfidic mine wastes are caused by the dissolution and precipitation of soluble mineral precipitates.

3.5.3 Coprecipitation

Coprecipitation refers to the removal of a trace constituent from solution which occurs at the same time as the precipitation of a major salt. This eventuates even when the solubility product of the trace constituent is not exceeded. The precipitating solid incorporates the minor constituent as an impurity into the crystal lattice. Various minerals can thereby host a wide variety of cations as "impurities". The cations can be incorporated into the crystal lattice of the minerals via single or coupled substitution. For example, a large number of ions have been reported to substitute for iron in the goethite crystal lattice (e.g. Al, Cr, Ga, V, Mn, Co, Pb, Ni, Zn, Cd) (Cornell and Schwertmann 1996). Also, jarosite has been found to incorporate various elements into its mineral structure (e.g. Cu, Zn, Pb, K, Na, Ca) (Levy et al. 1997).

3.5.4 Adsorption and Desorption

Trace elements move between dissolved and particulate phases. Adsorption is the term which refers to the the removal of ions from solution and their adherence to the surfaces of solids (Langmuir 1997). The attachment of the solutes onto the solid phases does not represent a permanent bond, and the adsorption is based on ionic attraction of the solutes and the solid phases (Smith 1999). The solid phases can be of organic or inorganic composition and of negative or positive charge attracting dissolved cations and anions, respectively. Adsorption reactions are an important control on the transport, concentration and fate of many elements in waters, including AMD waters.

Adsorption may occur in various AMD environments (Fuge et al. 1994; Bowell and Bruce 1995; Swedlund and Webster 2001). It may occur on iron- and aluminium-rich particulates and clay particles suspended in mine waters, on precipitates at seepage points, or on clayey sediments of stream beds and ponds. Different ions thereby exhibit different adsorption characteristics. Generally, solid compounds adsorb more anions at low pH and more cations at near neutral pH. In addition, the kind of metal adsorbed and the extent of metal adsorption is a function of: (*a*) the solution pH; (*b*) the presence of complexing ligands; and (*c*) the metal concentration of the AMD. Arsenic and lead are the most effectively adsorbed metals at acid pH values, whereas zinc, cadmium, and nickel are adsorbed at near-neutral pH values (Plumlee et al. 1999). Therefore, when AMD waters are gradually neutralized, various secondary minerals precipitate and adsorb metals. Adsorption is selective, and the chemical composition of the water changes as the pH increases. Ions are removed from solution by this process, and metal-rich sediment accumulates.

While sediment may remove ions from solution, it may also release adsorbed metals if the water is later acidified. In contrast, other elements such as arsenic and molybdenum may desorb at near-neutral or higher pH values to form oxyanions in the water (e.g. AsO_4^{3-}) (Jönsson and Lövgren 2000). Similarly, uranium, copper, and lead may desorb at near-neutral or higher pH values to form aqueous carbonate complexes. Sulfate may also be released from ferric precipitates if pH values raise to neutral or even alkaline values (Plumlee et al. 1999; Rose and Elliott 2000). As a result, sulfate, metal, and metalloid ions desorb and regain their mobility at near-neutral or alkaline pH values, and dissolved sulfate, metal, and metalloid concentrations of mine waters may in fact increase with increasing pH.

Sorption sites of particulates represent only temporary storage facilities for dissolved metals, metalloids and sulfate. In a worst case scenario, if excessive neutralization is used to treat AMD effected streams, sulfate, metals, and metalloids previously fixed in stream sediments may then be redissolved by the treated water. Thus, remediation of AMD waters should raise the pH only to values necessary to precipitate and adsorb metals.

3.5.5 Eh-pH Conditions

The solubility of many dissolved heavy metals is influenced by the pH of the solution. The generation of low pH waters due to sulfide oxidation, or the presence of process chemicals such as sulfuric acid, enhances the dissolution of many elements. This acidity significantly increases the mobility and bioavailability of elements, and the concentration of total dissolved solids in mine waters. Most of the metals have increasing ionic solubilities under acid, oxidizing conditions, and the metals are not adsorbed onto solids at low pH. In many cases, the highest aqueous concentrations of heavy metals are associated with oxidizing, acid conditions.

Precipitation of many of the dissolved metals occurs during neutralization of low pH drainage waters, for example, due to mixing with tributary streams or due to the movement of the seepage water over alkaline materials such as carbonate bedrocks. The metals are adsorbed onto solid phases, particularly precipitating iron-rich solids. Alternatively, the metals are incorporated into secondary minerals coating the seepage area or stream bed. Generally, as pH increases, aqueous metal species are inclined to precipitate as hydroxide, oxyhydroxide or hydroxysulfate phases (Berger et al. 2000; Munk et al. 2002). The resultant drainage water contains the remaining dissolved metals and products of the buffering reactions. Therefore, with increasing pH the dissolved metal content of mining influenced waters decreases.

While neutralization of AMD causes the removal of most metals, neutral to alkaline mine waters are known to contain elevated metal and metalloid concentrations. In fact, oxidized neutral to alkaline mine waters can have very high metal (Cd, Cu, Hg, Mn, Mo, Ni, Se, U, Zn) and metalloid (As, Sb) values (Carroll et al. 1998; Lottermoser et al. 1997b, 1999; Pettit et al. 1999; Plumlee 1999; Plumlee et al. 1999; Younger 2000; Schmiermund 2000; Scharer et al. 2000). Such waters are of environmental concern because the elements tend to remain in solution, despite pH changes. The elements can be carried for long distances downstream of their source, and they may adversely impact on the quality of receiving water bodies.

The ability of water to transport metals is not only controlled by pH but also by the Eh of the solution. The reduction-oxidation potential as measured by Eh affects the mobility of those metals which can exist in several oxidation states. Metals such as chromium, molybdenum, selenium, vanadium, and uranium are much more soluble in their oxidized states (e.g. U^{6+}, Cr^{6+}) than in their reduced states (e.g. U^{4+}, Cr^{3+}). Oxygenated water may oxidize metals present in their reduced, immobile state and allow mobility. These salient aspects of aqueous element chemistry are commonly described by Eh-pH diagrams. The diagrams illustrate the stability and instability of minerals under particular Eh-pH conditions and show the ionic element species present in solution (Brookins 1988).

3.5.6 Heavy Metals

The oxidation of various sulfide minerals will release their major and trace elements, including numerous heavy metals (Table 2.1). In some cases, the degradation of organic matter particularly in carbonaceous rocks (i.e. black shales) may release metals such as nickel to pore and drainage waters (e.g. Wengel et al. 2006; Falk et al. 2006). As a result, elevated concentrations of one or more heavy metals are characteristic of waters in contact with oxidizing sulfidic and carbonaceous rocks. The controls on heavy metal concentrations in mine waters are numerous, highly metal specific, and controlled by environmental conditions such as pH.

Heavy metals can occur in various forms in AMD waters. A metal is either dissolved in solution as ion and molecule, or it exists as a solid mass. Dissolved metal species include cations (e.g. Cu^{2+}), simple radicals (e.g. UO_2^{2+}), and inorganic (e.g. $CuCO_3$) and organic complexes (e.g. $Hg(CH_3)_2$). Metals may also be present in a solid form as substitutions in precipitates (e.g. Cu in eugsterite $Na_4Ca(SO_4)_3 \cdot 2H_2O$), as mineral particles (e.g. cerussite $PbCO_3$), and in living biota (e.g. Cu in algae) (Brownlow 1996; Smith and Huyck 1999). There is also a transitional state whereby very small particles, so-called "*colloids*", are suspended in water (Stumm and Morgan 1995). A colloid can be defined as a stable electrostatic suspension of very small particles (<10 µm) in a liquid (Stumm and Morgan 1995). The composition of colloids can be exceptionally diverse and includes organic and inorganic substances. Metals can be incorporated into organic (e.g. Pb fulvic acid polymers) or inorganic colloids (e.g. FeOOH), or are adsorbed onto them (e.g. Ni on clays). The stability of these colloids is influenced by a range of physical, chemical and biological changes of the solution (Brownlow 1996; Ranville and Schmiermund 1999). Upon such changes, colloids will aggregate into larger particles; that is, they undergo "*flocculation*" and occur as *suspended particles* in the water. Iron- and aluminium-rich colloids and suspended particles are especially common in AMD waters (Schemel et al. 2000; Zänker et al. 2002).

Metals may be transported in mine waters in various speciations. In AMD waters, most metals occur as simple metal ions or as sulfate complexes. In neutral and alkaline mine waters, elevated metal and metalloid concentrations are promoted by the formation of oxyanions (e.g. AsO_4^{3-}), aqueous metal complexes (e.g. U carbonate complexes, Zn sulfate and hydroxide complexes) as well as the lack of adsorption onto and coprecipitation with secondary iron hydroxides (Plumlee et al. 1999).

The size of the metal species progressively increases from cation to metal particle in living biota. The different size of the metal species and the common procedure to filter water prior to chemical analysis have a distinct implication on the analytical result. A common filter pore size used is 0.45 µm. Such filters will allow significant amounts of colloidal material to pass through, and analyses of these samples will reflect dissolved and colloidal constituents (Brownlow 1996; Ranville and Schmiermund 1999). For this reason, the US EPA has suggested the collection of both unfiltered and filtered water samples (Ranville and Schmiermund 1999). If significant differences are found in the metal concentrations, it is possible that the metals are transported via colloids. If detailed information on the speciation of metals is needed, other analytical methods need to be performed, including ultrafiltration and the use of exchange resins or Diffusion Gradient in Thin-Films (DGTs).

3.5.7 The Iron System

Elevated iron concentrations in mine waters are an obvious by-product of the oxidation of pyrite, pyrrhotite or any other iron-bearing sulfide. Dissolved iron is found in two oxidation states, ferrous (Fe^{2+}) and ferric (Fe^{3+}). Iron may also combine with organic and inorganic ions, so iron can be present in mine waters in several forms (e.g. Fe^{2+}, Fe^{3+}, $Fe(OH)^{2+}$, $Fe(OH)_2^+$, $Fe(SO_4)^+$, $Fe(SO_4)_2^-$).

Upon weathering of iron-bearing sulfides, iron enters the solutions as Fe^{2+}. Pore and drainage waters of sulfidic materials are commonly oxygen deficient, and reducing conditions are often prevalent. The rate of iron oxidation from Fe^{2+} to Fe^{3+} is now controlled by the pH of the mine water, the amount of dissolved oxygen, and the presence of iron oxidizing bacteria. Under reducing abiotic conditions and as long as the pH of the water remains less than approximately 4 to 4.5, the dissolved iron will remain in the ferrous state. Abiotic oxidation of Fe^{2+} to Fe^{3+} is relatively slow and strongly inhibited at a pH less than approximately 4.5 (Ficklin and Mosier 1999). However, in the presence of iron oxidizing bacteria, the oxidation rate of Fe^{2+} to Fe^{3+} is increased by five to six orders of magnitude over the abiotic rate (Singer and Stumm 1970). Therefore, AMD waters with bacteria, low dissolved oxygen concentrations, and acid to near neutral pH values can have elevated iron concentrations, with iron present as a mixture of Fe^{2+} and Fe^{3+}. Significant dissolved concentrations of Fe^{3+} only occur at a low pH; the exact pH value depends on the iron and sulfate contents of the mine water. The Fe^{2+} and Fe^{3+} ions participate in the oxidation of sulfides (Sec. 2.3.1). Alternatively, in the presence of abundant molecular oxygen and above pH values of approximately 3, the Fe^{2+} is oxidized to Fe^{3+} as illustrated in the following oxidation reaction:

$$4\,Fe^{2+}_{(aq)} + O_{2(g)} + 4\,H^+_{(aq)} \rightarrow 4\,Fe^{3+}_{(aq)} + 2\,H_2O_{(l)} \tag{3.14}$$

This Fe^{3+} will become insoluble and precipitates as ferric hydroxide, oxyhydroxide, and oxyhydroxysulfate colloids and particulates. The precipitation occurs as a result of the following hydrolysis reaction:

$$Fe^{3+}_{(aq)} + 3\,H_2O_{(l)} \leftrightarrow Fe(OH)_{3(s)} + 3\,H^+_{(aq)} \tag{3.15}$$

This reaction also generates hydrogen acidity. If appreciable amounts of Fe^{2+} are present in neutral mine drainage waters, oxidation of the Fe^{2+} to Fe^{3+} will result in precipitation of large amounts of Fe^{3+} hydroxides, and the neutral solution will become acid due to abundant hydrolysis reactions (Reaction 3.15). Oxidation of Fe^{2+} (Reaction 3.14) and hydrolysis of Fe^{3+} (Reaction 3.15) do not take place until the water is aerated. Nevertheless, further Fe^{2+} may be oxidized without the help of oxygen by oxidation at the surface of previously formed Fe^{3+} hydroxides. Such an iron removal process is referred to as autocatalytic iron oxidation.

The dissolved iron concentration and speciation (i.e. Fe^{2+} or Fe^{3+}) are strongly Eh and pH dependent. In addition, the dissolved iron concentration of AMD waters is influenced by factors other than the presence of iron oxidizing bacteria. For example, solar radiation and associated photolytic processes increases the dissolved Fe^{2+} and reduces the dissolved Fe^{3+}. Iron photoreduction involves the absorption of UV radiation by Fe^{3+} species, resulting in Fe^{2+} and OH^- ions. As a consequence, the colloidal Fe^{3+} hydroxide concentrations in oxygenated surface waters can be reduced during daytime or summer (Nordstrom and Alpers 1999a). While seasonal variations in the composition of AMD waters are typically controlled by climatic factors (e.g. evaporation, precipitation, runoff events and volumes) (Herbert 2006), other factors such as the water temperature can also impact indirectly on the chemistry of mine waters. Higher water temperatures favour the optimum rate of bacterially mediated iron oxidation (Butler and Seitz 2006).

AMD waters typically precipitate iron hydroxides, oxyhydroxides or oxyhydroxysulfates (Reaction 3.17) which are collectively termed "*ochres*", "*boulder coats*", or with the rather affectionate term "*yellow boy*". The iron solids commonly occur as colourful bright reddish-yellow to yellowish-brown stains, coatings, suspended particles, colloids, gelatinous flocculants, and precipitates in AMD affected waters, streams and seepage areas (e.g. Zänker et al. 2002; Kim and Kim 2004; Lee and Chon 2006; España et al. 2005). The poor crystallinity of ochre precipitates has led some authors to the conclusion that these substances should be referred to as "amorphous ferric hydroxides" or "hydrous ferric oxides" (i.e. HFO). The iron precipitates, in fact, consist of a variety of amorphous, poorly crystalline and/or crystalline Fe^{3+} hydroxides, oxyhydroxides and oxyhydroxy-sulfate minerals. Moreover, the ochres may contain other crystalline solids including sulfates, oxides, hydroxides, arsenates, and silicates (Table 2.5).

Iron minerals such as jarosite ($KFe_3(SO_4)_2(OH)_6$), ferrihydrite ($Fe_5HO_8 \cdot 4H_2O$), schwertmannite ($Fe_8O_8(SO_4)(OH)_6$), and the FeOOH polymorphs goethite, feroxyhyte, akaganéite, and lepidocrocite are very common. Different iron minerals appear to occur in different AMD environments (Bigham 1994; Bigham et al. 1996; Carlson and Kumpulainen 2000). Low pH (<3), high sulfate concentrations (>3 000 mg l^{-1}) and sustained bacterial activity cause the formation of jarosite. Schwertmannite is most commonly associated with mine effluents with pH from 2 to 4 and medium dissolved sulfate concentrations (1 000 to 3 000 mg l^{-1}), whereas ferrihydrite is associated with mine drainage with a pH of about 6 and higher (Bigham 1994; Bigham et al. 1996; Carlson and Kumpulainen 2000; Bigham and Nordstrom 2000; Lee et al. 2002; Murad and Rojik 2003; España et al. 2005). Goethite (α-FeOOH) may be formed at near neutral conditions, or when low pH ($pH < 4$), low sulfate (<1 000 mg l^{-1}) solutions are neutralized by carbonate-rich waters. Whether such a simplified iron mineral occurrence is valid remains to be confirmed with further field and laboratory studies. The mineralogy of secondary iron precipitates is complex and depends on solution composition, pH, temperature, redox conditions, and the rate of Fe^{2+} oxidation (Alpers et al. 1994; Jönsson et al. 2005, 2006).

Various soluble Fe^{2+} sulfates such as melanterite precipitate from AMD waters. These secondary salts can be regarded as intermediate phases. Melanterite may dehydrate to less hydrous Fe^{2+} sulfates. The Fe^{2+} of these reduced minerals will eventually be oxidized and hydrolyzed to form one or more of the FeOOH polymorphs. Also, when iron is precipitated from solutions enriched in sulfate, these anions often combine with hydroxyl (OH^-) to form metastable schwertmannite. Schwertmannite may convert to goethite as it is metastable with respect to goethitegoethite (Schroth and Parnell 2005; Acero et al. 2006). Similarly, ferrihydrite and the goethite polymorphs feroxyhyte, akaganéite, and lepidocrocite are thought to be metastable. Over time, they may ultimately convert and recystallize forming goethite and hematite, respectively (Murad et al. 1994; Bigham et al. 1996; Rose and Cravotta 1999). Therefore, a distinct paragenetic sequence of secondary iron minerals may occur (Jerz and Rimstidt 2003).

The formation of secondary iron minerals also impacts on the behaviour of other elements. Freshly precipitated iron minerals have a fine particle size and a large surface area which favours the adsorption of metals. In addition, coprecipitation of metals occurs with the formation of the secondary solids. As a result, the iron ochre minerals can contain significant concentrations of metals through coprecipitation and adsorption. The precipitates may contain apart from iron and sulfur a number of other elements (e.g. Al, Cr, Co, Cu, Pb, Mn, Ni, REE, Sc, U, Y, Zn) due to coprecipitation and adsorption processes (Rose and Ghazi 1998; Dinelli et al. 2001; Lee et al. 2002; Swedlund et al. 2003; Schroth and Parnell 2005; Regenspurg and Pfeiffer 2005; Sidenko and Sherriff 2005; Lee and Chon 2006). In particular, arsenic readily adsorbs to and is incorporated into precipitated iron minerals (Foster et al. 1998). These metal-rich suspended particles and colloidal materials may be deposited in stream sediments or transported farther in ground and surface waters. Colloidal iron precipitates are exceptionally small. Therefore, such materials with adsorbed and incorporated trace elements can represent important transport modes for metals and metalloids in mine environments and streams well beyond the mine site (Schmiermund 1997; Smith 1999).

3.5.8 The Aluminium System

High aluminium and silicon concentrations in acid waters derive from the weathering of aluminosilicate minerals such as clays, or from the dissolution of secondary minerals such as alunite ($KAl_3(SO_4)_2(OH)_6$). Aluminium is least soluble at a pH between 5.7 and 6.2; above and below this range aluminium may be solubilized. Dissolved aluminium is found in only one oxidation state as Al^{3+}. Aluminium may combine with organic and inorganic ions; hence, it can be present in mine waters in several forms (e.g. Al^{3+}, $Al(OH)^{2+}$, $Al(OH)_2^+$, $Al_2(OH)_2^{4+}$, $Al(SO_4)^+$, $Al(SO_4)_2^-$) (Nordstrom and Alpers 1999a). Aluminium is similar to iron in its tendency to precipitate as hydroxides, oxyhydroxides, and oxyhydroxysulfates in waters which have increased their pH from acid to near neutral conditions. These precipitated phases are predominantly amorphous, colloidal substances. Aggregation of these phases may eventually form microcrystalline gibbsite ($Al(OH)_3$) and other solids (Schemel et al. 2000; Munk et al. 2002). Dissolved aluminium concentrations are strongly pH dependent, and the formation of secondary aluminium minerals, colloids, and amorphous substances controls the aqueous aluminium concentrations (Nordstrom and Alpers 1999a). While a change to more neutral pH conditions results in the precipitation of aluminium hydroxides, the formation of aluminium hydroxides such as gibbsite also generates acid. The dissolved trivalent aluminium thereby hydrolyses in a manner similar to ferric iron:

$$Al^{3+}_{(aq)} + 3\,H_2O_{(l)} \leftrightarrow Al(OH)_{3(s)} + 3\,H^+_{(aq)} \quad (3.16)$$

The solid phase resulting from Reaction 3.16 typically forms a white precipitate, which is commonly amorphous and converts to gibbsite upon ageing. In aqueous environments with turbulence, the phase may occur as white foam floating on the water surface. As in the case of dissolved iron, flocculation and precipitation of dissolved aluminium will add colloidal and suspended matter to the water column, causing increased turbidity. In some mine waters, the aluminium concentrations are limited by the precipitation of aluminium-bearing sulfate minerals such as jarosite. Jarosite ($KFe_3(SO_4)_2(OH)_6$) forms a solid solution with alunite ($KAl_3(SO_4)_2(OH)_6$), and alunite-jarosite minerals commonly form because of evaporation of AMD seepage and pore waters (Alpers et al. 1994). Jarosite is a diagnostic yellow precipitate and occurs in mine drainage waters at pH values of less than 2.5 (Bigham 1994). The most prevalent type of jarosite is a potassium-type formed with available dissolved K^+ in the system. Other jarosite-type phases include the sodium-rich natrojarosite and the lead-rich plumbojarosite. The Al^{3+}, K^+ and Na^+ derive from dissolved ions in solution or from the decomposition of alkali feldspars, plagioclase, biotite, and muscovite. Jarosite-type phases are a temporary storage for acidity, sulfate, iron, aluminium, alkalis, and metals. The minerals release these stored components upon redissolution in a strongly acid environment and form solid Fe^{3+} hydroxides, according to the following equilibrium reactions (Hutchison and Ellison 1992; Levy et al. 1997):

$$KFe_3(SO_4)_2(OH)_{6(s)} + 6\,H^+_{(aq)} \leftrightarrow K^+_{(aq)} + 3\,Fe^{3+}_{(aq)} + 6\,H_2O_{(l)} + 2\,SO^{2-}_{4(aq)} \quad (3.17)$$

$$3\,Fe^{3+}_{(aq)} + 9\,H_2O_{(l)} \leftrightarrow 3\,Fe(OH)_{3(s)} + 9\,H^+_{(aq)} \quad (3.18)$$

(Reaction 3.17 + Reaction 3.18 = Reaction 3.19)

$$KFe_3(SO_4)_2(OH)_{6(s)} + 3\,H_2O_{(l)} \leftrightarrow K^+_{(aq)} + 3\,Fe(OH)_{3(s)} + 2\,SO^{2-}_{4(aq)} + 3\,H^+_{(aq)} \quad (3.19)$$

3.5.9 The Arsenic System

Elevated arsenic concentrations are commonly found in tailings and sulfidic mine wastes of gold, copper-gold, tin, lead-zinc, and some uranium ores. The common occurrence of arsenic in gold deposits is explained by the similar solubility of arsenic and gold in the ore forming fluids. Consequently, mine waters of many gold mining operations are enriched in arsenic (Marszalek and Wasik 2000; Lazareva et al. 2002; Craw and Pacheco 2002; Gieré et al. 2003). Arsenic in mine waters generally originates from the oxidation of arsenopyrite (FeAsS), orpiment (As_2S_3), realgar (AsS), enargite (Cu_3AsS_4), and arsenical pyrite and marcasite (FeS_2) (Foster et al. 1998; Roddick-Lanzilotta et al. 2002). Oxidation of these sulfides results in the release of arsenic, sulfate, and metals.

The aqueous chemistry of arsenic differs significantly from that of heavy metals. Mobilization of heavy metals is controlled by pH and Eh conditions and occurs primarily in low pH, oxidizing environments. In contrast, arsenic is mobile over a wide pH range (i.e. extremely acid to alkaline), and mine waters of an oxidized, neutral to alkaline pH nature can contain several mg l^{-1} of arsenic (Marszalek and Wasik 2000; Williams 2001; Roddick-Lanzilotta et al. 2002). Thus, contamination of mine waters by arsenic is not exclusive to AMD waters.

Arsenic exists in natural waters in two principal oxidation states, as As^{3+} in arsenite (AsO_3^{3-}) and as As^{5+} in arsenate (AsO_4^{3-}) (Yamauchi and Fowler 1994). In oxygenated environments, As^{5+} is the stable species. In more reduced environments, As^{3+} is the dominant form. The more reduced species As^{3+} is more soluble, mobile and toxic than As^{5+} (Yamauchi and Fowler 1994). The oxidation of As^{3+} to As^{5+} is relatively fast and increases with pH and salinity in the presence of particular bacteria and protozoa (Casiot et al. 2003; Casiot et al. 2004).

Iron exerts an important control on the mobility of arsenic in water (Bednar et al. 2005). In an oxidizing environment with a pH greater than 3, hydrous ferric oxides (HFO) are abundantly precipitated. Dissolved arsenic species are adsorbed by and coprecipitated with these ferric hydroxides, and As^{5+} is thereby more strongly sorbed than As^{3+} (Manceau 1995; Roddick-Lanzilotta et al. 2002). Adsorption onto and coprecipitation with Fe^{3+} hydroxides are very efficient removal mechanisms of arsenic from mine waters. The formation of jarosite, schwertmannite and ferrihydrite may also remove arsenic from solution (Fukushi et al. 2003; Gault et al. 2005; Courtin-Nomade et al. 2005). In general, precipitation of Fe^{3+} from mine waters is accompanied by a reduction in the concentration of dissolved arsenic.

The solubility of arsenic is also limited by: (*a*) the adsorption of arsenic onto clays; (*b*) the formation of amorphous iron sulfoarsenates and secondary arsenic minerals such as scorodite ($FeAsO_4 \cdot 2H_2O$), arsenolite (As_2O_3), or iron-calcium arsenates such as pharmacolite ($Ca(AsO_3OH) \cdot 2H_2O$); and (*c*) the substitution of arsenic for sulfate in jarosite and gypsum, and for carbonate in calcite (Foster et al. 1998; Savage et al. 2000; Gieré et al. 2003; Lee et al. 2005). In turn, the dissolution of arsenic salts will lead to arsenic release and mobilization. For instance, arsenolite (As_2O_3) is a high solubility phase that readily liberates arsenic into waters (Williams 2001). Also, scorodite is a common arsenic mineral which is formed during the oxidation of arsenopyrite-rich wastes. Scorodite solubility is strongly controlled by pH (Krause and Ettel 1988). It is soluble at very low pH; its solubility is at its minimum at approximately pH 4; and the solubility increases above pH 4 again. Hence, scorodite leads to the fixation of arsenic at approximately pH 4 whereas waters of low pH (<pH 3) and high pH (>pH 5) can contain significant amounts of arsenic.

While precipitation of secondary arsenic minerals and adsorption can limit the mobility of arsenic, the mobilization of arsenic from minerals back into mine waters may be triggered through various processes. Important processes include: (*a*) desorption at high pH (pH > 8.5) under oxidizing conditions; (*b*) desorption and Fe^{3+} hydroxide dissolution due to a change to reducing conditions; and (*c*) arsenic mineral dissolution (Smedley and Kinniburgh 2002; Salzsauler et al. 2005). In particular, reducing conditions can lead to the desorption of arsenic from Fe^{3+} hydroxides and to the reductive dissolution of Fe^{3+} hydroxides, also leading to an arsenic release (Pedersen et al. 2006). Therefore, the reduction of Fe^{3+} to Fe^{2+} increases the mobility of arsenic. However, strongly reducing conditions do not favour arsenic mobility because both iron and hydrogen sulfide would be present, leading to the coprecipitation of arsenic sulfide with iron sulfide. In contrast, mildly reducing environments that lack hydrogen sulfide can allow the dissolution of arsenic. In such environments, iron is in the soluble Fe^{2+} state, and arsenic is present as As^{3+} in the arsenite form (AsO_3^{3-}). In mildly reducing environments such as saturated tailings, precipitated Fe^{3+} hydroxides and oxides can be reduced with the help of microorganisms to form dissolved Fe^{2+} and As^{3+} (McCreadie et al. 2000; Macur et al. 2001). Consequently, pore and seepage waters of such tailings repositories may contain strongly elevated iron and arsenic concentrations. When these seepage waters reach the surface, oxidation of the waters will result in the precipitation of iron and coprecipitation of arsenic, forming arsenic-rich yellow boys.

3.5.10 The Mercury System

The determination of mercury speciation in mine waste requires the application of appropriate methods (Sladek et al. 2002; Sladek and Gustin 2003; Kim et al. 2004). Mercury in mine waters is sourced from the weathering of cinnabar (HgS), metacinnabar (HgS), calomel (HgCl), quicksilver ($Hg_{(l)}$), livingstonite ($HgSb_4S_7$), and native mercury (Hg). It may also be released from mercury amalgams present in historic alluvial gold mines. While cinnabar weathers slowly under aerobic conditions (Barnett et al. 2001), the slow oxidation of mercury-bearing sulfides can still provide elevated mercury levels to mine waters. Mercury exists in natural waters as elemental mercury (Hg^0) and ionic mercury (Hg^+ and Hg^{2+}), and it is prone to be adsorbed onto organic matter, iron hydroxides, and clay minerals (Covelli et al. 2001; Domagalski 1998, 2001). As a result, mercury can be transported in natural waters as dissolved species and adsorbed onto suspended particles and colloids. Furthermore, mercury is transformed by bacteria into organic forms, notably monomethyl mercury (CH_3Hg^+) and dimethyl mercury ($(CH_3)_2Hg$) (Gray et al. 2002b; Bailey et al. 2002). These organic forms are highly toxic, fat-soluble compounds and tend to bioaccumulate in the foodchain (Ganguli et al. 2000; Hinton and Veiga 2002). Factors encouraging mercury methylation include high concentrations of dissolved carbon and organic matter, abundant bacteria and acidic water. Consequently, AMD waters are especially susceptible to mercury methylation.

3.5.11 The Sulfate System

Upon oxidation of sulfides, the sulfur S^{2-} (S: 2–) in the sulfides will be oxidized to elemental sulfur (S: 0), and more commonly to sulfate SO_4^{2-} (S: 6+). The sulfate may remain in solution or precipitate to form secondary minerals (e.g. melanterite $FeSO_4 \cdot 7H_2O$). However, sulfides may not be completely oxidized to form dissolved sulfate ions or sulfate minerals. The sulfur may be oxidized to metastable, intermediate sulfur oxyanions. These include sulfite SO_3^{2-} (S: 4+), thiosulfate $S_2O_3^{2-}$ (S: 2+), and polythionates ($S_nO_6^{2-}$), which are then subsequently oxidized to sulfate (Moses et al. 1987; Descostes et al. 2004). The occurrence of these intermediate sulfur species in mine waters is controversial, yet such reactions are supported by the occurrence of sulfite and thiosulfate minerals as natural weathering products (Braithwaite et al. 1993).

AMD waters carry significant concentrations of sulfate which exceed those of iron and heavy metals. Strongly elevated sulfate concentrations are prevalent because relatively few natural processes remove sulfate from ground and surface waters. Only the precipitation of secondary sulfate minerals influences the concentration of sulfate in solution. The formation of secondary sulfates generally occurs in response to evaporation or neutralization reactions. Gypsum and other sulfates such as epsomite ($MgSO_4 \cdot 7\,H_2O$) and jarosite ($KFe_3(SO_4)_2(OH)_6$) are such precipitates in AMD affected seepages, streams, and ponds. Gypsum is the most common sulfate salt in AMD environments. The Ca^{2+} for gypsum formation is released by the acid weathering of carbonate and silicate minerals such as dolomite, calcite, and plagioclase. The concentration of calcium sulfate in mine waters may rise to a level at which gypsum precipitates. This level is not influenced by pH and is dependent on the detailed chemical conditions of the water such as the amount of magnesium in solution. Gypsum formation may also be due to neutralization of AMD waters. Neutralization reactions between AMD waters and calcite or dolomite result in gypsum (Reaction 3.20) and epsomite precipitation (Reaction 3.21). The reactions can be written as follows:

$$CaCO_{3(s)} + H_2SO_{4(aq)} + 2\,H_2O_{(l)} \rightarrow CaSO_4 \cdot 2\,H_2O_{(s)} + H_2CO_{3(aq)} \tag{3.20}$$

$$CaMg(CO_3)_{2(s)} + 2\,H_2SO_{4(aq)} + 9\,H_2O_{(l)} \rightarrow MgSO_4 \cdot 7\,H_2O_{(s)} + CaSO_4 \cdot 2\,H_2O_{(s)} + 2\,H_2CO_{3(aq)} \tag{3.21}$$

While the formation of gypsum and other sulfates reduces the dissolved sulfate concentration, the minerals' solubility in water is also high. The major chemical mechanism that removes sulfate from solution also causes elevated sulfate concentrations in water. In addition, many oxidized ores may contain gypsum as a pre-mining mineral. Thus, not all high sulfate concentrations of mine waters are caused by sulfide oxidation; they can also be the result of the dissolution of gypsum and other sulfates.

AMD processes lead to high concentrations of dissolved sulfate at the AMD source. Once released into solution, the sulfate ion has the tendency to remain in solution. Sulfate concentrations in AMD waters are exceptionally high when compared to those of uncontaminated streams. Therefore, the sulfate ion can be used to trace the behaviour of contaminant plumes impacting on streams and aquifers. For example, sulfate-rich mine waters discharge into a surface stream with little organic activity, and there is a decrease in sulfate concentration downstream from the discharge point. This can only be ascribed to dilution by non-contaminated streams (Schmiermund 1997; Ghomshei and Allen 2000). If other mine derived constituents such as metals decrease to a greater extent in the same reach of the stream, then they must have been removed from the water by geochemical processes such as adsorption or coprecipitation. The behaviour of sulfate helps to trace and assess the fate of other mine water constituents.

3.5.12 The Carbonate System

The so-called "carbonic acid system" or "carbonate system" greatly affects the buffer intensity and neutralizing capacity of waters (Brownlow 1996; Langmuir 1997). The system comprises a series of reactions involving carbon dioxide (CO_2), bicarbonate (HCO_3^-), carbonate (CO_3^{2-}), and carbonic acid (H_2CO_3). The reactions affecting these different species are very important in ground and surface waters and involve the transfer of carbon among the solid, liquid and gas phase. This transfer of carbon also results in the production of carbonic acid. Carbonic acid in water can be derived from several sources, the most important of which are the weathering of carbonate rocks (Reactions 3.22–3.24) and the uptake of carbon dioxide from the atmosphere (Reaction 3.25):

$$CaCO_{3(s)} \leftrightarrow Ca^{2+}_{(aq)} + CO^{2-}_{3(aq)} \tag{3.22}$$

$$CO^{2-}_{3(aq)} + H^+_{(aq)} \leftrightarrow HCO^-_{3(aq)} \tag{3.23}$$

$$HCO^-_{3(aq)} + H^+_{(aq)} \leftrightarrow H_2CO_{3(aq)} \tag{3.24}$$

$$CO_{2(g)} + H_2O_{(l)} \leftrightarrow H_2CO_{3(aq)} \tag{3.25}$$

Contribution of carbonic acid from weathering processes of carbonate rocks is far more important than the uptake of carbon dioxide from the atmosphere. Which carbonate species will be present in the water is determined by the pH of the water, which in turn is controlled by the concentration and ionic charge of the other chemical compounds in solution. Bicarbonate is the dominant species found in natural waters with a pH greater than 6.3 and less than 10.3; carbonate is dominant at pH greater than 10.3; carbonic acid is the dominant species below pH 6.3 (Sherlock et al. 1995; Brownlow 1996; Langmuir 1997).

The distinction between bicarbonate and carbonic acid is important for the evaluation of AMD chemistry. Firstly, bicarbonate is a charged species whereas carbonic acid does not contribute any electrical charge or electrical conductivity to the water. In other words, in a low pH AMD water, the carbonic acid does not contribute a significant amount of anionic charge or conductivity to the water. With increasing pH value of the AMD water, the proportions of carbonic acid and bicarbonate will change. This alters the amount of negative charge and conductivity because bicarbonate ions will contribute to the negative charge. Secondly, dissolved bicarbonate ions consume hydrogen ions; hence, bicarbonate ions provide neutralizing capacity to the water as illustrated by the following reaction:

$$HCO^-_{3(aq)} + H^+_{(aq)} \leftrightarrow H_2CO_{3(aq)} \tag{3.26}$$

Bicarbonate removes free hydrogen from the solution, lowering the solution's acidity. Thus, the greater the total concentration of the bicarbonate species, the greater the buffering capacity and alkalinity of the AMD water. The alkalinity of a water is a measure of the bicarbonate and carbonate concentration, indicating the buffering capacity of the water (Table 3.1). The greater the alkalinity, the greater the hydrogen concentration that can be balanced by the carbonate system.

The reaction of free hydrogen with bicarbonate is easily reversible (Reaction 3.26). Consequently, carbonic acid formation does not cause a permanent reduction in acidity of AMD waters. The consumed hydrogen may be released back into the mine water. In fact, Reaction 3.26 is part of a series of equilibrium reactions (Reaction 3.27): bicarbonate reacts with hydrogen ions to form carbonic acid; carbonic acid then reacts to dissolved carbon dioxide and water, and finally to gaseous carbon dioxide and water:

$$HCO^-_{3(aq)} + H^+_{(aq)} \leftrightarrow H_2CO_{3(aq)} \leftrightarrow CO_{2(aq)} + H_2O_{(l)} \leftrightarrow CO_{2(g)} + H_2O_{(l)} \quad (3.27)$$

These equilibrium reactions can be forced to react towards the production of gaseous carbon dioxide. For example, if AMD water is neutralized with limestone and stirred at the same time, the carbon dioxide exsolves as a gas phase; the dissolved carbon dioxide content is lowered. As a result, the degassing of carbon dioxide does not allow the equilibrium reactions to proceed back to the production of hydrogen ions. Carbon dioxide degassing supports the permanent consumption of hydrogen by bicarbonate ions, and the acidity of AMD waters can be permanently lowered (Carroll et al. 1998).

3.5.13 pH Buffering

At mine sites, water reacts with minerals of rocks, soils, sediments, wastes, and aquifers. Different minerals possess different abilities to buffer the solution pH (Blowes and Ptacek 1994). Figure 2.3 shows a schematic diagram of AMD production for a hypothetical sulfidic waste dump. The initial drainage stage involves the exposure of sulfide to water and oxygen. The small amount of acid generated will be neutralized by any acid buffering minerals such as calcite in the waste. This maintains the solution pH at about neutral conditions. As acid generation continues and the calcite has been consumed, the pH of the water will decrease abruptly. As shown in Fig. 2.3, the pH will proceed in a step-like manner. Each plateau of relatively steady pH represents the weathering of specific buffering materials at that pH range. In general, minerals responsible for various buffering plateaus are the calcite, siderite, silicate, clay, aluminium hydroxysulfate, aluminium/iron hydroxide, and ferrihydrite buffers (Sherlock et al. 1995; Jurjovec et al. 2002). Theoretically, steep transitions followed by pH plateaus should be the result of buffering by different minerals. Such distinct pH buffering plateaus may be observed in pore and seepage waters of sulfidic tailings, waste rock piles, spoil heaps or in ground waters underlying sulfidic materials. However, in reality, such distinct transitions and sharp plateaus are rarely observed as many different minerals within the waste undergo kinetic weathering simultaneously and buffer the mine water pH.

The buffering reactions of the various minerals operate in different pH ranges. Nonetheless, there are great discrepancies in the literature about the exact pH values of these zones (Blowes and Ptacek 1994; Ritchie 1994b; Sherlock et al. 1995). Broad pH buffering of calcite occurs around neutral pH (pH 6.5 to 7.5) in an open or closed system (Sec. 2.4.2):

$$CaCO_{3(s)} + CO_{2(g)} + H_2O_{(l)} \leftrightarrow Ca^{2+}_{(aq)} + 2\,HCO^-_{3(aq)} \quad (3.28)$$

$$CaCO_{3(s)} + H^+_{(aq)} \leftrightarrow Ca^{2+}_{(aq)} + HCO^-_{3(aq)} \quad (3.29)$$

The presence of bicarbonate is influenced by the pH of the solution. Below pH 6.3, the dominant carbonate species in solution is carbonic acid. Hence, bicarbonate may form carbonic acid as follows:

$$HCO_{3(aq)}^{-} + H_{(aq)}^{+} \leftrightarrow H_2CO_{3(aq)} \quad (3.30)$$

If all of the calcite has been dissolved by acid, or the mineral is absent, then siderite provides buffering between pH values of approximately 5 and 6 (Blowes and Ptacek 1994; Sherlock et al. 1995):

$$FeCO_{3(s)} + H_{(aq)}^{+} \leftrightarrow HCO_{3(aq)}^{-} + Fe_{(aq)}^{2+} \quad (3.31)$$

The silicate minerals provide neutralizing capacity between pH 5 and 6. Their chemical weathering can be congruent (Reaction 3.32) or incongruent (Reaction 3.33) (Sec. 2.4.1). Either reaction pathway results in the consumption of hydrogen ions:

$$MeAlSiO_{4(s)} + H_{(aq)}^{+} + 3\,H_2O \rightarrow Me_{(aq)}^{x+} + Al_{(aq)}^{3+} + H_4SiO_{4(aq)} + 3\,OH_{(aq)}^{-} \quad (3.32)$$

$$2\,MeAlSiO_{4(s)} + 2\,H_{(aq)}^{+} + H_2O \rightarrow Me_{(aq)}^{x+} + Al_2Si_2O_5(OH)_{4(s)} \quad (3.33)$$

(Me = Ca, Na, K, Mg, Mn or Fe)

Exchange buffering of clay minerals is dominant between pH 4 and 5 and causes alkali and alkali earth cation release:

$$clay\text{-}(Ca^{2+})_{0.5(s)} + H_{(aq)}^{+} \rightarrow clay\text{-}(H^{+})_{(s)} + 0.5\,Ca_{(aq)}^{2+} \quad (3.34)$$

Aluminium and iron hydroxide buffering of minerals (e.g. ferrihydrite, goethite, gibbsite, hydroxysulfates, and amorphous iron and aluminium hydroxides) occurs at a lower pH than all other minerals; that is, between pH values of approximately 3 and 5. Their buffering results in the release of aluminium and iron cations:

$$Al(OH)_{3(s)} + 3\,H_{(aq)}^{+} \leftrightarrow Al_{(aq)}^{3+} + 3\,H_2O_{(l)} \quad (3.35)$$

$$Fe(OH)_{3(s)} + 3\,H_{(aq)}^{+} \leftrightarrow Fe_{(aq)}^{3+} + 3\,H_2O_{(l)} \quad (3.36)$$

3.5.14 Turbidity

Turbidity is the ability of a water to disperse and adsorb light. It is caused by suspended particles floating in the water column. The suspended particles of AMD affected streams and seepages are diverse in composition. Firstly, they may include flocculated colloids due to hydrolysis and particulate formation. Such flocculants are typically composed of poorly crystalline iron hydroxides, oxyhydroxides, and oxyhydroxysulfates (e.g. schwertmannite), and less commonly of aluminium hydroxides (Sullivan and

Drever 2001; Mascaro et al. 2001). Secondly, suspended particles may also originate from the overflow of tailings dams and from the erosion of roads, soils, and fine-grained wastes during periods of heavy rainfall. These particulate may consist of clays and other inorganic and organic compounds.

Regardless of their origin, suspended solids can be important in transporting iron, aluminium, heavy metals, metalloids, and other elements in solid forms far beyond the mine site (Schemel et al. 2000). The poorly crystalline nature of suspended particulates also allows the release of the incorporated or adsorbed elements back into the water column. The release may be initiated due to bacterial activity, reduction or photolytic degradation (McKnight et al. 1988).

3.6 Prediction of Mine Water Composition

The prediction of mine water quality is an important aspect of mining and mineral processing activities. Static and kinetic test data on sulfidic wastes provide information on the potential of wastes to generate acid (Sec. 2.7.4). However, the prediction of mine water composition is a very complex task and remains a major challenge for scientists and operators (Younger et al. 2002). Nevertheless, the mine operator would like to know in advance: (*a*) the composition of the mine water at the site; (*b*) whether or not mine water can be discharged without treatment; (*c*) whether or not mine water will meet effluent limits; and (*d*) whether or not the drainage water will turn acid, and if so, when.

3.6.1 Geological Modeling

The geological approach is an initial step in assessing the mine water quality of a particular ore deposit. Similar to the geological modeling of sulfidic wastes (Sec. 2.7.1), geological modeling of mine waters involves the classification of the deposit and the deduction of water quality problems (Plumlee et al. 1999). The reasoning behind this method is that the same types of ore deposits have the same ore and gangue minerals, meaning the same acid producing and acid buffering materials. Consequently, the mine waters should be similar in terms of pH and combined metal contents. This empirical classification constrains the potential ranges in pH and ranges in metal concentrations of mine waters that may develop. However, the technique cannot be applied to predict the exact compositions of mine waters (Plumlee et al. 1999).

3.6.2 Mathematical and Computational Modeling

There are simple mathematical models and computational tools which help to predict the chemistry of water at a mine site. All presently available mathematical and computational models have limitations and rely on good field and laboratory data obtained from solid mine wastes and mine waters. In other words, any modeling will only be as good as the data used to generate the model.

A simple mathematical model for predicting the chemistry of water seeping from waste rock piles has been presented by Morin and Hutt (1994). This empirical model provides rough estimates of future water chemistries emanating from waste rock piles. It considers several factors including:

1. the production rates of metals, non-metals, acidity, and alkalinity under acid and pH neutral conditions from a unit weight of rock;
2. the volume of water flow through the waste pile based on the infiltration of precipitation;
3. the elapsed time between infiltration events;
4. the residence time of the water within the rock pile; and
5. the percentage of mine rock in the pile flushed by the flowing water.

In Morin and Hutt's (1994) example, a hypothetical waste rock pile is 600 m long, 300 m wide, and 20 m high, and contains 6.5 Mt of rock. The long-term production rate of zinc has been obtained from kinetic test data at 5 mg kg^{-1} per week (factor 1). Rainfall occurs every second day and generates 1 mm of infiltration. Such infiltration converts to 180 000 l of water over the surface of the waste pile (factor 2). The elapsed time between rainfall and infiltration events is assumed to be equal to the infiltration events of every two days (factor 3). As a result, the waste pile accumulates 1.4 mg kg^{-1} zinc between events (5 mg kg^{-1} Zn per 7 days $\times$ 2 days = 1.4 mg kg^{-1} Zn). The residence time of the water within the waste is also assumed to be two days (factor 4), and the percentage of the total waste flushed by the pore water is assumed to be 10% (factor 5). Accordingly, the predicted zinc concentration in the waste rock seepage amounts to 506 mg l^{-1} (1.4 mg kg^{-1} Zn $\times$ 6.5 Mt $\times$ 10% / 180 000 l = 506 mg l^{-1} Zn).

While such simple mathematical models provide some insight into seepage chemistry, complex geochemical processes occurring in mine waters need to be modelled using computational software (Gerke et al. 2001; Fala et al. 2005; Accornero et al. 2005). There are computer programs which model geochemical databases, mass balances, secondary mineral saturations, phase diagrams, speciations, equilibria, reactions paths, and flows (Perkins et al. 1997). Each computational tool has been developed for slightly different purposes. Each geochemical model relies on accurate and complete data sets. Input parameters may include water composition, mineralogy, bacterial activity, reactive surface area, temperature, oxygen availability, water balance, waste rock pile structure and composition, humidity cell and leach column test data, and thermodynamic data (Perkins et al. 1997).

The predicted concentrations of individual metals, metalloids, and anions in mine waters obtained from computational geochemical models should be compared with actual mine water chemistries measured in the field or obtained through kinetic test work. Geochemical modeling programs of waters are also able to calculate the mineral saturation indices and to identify minerals that might be forming and limiting solution concentrations of these constituents. In low pH environments, many metals are mobilized and present at concentrations which cause precipitation of secondary minerals. Adsorption is also an important geochemical process operating in these waters. Precipitation and adsorption capabilities of an acidic system need to be evaluated using computational software (Smith 1999). The computational programs are used to predict the precipitation of secondary minerals from mine waters. These predicted mineral precipitates have to be verified by comparing them with those secondary minerals actually identified in the AMD environment.

Modeling is not an exact science; its application has numerous pitfalls, uncertainties, and limitations, and the calculations are at best well-educated guesses (Nordstrom and Alpers 1999a; Alpers and Nordstrom 1999). None of the programs should be used to predict the exact water composition even though they can be used to improve the understanding of geochemical processes and to perform comparisons between possible mine water scenarios (Perkins et al. 1997).

3.7
Field Indicators of AMD

Any seepage water flowing from a mine, mine waste pile, tailings dam or pond may be acid. The most common indicators in the field for the presence of AMD waters are:

- *pH values less than 5.5.* Many natural surface waters are slightly acidic (pH ~5.6) due to the dissolution of atmospheric carbon dioxide in the water column and the production of carbonic acid. Waters with a pH of less than 5.5 may have obtained their acidity through the oxidation of sulfide minerals.
- *Disturbed or absent aquatic and riparian fauna and flora.* AMD waters have low pH values and can carry high levels of heavy metals, metalloids, sulfate, and total dissolved solids. This results in the degradation or even death of aquatic and terrestrial ecosystems.
- *Precipitated mineral efflorescences covering stream beds and banks.* The observation of colourful yellow-red-brown precipitates, which discolour seepage points and stream beds, is typical for the AMD process. The sight of such secondary iron-rich precipitates (i.e. yellow boy) is a signal that AMD generation is well underway.
- *Discoloured, turbid or exceptionally clear waters.* AMD water can have a distinct yellow-red-brown colouration, caused by an abundance of suspended iron hydroxides particles. The turbidity of the AMD water generally decreases downstream as the iron and aluminium flocculate, and salts precipitate with increasing pH. As a result, acid waters can also be exceptionally clear and may give the wrong impression of being of good quality.
- *Abundant algae and bacterial slimes.* Elevated sulfate levels in AMD waters favour the growth of algae, and acid waters may contain abundant slimy streamers of green or brown algae.

3.8
Monitoring AMD

Mine water monitoring is largely based on the analysis and measurement of ground, pore and surface waters over a significant time period because their chemistry commonly changes over time (Scientific Issue 3.1). The monitoring of waters in and around mine sites is designed: (*a*) to define natural baseline conditions; (*b*) to identify the early presence of or the changes to dissolved or suspended constituents; (*c*) to ensure that discharged water meets a specified water quality standard; (*d*) to protect the quality of the region's water resources; and (*e*) to provide confirmation that AMD control measures on sulfide oxidation are operating as intended. The acquisition of baseline data prior to mining is particularly important as some sulfide orebodies may have

Scientific Issue 3.1. Seasonal and Diel Factors Controlling Water Composition

Seasonal Factors

The chemistry of rivers, streams and lakes as well as mine waters and natural waters draining mined lands may change over time (Kim and Kim 2004; Herbert 2006). Such changes are largely a function of local climatic, hydrological and geochemical factors. For example, seasonal variations in the composition of AMD waters are typically controlled by evaporation, precipitation as well as runoff events and volumes. Increased evaporation will lead to concentration of solutes and hence higher concentrations, whereas high rainfall, associated runoff events and greater stream flow volumes cause dilution of chemical constituents. Significant increases in solute concentrations have been reported during the melting of snow covers and during flushing events in semi-arid regions, when sulfide oxidation products are washed from waste repositories (Harris et al. 2003; Herbert 2006). Other factors such as the water temperature can impact indirectly on the chemistry of mine waters. Higher water temperatures favour the proliferation of bacteria. This in turn leads to an optimum rate of bacterially mediated iron oxidation, increased sulfide oxidation and greater trace metal release to waters. In addition, the dissolved iron concentration of AMD waters is influenced by solar radiation and associated photolytic processes, which increase the dissolved Fe^{2+} and reduce the dissolved Fe^{3+}. Iron photoreduction involves the absorption of UV radiation by Fe^{3+} species, converting Fe^{3+} present in either dissolved or suspended particle form into soluble Fe^{2+} (McKnight et al. 1988; Butler and Seitz 2006). As a consequence, the colloidal Fe^{3+} hydroxide concentrations in oxygenated surface waters can be reduced during daytime or summer, and such waters contain higher Fe^{2+} and total Fe concentrations.

Diel Factors

Rivers and lakes are not only known to possess seasonal but also daily cycles in pH, dissolved oxygen and water temperature. These cycles occur in response to the daily cycles of photosynthesis and respiration and to weather changes. For example, the pH of natural waters varies with the amount of amount of carbon dioxide present because carbon dioxide combines with water to form carbonic acid. During the day, plants consume carbon dioxide and the pH generally increases. At night, photosynthesis ceases and the increase in carbon dioxide due to respiration and decay of organic matter is most evident, causing pH to decrease. As a result, stream waters possess a photosynthesis-induced diel (24-hour) cycle in pH. In addition, the concentrations of several elements in acid stream waters draining mined land have been reported to change substantially during a 24-hour period, irrespective of stream flow. Diel fluctuations in the concentration of total iron and the Fe^{2+}/Fe^{3+} ratio have been related to: (*a*) daytime photoreduction of Fe^{3+} to Fe^{2+}, and (*b*) subsequent reoxodiation of Fe^{2+} to Fe^{3+} and temperature-dependent hydrolysis and precipitation of hydrous ferric oxides at night (McKnight et al. 1988; Butler and Seitz 2006).

Diel fluctuations have also been reported for dissolved arsenic, cadmium, copper, manganese, nickel and zinc. At some sites, the reported concentrations of zinc increased as much as 500% during the night (Nimick et al. 2003; Shope et al. 2006). The cyclic changes of arsenic, cadmium, copper, manganese, nickel and zinc contents have been related to adsorption/desorption equilibria onto secondary ferric iron and manganese oxide precipitates in the streambed. Sorption behaviour is thereby influenced by the pH and temperature of the stream water, and the characteristics and abundance of the precipitate (Gammons et al. 2005; Shope et al. 2006). Thus, the diel cycles in iron, pH and water temperature may induce a diel cycle in dissolved trace element concentrations, with large fluctuations possible during a 24-hour period. Consequently, a properly collected water sample does not necessarily provide an accurate assessment of trace metal and arsenic concentrations on a given day at a particular sampling location (Nimick et al. 2003).

undergone natural oxidation prior to mining. Ground and surface waters in these environments can be naturally enriched in sulfate, metals and metalloids. It is of critical importance to know the water, soil and sediment chemistry in a region prior to the development of a mining operation. Otherwise, pre-existing natural geochemical enrichments might be mistaken by the statutory authorities as being a result of mining and processing and could be thus subject to subsequent unnecessary (and unfair) remediation processes.

Effective monitoring of a mine site for its water composition requires the following.

- *Setup of monitoring system.* An effective monitoring system is site specific and fulfills the above mentioned monitoring aims. Most importantly, surface water sampling localities and ground water monitoring bores need to be established within and outside the mining area. Sites outside and upstream the mining area provide information on natural background conditions of waters whereas other sites serve as sampling points for mine waters as well as observation points for flow rates (Harries and Ritchie 1988). Monitoring of mine waters relies exclusively on chemical analyses obtained from surface and ground waters.
- *Monitoring climate.* Climate (i.e. precipitation, evapotranspiration, temperature) is an important factor that determines: (*a*) the quality and quantity of mine waters; and (*b*) the volume of surface water run-off vs. ground water recharge (Plumlee and Logsdon 1999; Younger et al. 2002). Meterological data, measured on a regular basis, are needed to understand site surface and ground water hydrology. Calculation of seasonal evapotranspiration rates and a net water balance at the site are crucial as they affect ground water recharge rates and water accumulation rates in mine water storage ponds.
- *Monitoring pore waters of sulfidic waste rock dumps.* Samples of water from the unsaturated zone of sulfidic waste rock dumps are taken using so-called "*lysimeters*". Lysimeters are small or large drums buried in waste piles, in some cases under dry covers, and connected with tubes to the surface enabling the collection of the leachate. Meteoric data and analyses of these leachates provide information on water infiltration rate into the dump and contaminant production rates (Ritchie 1998; Bews et al. 1999). The analysis of lysimeter leachates also allows an evaluation as to what depth water infiltrates into waste rock piles, and whether water infiltrates into sulfidic material. Consequently, lysimeters allow a performance evaluation of dry covers.
- *Monitoring ground waters.* Major fractures are commonly present within sulfidic ore deposits, and the hydrology of many metal mines is structurally controlled. These fractures represent permeability zones for ground water, and the interaction between water and the sulfide-bearing rocks along such fractures can lead to AMD. Highly metalliferous and strongly acidic solution may develop and flow along permeable fractures. Some of the waters may surface in open pits as distinct seepages. Therefore, it is important to ensure that ground water monitoring bores are sunk into fracture zones.

  AMD impacts more frequently on ground water quality than on the surface drainage from a mine. In particular, in low rainfall arid areas, most of the drainage is likely to move into the aquifer. Ground water samples from the saturated zone immediately beneath sulfidic waste dumps or tailings can be obtained using piezometers. Chemical analyses of such waters will indicate the contaminant transport into the local aquifer.

  Electrical and electromagnetic techniques can be used to map ground water contamination and pore waters in waste dumps, watersheds and mined areas (Benson 1995; Campbell and Fitterman 2000; Hammack et al. 2003a,b; Ackman

2003). The geophysical surveys measure the electrical conductivity, which increases with increasing TDS concentrations that are particularly found in acidic, metal-rich ground waters. However, geophysical techniques only provide indirect information about the ground water composition, and the data can be difficult to interpret as changing rock types cause pronounced resistivity or conductivity variations. The detection and mapping of contaminant plumes in areas of diverse lithologies are only possible in highly contaminated areas where other factors have a relatively small effect on the data variations.

- *Monitoring open pit lakes.* Acid waters may accumulate within final mining voids, and these acid pit lakes require detailed monitoring and hydrological studies (Scientific Issue 3.2).
- *Monitoring the ecosystem health.* The ecosystem health of waterways surrounding a mine site can be measured using indicator species (e.g. mussels; Martin et al. 1998). Such indicator species are sensitive to contaminants which may be released from a mining operation. The chemistry, behaviour, breeding cycle, population size, and health of indicator species in ecosystems immediately downstream of the mine site may then be compared with those upstream. This direct biological monitoring can reveal any significant impacts on the health of ecosystems surrounding the mine.
- *Monitoring surface waters.* The monitoring of surface waters is based on the analysis and measurement of surface waters over a significant time period. Changes to pH, conductivity, and sulfate and metal content over time are good indicators of AMD generation. In particular, the SO_4/Cl ratio is a good indicator of an input of sulfate from the oxidation of sulfides. Nonetheless, high sulfate concentrations, elevated conductivities up to 100 times higher than the local ground and surface water, and extreme salinities in pore and seepage waters do not necessarily indicate sulfide oxidation. They may also be the result of dissolution of soluble secondary sulfates within the wastes. Furthermore, changes in pH, conductivity and sulfate, metal and major cation concentrations may be the result of changes in hydrological conditions such as evaporation or dilution. These latter processes lead to increases or decreases in the contaminant's concentration and do not reflect changing chemical processes at the contamination source. Therefore, sites should measure or estimate flow rates or periodic flow volumes as they are additional important parameters of mine waters. The measurement of the solute concentration (mg l^{-1}) and of the flow rate or flow volume (l s^{-1}) will allow the calculation of the contaminant load (mg s^{-1}) in the mine water. The load of a chemical species is defined as the rate of output of the species over time (mg s^{-1}). Monitoring solute concentration and flow rate over time will allow the calculation of the output rate of the species over time. As a result, chemical processes within the contaminant source can be recognized. For example, seepage water appears at the toe of a sulfidic waste pile. The changing water flow rate (l s^{-1}) is measured using a calibrated flow gauging structure instrumented with a water sampler (Younger et al. 2002). In addition, contaminant concentrations (mg l^{-1}) are determined in the water samples over time. The knowledge of flow rates and contaminant concentrations over time allows the calculation of loads (mg s^{-1}) of individual chemical species released from the waste pile. Increases in metal and sulfate loads (mg s^{-1}) over time indicate the onset of AMD.

Scientific Issue 3.2. Acid Pit Lakes

Formation of Pit Lakes

The mining industry extracts a number of mineral commodities and coal from open pits. The large increase in open pit mining of metal ores since the 1970s has resulted in the ever increasing creation of mining voids worldwide. If open pit mines extend below the water table, ground water must be removed to permit mining operations. This is achieved through water pumping and extraction and temporarily lowering of the water table. The effect is generally limited to the immediate vicinity of the mining operation. The mining method leaves an open pit (i.e. final mining void) at the end of the operations.

Final mining voids may be used as water storage ponds, wetland/wildlife habitats, recreational lakes, engineered solar ponds, or waste disposal sites. However, the range of realistic end-use options is limited for a specific mine void, particularly in remote areas, and many open pits are left to fill with water. If a pit remains unfilled, the water table usually returns to its pre-mining level after pumping is discontinued, and the large void will flood. Open pit water can percolate into ground water or ground water can flow into the void, coming in contact with the atmosphere. As a result, changes in the water quality and quantity of the final mining void occur, and pit lake properties change over time. Thus, water balance and water quality of a final void are highly site specific and may evolve for many years before approaching a steady state. The final use of the pit lake will be determined largely by long-term trends in the quantity and quality of water within the void.

General Characteristics of Pit Lakes

The key processes which control the quantity of water in a final void are direct rainfall into the pit, evaporation from the standing water, and surface and ground water inflows and outflows (Doyle and Runnels 1997; Warren et al. 1997). Generally, pit lakes represent – similar to natural lakes – ground water discharge lakes, recharge lakes, or flow-though lakes. Evaporation of water and resulting concentration of solutes in pit lakes is less significant at most pit lakes than at natural lakes. This is because pit lakes are much smaller and deeper than natural lakes; therefore, the surface area to depth ratio of pit lakes is several magnitudes smaller than that of natural lakes.

The key processes which determine the water chemistry of final voids include: (*a*) weathering reactions in pit walls and wastes draining into the pit; (*b*) concentration as a result of evaporation; (*c*) stratification of the water column due to temperature or chemical gradients; (*d*) adsorption of dissolved elements on and coprecipitation with particulates; (*e*) precipitation of secondary salts from the pit water; and (*f*) biochemical and redox reactions in the water (Davis and Ashenberg 1989; Miller et al. 1996; Meyer et al. 1997; Eary 1998, 1999; Shevenell et al. 1999; Shevenell 2000; Tempel et al. 2000; Ramstedt et al. 2003; Denimal et al. 2005; Castendyk et al. 2005).

3.9 AMD from Sulfidic Waste Rock Dumps

Sulfidic waste rock dumps and spoil heaps are, because of their sheer volume, the major sources of AMD. The chemistry and volume of AMD seepage waters emanating from sulfidic piles are largely influenced by the properties of the waste materials. AMD development in waste heaps occurs via complex weathering reactions (Sec. 2.2). The different rates of the various weathering reactions within the waste may cause temporal changes to the drainage chemistry. Thus, the composition of drainage waters from waste rock piles depends on three factors:

Scientific Issue 3.2. ***Continued***

Acid Pit Lakes

Open pit lakes may be acidic because they receive drainage from adjacent mine workings or sulfidic wastes, soluble salts are leached from the weathered pit walls, or subaqueous oxidation of sulfides occurs via dissolved Fe^{3+} on the submerged pit walls (Pellicori et al. 2005). If any sulfidic pit walls are exposed to atmospheric conditions, the remaining sulfides will undergo oxidation to a particular depth (Fennemore et al. 1998). A vuggy texture and intense fracturing of the wall rocks will greatly aid sulfide oxidation. The oxidation of sulfide minerals and the dissolution of secondary minerals contribute to the generation of acid solutes into pit waters, and surface run-off waters draining into the pit can be saline, acid and metal-rich. Natural attenuation of any acid formed may occur in pit lakes located in limestone or marble units. Such voids have alkaline ground water inflows which can react with acid mine waters, forming precipitates and resulting in metal-rich sludges accumulating at the bottom of the voids. In regions with distinct wet and dry seasons, exposed, reactive pit walls can become invariably coated with oxidation products, including soluble and insoluble secondary mineral salts. The alternate wetting and drying of the walls will favour the oxidation of pyrite, the dissolution of secondary phases, and the periodic generation of acid, saline surface run-off waters into the pit. In arid environments, the oxidation rate of pyrite is retarded due to the lack of moisture, and the thickness of oxidized pyritic wall rocks is reduced. This in turn reduces the solute loads entering the incipient acid pit lake (Fennemore et al. 1998).

Remediation practices of acid pit lakes include the addition of organic waste such as wood sawdust, hay, straw, manure, mushroom compost, and sewage sludge, or of neutralizing materials such as lime, caustic soda, and limestone (Castro and Moore 2000). Organic matter accelerates the reduction of dissolved metals and sulfate and leads to their precipitation as metal sulfides (Levy et al. 1996; Costa and Duarte 2005). Also, rapid filling of a pit with water can suppress sulfide oxidation and result in good water quality (Parker et al. 1996; Castro and Moore 2000).

- The hydrology of the waste pile
- The presence of different weathering zones within the pile
- The rate of weathering reactions (i.e. weathering kinetics), causing temporal changes to the composition of the drainage waters

3.9.1 Hydrology of Waste Rock Dumps

Waste rock piles have physical and hydrological properties unlike the unmined, in situ waste. Mining and blasting increase the volume and porosity of waste rocks and create large pores and channels through which atmospheric gases and water can be transported.

Waste rock dumps frequently contain "perched aquifers" located well above the underlying bedrock (Younger et al. 2002). The dumps generally contain an unsaturated and a saturated zone separated by a single continuous water table with a moderate hydraulic gradient (Blowes and Ptacek 1994; Hawkins 1999; Younger et al. 2002) (Fig. 3.6). The water table tends to reflect the waste dump surface topography. Within the unsaturated part, water typically fills small pores and occurs as films on particle surfaces. Flow rates of the water vary from relatively rapid movements through inter-

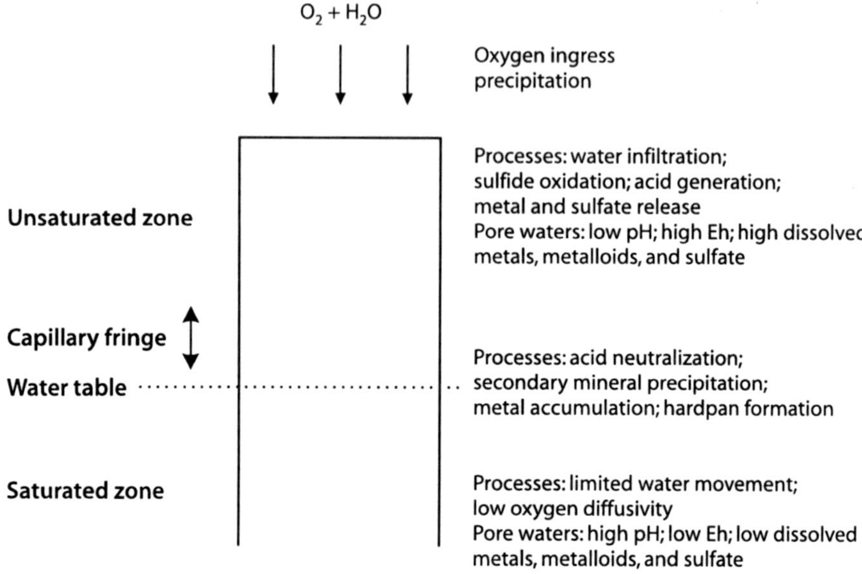

Fig. 3.6. Generalized profile of a sulfidic waste rock dump undergoing sulfide oxidation and AMD development. Hydrological subdivisions as well as hydrological, hydrochemical and geochemical processes are also shown (after Blowes and Ptacek 1994). The profile and processes of a sulfidic tailings pile are analogous

connected large pores, fractures and joints, to slow movements or nearly stagnant conditions in water films or small pores (Rose and Cravotta 1999). Within the saturated part, flow rates depend on the hydraulic properties of the waste material. Water movement is thought to be highly channellized, similar to karst environments, where water flows preferentially through randomly located channels, voids and conduits (Hawkins 1999; Younger et al. 2002). The flow of water is also influenced by the physical properties of the dump material. For instance, clay-bearing rocks tend to break to small fragments during mining and weather readily to release small mineral particles, decreasing the hydraulic conductivity (Hawkins 1999). Also, many dumps are constructed by end-dumping. This leads to some segregation of dump material down the slope at the end of the dump and causes some layering in the dump. Where large rock fragments are present, a significant volume of interstitial pores is created. Consequently, the hydraulic properties of waste rock are influenced by the dump structure, particularly the propensity of coarse material to collect at the bottom of the dump end-slope, and the tendency of fine material to remain on the sides and top. Differential settling and piping of finer material will occur shortly after dumping of waste materials. The shifting and repositioning of dump fragments are further facilitated by infiltrating meteoric water or surface run-off. Fine-grained materials migrate towards the base of the dump, and the settling of dump fragments may cause decreasing hydraulic conductivities (Hawkins 1999). Alternatively, ground water flow and infiltration of meteoric water may result in the interconnection of voids and increasing hydraulic conductivity.

The shear strength of a waste dump and its stability are influenced by the pore water pressure. Increasing pore water pressures may develop due to the increasing weight and height of the waste dump, or due to increasing seepage through the dump. Excess pore water pressures are usually associated with fine-grained materials since they possess lower permeabilities and higher moisture contents than coarse-grained wastes. Fine-grained wastes may, therefore, become unstable and fail at lower pore water pressures than coarse-grained wastes.

The hydraulic properties of wastes (e.g. hydraulic conductivity, transmissivity, porosity, pore water velocity, recharge) vary greatly. As a result, the flow direction and paths of pore waters, as well as the location and elevation of saturated zones are often difficult to predict. Detailed hydrological models are required to understand water storage and transport in waste rock dumps (Moberly et al. 2001).

3.9.2 Weathering of Waste Rock Dumps

Perkins et al. (1997) have provided a simplified model for the generation of drainage waters from sulfidic waste rock piles. The production and flow of drainage from waste rock piles is controlled by wetting and drying cycles. The waste piles are intermittently wetted by meteoric water and seasonal run-off; they are dried by drainage and evaporation. The time it takes to complete the entire wetting-drying cycle is dependent upon porosity, permeability, and climatic factors. A complete wetting-drying cycle, for a waste rock pile located in a region of moderate to high rainfall with distinct seasons, consists of four sequential stages (Perkins et al. 1997):

1. Sulfide oxidation and formation of secondary minerals
2. Infiltration of water into the dump
3. Drainage of water from the dump
4. Evaporation of pore water

The first stage represents the atmospheric oxidation of sulfides which results in the destruction of sulfides and the formation of secondary minerals. The second stage is the infiltration of meteoric water and seasonal run-off. Pores are wetted to the extent that weathering of minerals occurs. The third stage involves drainage of water from the pore spaces. Solutes dissolved in the pore water are transported to the water table or are channelled to surface seepages. Air replaces the pore water during drainage, and a thin pore water film is left behind, coating individual grains. The fourth stage is the evaporation of the water film during the drying cycle. During drying, the relative importance of drainage compared to evaporation is determined by the physical properties of the waste rock pile such as hydraulic conductivity. The drying results in the precipitation of secondary minerals that may coat the sulfide mineral surfaces. If drying continues, some of these minerals may dehydrate, crack, and spall from the sulfide surfaces, exposing fresh sulfides to atmospheric oxygen (Perkins et al. 1997). In an arid climate, there are no percolating waters present, and the flow of water through a waste rock pile is greatly reduced. In such locations, sulfide oxidation occurs, and the secondary salts generated from the limited available moisture reside within the waste. As a result, the first (i.e. sulfide oxidation and formation of secondary minerals) and fourth (i.e. evaporation of pore water) stages of the wetting-drying cycle may only be important (Perkins et al. 1997). In an arid environment, sulfide destruction does not necessarily lead to drainage from waste rock piles. However, during high rainfall events, excess moisture is present, and the secondary weathering products are dissolved and transported with the water moving through the material to the saturated zone or surface seepages.

The position of the water table in mine wastes has an important role in influencing the composition of drainage waters (Fig. 3.6). This is because the water table elevation fluctuates in response to seasonal conditions, forming a zone of cyclic wetting and drying. Such fluctuations provide optimal conditions for the oxidation of sulfides in the unsaturated zone and subsequent leaching of sulfides and associated secondary weathering products.

Ritchie (1994b) and Paktunc (1998) provide a model for the weathering of a hypothetical sulfidic waste rock dump. Weathering has proceeded for some time in the dump. Such a "mature" dump has three distinct domains (i.e outer unsaturated zone, unsaturated inner zone, saturated lower zone), reflecting the different distribution of oxidation sites and chemical reactions. This model implies that the types and rates of reactions and resulting products are different in the individual zones (Ritchie 1994b; Paktunc 1998). The outer zone of a mature waste pile is expected to have low levels of sulfide minerals. It is rich in insoluble primary and secondary minerals and can be depleted in readily soluble components. In contrast, the unsaturated inner zone is enriched in soluble and insoluble secondary minerals. In this zone, oxidation of sulfides should occur along a front slowly moving down towards the water table of the dump.

On the other hand, some authors reject the model of a stratified waste rock profile. They have argued: (*a*) that sulfidic waste dumps are heterogeneous; and (*b*) that any infiltrating rainwater would follow preferential flow paths acting as hydraulic conduits (Hutchison and Ellison 1992; Eriksson 1998; Hawkins 1999). Such discrete hydrogeological channels would limit water-rock interactions. In addition, local seeps from a single waste dump are known to have substantially different water qualities, which supports the hypothesis of preferential flow paths in waste piles. Also, the abundance and distribution of acid producing and acid buffering minerals vary from one particle to another. Waste parcels with abundant pyrite, free movement of air, and impeded movement of water are expected to develop higher acidities than equal volumes that contain less pyrite or that are completely saturated with water (Rose and Cravotta 1999). Chemical and physical conditions within waste dumps vary even on a microscopic scale. The resulting drainage water is a mixture of fluids from a variety of dynamic micro-environments within the dump. Consequently, the water quality in different parts of waste dumps exhibits spatial and temporal variations. One could conclude that prediction of drainage water chemistry from waste dumps is difficult and imprecise.

3.9.3 Temporal Changes to Dump Seepages

When mine wastes are exposed to weathering processes, some soluble minerals go readily into solution whereas other minerals take their time and weather at different rates (Morin and Hutt 1997). The drainage chemistry of readily soluble minerals remains constant over time as only a limited, constant amount of salt is able to dissolve in water. Such a static equilibrium behaviour is commonly found in secondary mineral salts such as sulfates and carbonates. Secondly, there are other minerals such as silicates and sulfides which weather and dissolve slowly over time. Their reactions are strongly time dependent (i.e. kinetic); hence, the drainage chemistry of these minerals changes through time.

Kinetic or equilibrium chemical weathering and dissolution of different minerals within mine wastes have an important influence on the chemistry of mine waters. The different weathering processes cause or contribute to the chemical load of waters draining them. In particular, kinetic weathering processes determine changes to mine water chemistries over time because acid producing and acid neutralizing minerals have different reaction rates. These different weathering and dissolution behaviours of minerals have an influence on the temporal evolution of mine water chemistries. The drainage water chemistry of a dump or tailings dam evolves with time as different parts of the material start to contribute to the overall chemical load. Generally, the chemical load reaches a peak, after which the load decreases slowly with time (Fig. 3.7).

When altered, weathered or oxidized wastes are subjected to rinsing and flushing, the pore water will be flushed first from the waste. Then easily soluble alteration minerals, weathering and oxidation products, and secondary efflorescences will dissolve and determine early rates of metal release and seepage chemistry. In particular, the soluble and reactive minerals will contribute to equilibrium dissolution at an early stage. Finally, weathering kinetics of sulfides and other acid neutralizing minerals will take over and determine the drainage chemistry.

Mine drainage quality prediction cannot be based on the assumption that 100% of the waste material experiences uniform contact with water (Hawkins 1999). Water moving through the unsaturated portion of the waste contacts waste briefly whereas water of the saturated zone has a longer contact time with the waste. In addition, some material may have a very low permeability, allowing very little ground water to flow through it. These waste portions contribute little to the chemistry of drainage waters. In order to understand the chemistry of drainage waters emanating from waste rock

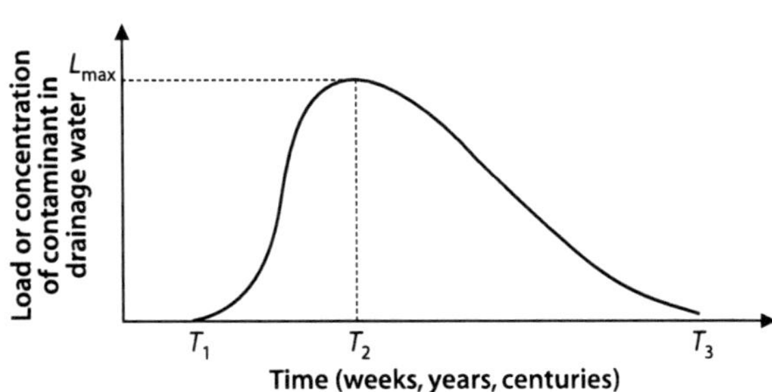

Fig. 3.7. Schematic evolution of contaminants in AMD waters emanating from sulfidic mine wastes (after Ritchie 1995)

T_1 time to first appearance of AMD (weeks to years)
T_2 time when pollutant concentration/load reaches a maximum (years to tens of years)
T_3 time when drainage water is no longer contaminated (tens to thousands of years)
L_{max} maximum contaminant load/concentration (e.g. Cu, Zn or SO_4 in mg l^{-1})

dumps, it is important to determine what waste portions are contacted by water and what is the nature of this contact (Hawkins 1999).

3.10 Environmental Impacts of AMD

AMD water from tailings dams, mine waste dumps, heap leach pads, and ore stockpiles should not be released from the mine site due to the presence of suspended solids and dissolved contaminants such as acid, salts, heavy metals, metalloids, and sulfate. The uncontrolled discharge of AMD waters into the environment may impact on surface waters, aquatic life, soils, sediments, and ground waters.

- *Surface water contamination.* The release of AMD waters with their high metal and salt concentrations impacts on the use of the waterways downstream for fishing, irrigation, and stock watering (Table 3.4). Potable water supplies can be affected when national drinking water quality guidelines are not met (Cidu and Fanfani 2002). Poor water quality also limits its reuse as process water at the mine site and may cause corrosion to and encrustation of the processing circuit. Seasonally high concentrations of acidity and metals and increased conductivity, total dissolved and suspended solids, and turbidity can be observed in AMD waters at the beginning of the wet season or spring (e.g. Gray 1998). Specifically, the first flush can cause distinct impacts on downstream ecosystems with potentially severe effects on biota.
- *Impact on aquatic life.* The high acidity of AMD waters can destroy the natural bicarbonate buffer system which keeps the pH of natural waters within a distinct pH range. The destruction of the bicarbonate system by excessive hydrogen ions will result in the conversion of bicarbonate to carbonic acid and then to water and carbon dioxide (Reaction 3.27). Photosynthetic aquatic organisms use bicarbonate as their inorganic carbon source; thus, the loss of bicarbonate will have an adverse impact on these organisms. They will not be able to survive in waters below a pH value of less than 4.3 (Brown et al. 2002). In addition, the bulk of the metal load in AMD waters is available to organisms and plants since the contaminants are present in ionic forms. Heavy metals and metalloids, at elevated bioavailable concentrations, are lethal to aquatic life and of concern to human and animal health (Gerhardt et al. 2004). Moreover, the methylation of dissolved mercury and other metals and metalloids is favoured by a low pH which turns the elements into more toxic forms. The impact on aquatic ecosystems and on downstream drainage channel plant and animal life can be severe (Gray 1998). A reduction of biodiversity, depletion of numbers of sensitive species, or even fish kills and death of other species are possible (Table 3.4).
- *Sediment contamination.* Improper disposal of contaminated water from mining, mineral processing, and metallurgical operations releases contaminants into the environment (Herr and Gray 1997; Gray 1997) (Table 3.4). If mine waters are released into local stream systems, the environmental impact will depend on the quality of the released effluent. Precipitation of dissolved constituents may result in abundant colourful mineral coatings (Fig. 3.8). This may cause soils, floodplain sediments, and

Table 3.4. Main characteristics of AMD waters and their environmental impact (after Ritchie 1994a)

Property	Chemical species	Concentration range in solution	Environmental impact
Acidity	H^+	pH < 4.5	Loss of bicarbonate to photosynthetic organisms; degradation and death to animals and plants; reduction in drinking water quality; mobilization of metal ions; corrosion of man-made structures
Iron precipitates	Fe^{3+}, Fe^{2+}, $Fe(OH)_{3(s)}$	100 to $1–9 \times 10^3$ mg l^{-1}	Discoloration and turbidity in receiving water as pH increases and ferric salts precipitate; smothering of benthic organisms and clogging up of fish gills; reduction in light penetrating the water column; encrustation of man-made structures
Dissolved heavy metals and metalloids	Cu, Pb, Zn, Cd, Co, Ni, Hg, As, Sb	0.01 to $1–9 \times 10^3$ mg l^{-1}	Degradation and death to animals and plants; bioaccumulation; reduction in drinking water quality; soil and sediment contamination
TDS	Ca, Mg, K, Na, Fe, Al, Si, Mn, sulfate	100 to more than $1–9 \times 10^4$ mg l^{-1}	Reduction in drinking water quality; reduction in stockwater quality; encrustation of man-made structures as TDS precipitate as salts; soil and sediment contamination

Fig. 3.8. Stream channel impacted by AMD, Rum Jungle uranium mine, Australia. The channel is devoid of plant life and encrusted with white sulfate efflorescences

stream sediments to become contaminated with metals, metalloids, and salts. The metals and metalloids may be contained in various sediment fractions. They may be present as cations: (*a*) on exchangeable sites; (*b*) incorporated in carbonates; (*c*) incorporated in easily reducible iron and manganese oxides and hydroxides; (*d*) incorporated in moderately reducible iron and manganese oxides and hydroxides; (*e*) incorporated in sulfides and organic matter; and (*f*) incorporated in residual silicate and oxide minerals.

- *Ground water contamination.* AMD impacts more frequently on the quality of ground waters than on that of surface waters. Ground water contamination may originate from mine workings, sulfidic tailings dams, waste rock piles, heap leach pads, ore stockpiles, coal spoil heaps, ponds, and contaminated soils (Paschke et al. 2001; Eary et al. 2003). Contaminated water may migrate from workings and waste repositories into aquifers, especially if the waste repository is uncapped, unlined and permeable at its base, or if the lining of the waste repository has been breached. At such sites, water may leak from the mine workings or the waste repository into the underlying aquifer. Significant concentrations of sulfate, metals, metalloids, and other contaminants have been found in ground water plumes migrating from mine workings and sulfidic waste repositories and impoundments. If not rectified, a plume of contaminated water will migrate over time downgradient, spreading beyond the mine workings and waste repositories, surfacing at seepage points, and contaminating surface waters (Lachmar et al. 2006). The migration rate of such a plume is highly variable and dependent on the physical and chemical characteristics of the aquifer or waste material. Generally, sulfate, metal, and metalloid concentrations in the ground water define a leachate plume extending downgradient of the AMD source (Lind et al. 1998; Johnson et al. 2000; Paschke et al. 2001). Contaminant levels depend on the interaction between the soil, sediment or rock through which the contaminated water flows and the contaminant in the water. Conservative contaminants (e.g. SO_4^{2-}) move at ground water velocities. However, reactive contaminants (e.g. heavy metals, metalloids) move more slowly than the ground water velocity, and a series of different pH zones may be present in the contaminant plume (Fig. 3.9). The occurrence of these zones is attributed to the successive weathering of different pH buffering phases in the aquifer. Such natural attentuation processes in the aquifer, including pH and Eh

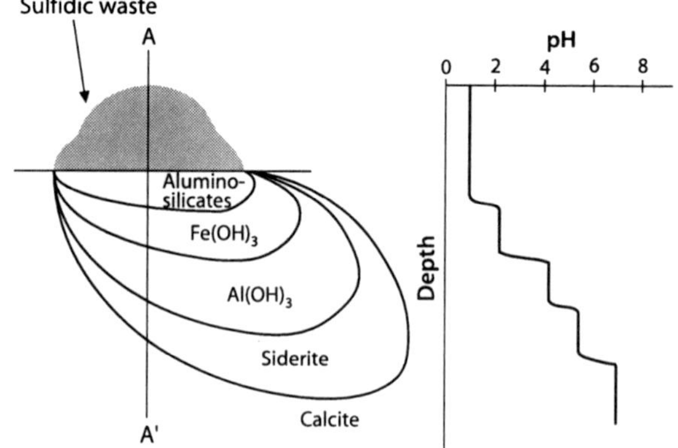

Fig. 3.9. Schematic cross-section of a sulfidic waste dump with a corresponding plume of acid water seeping into the ground. Various minerals buffer the acid ground water. The pH changes in the plume are shown for the cross-section *AA'* (Jurjovec et al. 2002). (Reprinted from Jurjovec J, Ptacek CJ, Blowes DW (2002) Acid neutralization mechanisms and metal release in mine tailings: A laboratory column experiment. Geochimica et Cosmochimica Acta 66:1511–1523, with permission from Elsevier Science)

changes, can reduce the constituent concentrations to background levels in the pathway of the subsurface drainage. Neutralizing minerals – such as carbonates – may be contained in the aquifers, and these minerals buffer acidic ground waters. Depending on the neutralization property of the aquifer through which this water moves, it could be many years before significant impact on ground and surface water quality is detected. In the worst case scenario, the neutralizing minerals are completely consumed before the acid generation is halted at the source. Then the acidic ground water plume will migrate downgradient and can eventually discharge to the surface.

3.11 AMD Management Strategies

At mine sites, containment of all contaminated water is to be ensured using water management strategies. These strategies aim to protect aquatic environments and to reduce the water volume requiring treatment. Depending on the location or climate of the mine site, different strategies are applied (SMME 1998; Environment Australia 1999). Various techniques can reduce mine water volumes: (*a*) interception and diversion of surface waters through construction of upstream dams; (*b*) diversion of runoff from undisturbed catchments; (*c*) maximization of recycling and reuse of water; (*d*) segregation of water types of different quality; (*e*) controlled release into nearby waters; (*f*) sprinkling of water over dedicated parts of the mine site area; (*g*) use of evaporative ponds; and (*h*) installation of dry covers over sulfidic wastes in order to prevent infiltration of meteoric water. These water management strategies will reduce the potential AMD water volume.

In coastal wet climates, the construction of pipelines and the discharge of AMD waters into the ocean may also be considered for the disposal of AMD waters (Koehnken 1997). Seawater has a strong buffering capacity due to the abundance of bicarbonate whereas ground and surface waters in a carbonate terrain have similarly a significant natural buffering capacity. Releasing waste waters during periods of high rainfall or peak river flow may also achieve dilution and reaction of the effluent to pollutant concentrations below water quality standards (i.e. dilution is the solution to pollution). However, in most cases such a disposal technique is not possible or politically and environmentally acceptable, and treatment of AMD waters is required prior to their discharge.

In many cases, mining operations have to discharge mine water to streams outside their operating licence areas. The release of water from mine sites has to conform with statutory directives; that is, the quality of discharged water has to meet a specified standard comprising a list of authorized levels of substances. Water quality standards list values for parameters such as pH, total suspended matter, and concentrations of sulfate, iron, metals, metalloids, cyanide, and radionuclides. National water quality guidelines are commonly used as a basis for granting a mining licence and allowing discharge of mine water. They are designed to protect downstream aquatic ecosystems, drinking water, and water for agricultural use. Water quality guidelines for metals in aquatic ecosystems are commonly based on total concentrations. However, the bioavailability of metals (i.e. the ability to pass through a biological cell membrane) and the toxicity of metals to aquatic organisms are dependent on the chemical form, that is, the speciation of these metals. Metals present as free ions are more bioavailable than metals adsorbed to colloids or particulate matter. Consequently, guidelines which are based on total metal concentrations are overprotective since only a fraction of the total metal concentration in water will be bioavailable.

3.12 Treatment of AMD

Once started, AMD is a persistent and potentially severe source of pollution from mine sites that can continue long after mining has ceased (Fig. 3.10). Abandoned historic mine sites still releasing AMD waters are a large liability for governments. Liabilities for historic AMD have been estimated around the world to include US$4 000 million in Canada, US$2 000 to 3 500 million in the United States, US$6 000 million for uranium mines in the former East Germany, US$300 million in Sweden, and US$500 million in Australia (Harries 1997; Brown et al. 2002). The total worldwide liability related to AMD is likely to be in excess of 10 000 million US dollars. In the United States alone, the mining industry spends over US$1 million every day to treat AMD water (Brown et al. 2002). The message is clear: it is always considerably more costly and more difficult to treat AMD problems after they have developed than to control the generation process through sulfide oxidation prevention technologies (Sec. 2.10). In other words, prevention or minimization of sulfide oxidation at the source is better than the treatment of AMD waters. Preventative measures applied to control sulfide oxidation will also help to control the volume of AMD waters (Sec. 2.10). A greater control of sulfide oxidation creates a smaller volume of AMD water requiring treatment.

Like the control techniques for sulfidic wastes, AMD treatment technologies are site specific, and multiple remediation strategies are commonly needed to achieve successful treatment of AMD waters (Skousen and Ziemkiewicz 1996; Environment Australia 1997; Evangelou 1998; SMME 1998; Taylor et al. 1998; Mitchell 2000; Brown et al. 2002; Younger et al. 2002). Collection and treatment of AMD can be achieved using established and sophisticated treatment systems.

Established treatment processes include evaporation, neutralization, wetlands, and controlled release and dilution by natural waters. More technologically advanced processes involve osmosis (i.e. metal removal through membranes), electrodialysis (i.e. selective metal removal through membranes), ion exchange (i.e. metal removal using various ion exchange media such as resins or polymers), electrolysis (i.e metal recovery with electrodes), biosorption (i.e. metal removal using biological cell material), bioreactor tanks (i.e. vessels that contain colonies of metal immobilizing bacteria or contain sulfate reducing bacteria causing the metal to precipitate as sulfides) aerated bioreactors and rock filters (i.e. removal of manganese from mine waters), limestone reactors (i.e. enhanced limestone dissolution in a carbon dioxide pressurized reactor), and solvent extraction (i.e. removal of particular metals with solvents) (e.g. Shelp et al. 1996; Sibrell et al. 2000; Brown et al. 2002; Greben and Maree 2005; Johnson and Younger 2005; Watten et al. 2005).

Many of the innovative treatment techniques are not standard industry practices, are used only at some individual mine sites, or are still at the exploratory stage (Scientific Issue 3.3). Both established and innovative AMD treatment techniques are generally designed:

Fig. 3.10. Unvegetated waste rock dump at the Mt. Lyell copper mine, Queenstown, Australia. Waste rock dumps and mine workings are significiant sources of AMD into the Queen River. It has been estimated that AMD will continue for another 600 years with the present copper load being 2 000 kg per day (Koehnken 1997)

- to reduce volume;
- to raise pH;
- to lower dissolved metal and sulfate concentrations;
- to lower the bioavailability of metals in solution;
- to oxidize or reduce the solution; or
- to collect, dispose or isolate the mine water or any metal-rich sludge generated.

AMD treatment techniques can also be classified as active or passive (Walton-Day 2003; Johnson and Hallberg 2005b):

- *Active treatment.* Active treatment systems such as lime neutralization require continued addition of chemical reagents, active maintenance and monitoring, and mechanical devices to mix the reagent with the water.
- *Passive treatment.* Passive methods like wetlands, bioreactors or anoxic limestone drains use chemical and biological processes to reduce dissolved metal concentrations and to neutralize acidity. Such methods require little or no reagents, active maintenance and monitoring, or mechanical devices.

Active treatment techniques such as neutralization and passive treatment methods such as abiotic ponds result in the precipitation of heavy metals from AMD waters and produce voluminous sludge (Dempsey and Jeon 2001). This sludge needs to

Scientific Issue 3.3. The Use of Red Mud from Bauxite Refineries to Treat AMD

Production of Bauxite Refinery Waste

More than 95% of bauxite is refined worldwide by using concentrated sodium hydroxide (NaOH) to dissolve the aluminium ore minerals. Materials that are insoluble in sodium hydroxide are filtered out of the alkaline aluminate solution and removed from the hydrometallurgical processing circuit. This waste product is known as "bauxite residue". Alumina refineries generate significant volumes of extremely alkaline (pH 10–13), saline, sodic bauxite residues. For example, Australia alone produces up to 60 Mt of residue each year. The bauxite residue is frequently disposed of in large ponds or dry stacks. In the dry stacking disposal technique, the finer residue fraction (i.e "*red mud*") is separated from the coarser fraction (i.e "red sand"). Washing of the coarser fraction produces a benign sand which can be safely disposed of. In contrast, red mud cannot be washed, is saturated with saline, alkaline, sodic pore waters, and requires secure disposal. The red mud is dewatered and dry stacked in lined waste repositories. Dry stacking involves discharging and spreading the red mud in layers, and allowing each layer to dry before adding another layer. Red mud waste repositories require careful capping, soil management and revegetation as long-term leaching of the waste may contaminate local ground waters. Untreated red mud has a pH value (10–13) which does not allow plant growth. Neutralizing the red mud with seawater lowers the pH value to approximately 8.5 and converts soluble, alkaline sodium salts into less soluble, weakly alkaline solids (Harrison 2002). As a result, seawater-neutralized red mud has a reduced pH and alkalinity as well as lower sodium concentrations. These properties improve settling characteristics, simplify handling and allow plant growth.

Use of Bauxite Refinery Waste

Several alternatives to the disposal of bauxite residues have been proposed such as using the metallurgical waste as a raw material for making glass, ceramics or bricks. In addition, red mud can be used to treat dairy wastewaters, to remove dyes from wastewater, and to remove metals from solution (e.g. Zouboulis and Kydros 1993; Soner Altundogan et al. 2000). Seawater-neutralized red mud consists of a mixture of amorphous phases and finely crystalline minerals, including anhydrite/gypsum, quartz, hematite, goethite, anatase, rutile, calcite, aragonite ($CaCO_3$), boehmite ($AlO(OH)$), brucite ($Mg(OH)_2$), cancrinite ($(Na,Ca)_8(Si_6Al_6)O_{24}(CO_3)_2$), gibbsite ($Al(OH)_3$), hydrocalumite ($Ca_4Al_2(OH)_{12}(Cl,CO_3,OH,H_2O)_{2.5} \cdot 4H_2O$), hydrotalcite ($Mg_4Al_2(OH)_{12}CO_3 \cdot 3H_2O$), portlandite ($Ca(OH)_2$), sodalite ($Na_4(Si_3Al_3)O_{12}Cl$), and whewellite ($CaC_2O_4 \cdot H_2O$) (McConchie et al. 1998, 2000; Harrison 2002). These amorphous and finely crystalline phases give the red mud very high acid neutralization and metal binding capacities. The injection of seawater-neutralized red mud into AMD waters not only increases the pH but also reduces the aqueous metal and metalloid concentrations. Thus, seawater-neutralized red mud has successfully been applied to neutralize AMD waters and to strip AMD waters of their dissolved metals and metalloids (McConchie et al. 1998, 2000; Harrison 2002; Munro et al. 2004; Lapointe et al. 2006). Regardless of whether red mud is used to neutralize AMD waters, to treat agricultural or industrial wastewaters, or to manufacture bricks, any recycling will reduce the amount of bauxite residues stacked in waste repositories.

be removed from the treatment system on a regular basis. The sludge is either disposed of in appropriate impoundments or treated further for metal recovery. In fact, the recovery of metals from sludge may partly pay for the costs of water treatment (Miedecke et al. 1997).

3.12.1 Neutralization

Neutralization involves collecting the leachate, selecting an appropriate chemical, and mixing the chemical with the AMD water. In the process, acid is neutralized, and metals and sulfate are removed from solution and precipitated as sludge. Neutralization treatment systems for AMD include:

1. *A neutralizing agent.* A large variety of natural, by-product or manufactured chemicals are used for AMD treatment including local waste rock with high ANC (Table 3.5). Each of these neutralizing agents has different advantages and disadvantages. For example, the advantages of using limestone (largely $CaCO_3$) include low cost, ease of use, and formation of a dense, easily handled sludge. Disadvantages include slow reaction times and coating of the limestone particles with iron precipitates. In the reaction of limestone with AMD waters, hydrogen ions are consumed, bicarbonate ions generated, and dissolved metals are converted into sparingly soluble minerals such as sulfates, carbonates, and hydroxides:

$$CaCO_{3(s)} + H^+_{(aq)} + SO^{2-}_{4(aq)} + Pb^{2+}_{(aq)} \rightarrow PbSO_{4(s)} + HCO^-_{3(aq)} \quad (3.37)$$

$$CaCO_{3(s)} + Pb^{2+}_{(aq)} \rightarrow PbCO_{3(s)} + Ca^{2+}_{(aq)} \quad (3.38)$$

$$CaCO_{3(s)} + Zn^{2+}_{(aq)} + 2\,H_2O_{(l)} \rightarrow Zn(OH)_{2(s)} + Ca^{2+}_{(aq)} + H_2CO_{3(aq)} \quad (3.39)$$

Also, hydrated lime ($Ca(OH)_2$) is easy and safe to use, effective, and relatively inexpensive. The major disadvantages are the voluminous sludge produced and high initial costs for the establishment of the active treatment plant (Zinck and Griffith 2000; Brown et al. 2002). In this process, acid is neutralized (Reaction 3.40), metals (Me^{2+}/Me^{3+}) are precipitated in the form of metal hydroxides (Reaction 3.41), and gypsum is formed, if sufficient sulfate is in solution (Reaction 3.42):

$$Ca(OH)_{2(s)} + 2\,H^+_{(aq)} \rightarrow Ca^{2+}_{(aq)} + 2\,H_2O_{(l)} \quad (3.40)$$

$$Ca(OH)_{2(s)} + Me^{2+} / Me^{3+}_{(aq)} \rightarrow Me(OH)_{2(s)} / Me(OH)_{3(s)} + Ca^{2+}_{(aq)} \quad (3.41)$$

$$Ca^{2+}_{(aq)} + SO^{2-}_{4(aq)} + 2\,H_2O_{(l)} \rightarrow CaSO_4 \cdot 2H_2O_{(s)} \quad (3.42)$$

Lime neutralization is efficient for removing metals such as cadmium, copper, iron, lead, nickel, and zinc from solution. Nonetheless, the solubility of metals varies with pH, and the lowest dissolved metal concentration is not achieved at the same pH (Kuyucak 2000; Brown et al. 2002). Not all metals can be precipitated at the same pH, and a combination of neutralizing agents (e.g. lime plus limestone) and other chemical additives may be needed to achieve acceptable water quality. Caustic soda (NaOH) is especially effective for treating AMD waters having a high manganese content. Manganese is difficult to remove from mine waters because the pH must be raised to above 10 before manganese will precipitate. Caustic soda raises the pH to above 10. Major disadvantages of caustic soda are its high costs, the dangers in handling the chemical, and poor sludge properties. Other rock types (serpentinite; Bernier 2005) and unconventional industrial waste or byproducts such as fly ash from coal power stations or kiln dust from cement factories have been suggested for the treatment of AMD waters. However, fly ash commonly contains elevated metal and metalloid concentrations, and its reaction rate is slow compared to lime (Kuyucak 2000).

Table 3.5. Chemical compounds commonly used in AMD treatment (after Skousen and Ziemkiewicz 1996; Environment Australia 1997)

Common name	Chemical name	Chemical formula
Limestone	Mainly calcium carbonate	$CaCO_3$
Quicklime or caustic lime	Calcium oxide	CaO
Hydrated lime	Calcium hydroxide	$Ca(OH)_2$
Dolomite	Calcium-magnesium carbonate	$CaMg(CO_3)_2$
Caustic magnesia or brucite	Magnesium hydroxide	$Mg(OH)_2$
Magnesite	Magnesium carbonate	$MgCO_3$
Soda ash	Sodium carbonate	Na_2CO_3
Caustic soda	Sodium hydroxide	$NaOH$
Ammonia	Anhydrous ammonia	NH_3
Kiln dust		Largely CaO and $Ca(OH)_2$
Coal fly ash		Largely $CaCO_3$ and CaO

The addition of neutralizing agents reduces the acidity and dissolved heavy metal concentrations of mine waters. Excessive neutralization can also lead to the enhanced dissolution of metals and metalloids and to waters with high metal and metalloid concentrations. Neutral to alkaline oxidizing conditions favour increased concentrations of some metals and metalloids (e.g. As, Sb, U) as ions or complexes in solution. Consequently, neutralization of AMD waters should raise the pH only to values necessary to precipitate and adsorb metals.

2. *Means for mixing AMD and the agents.* The chemicals are dispensed as a slurry by a range of neutralization plants or dosing systems. Active mixing of the chemicals is essential in order to prevent armouring of the reagent particles with reaction products such as metal hydroxides. These precipitates inhibit the neutralization reactions and cause excessive reagent consumption (Mitchell 2000).
3. *Procedures to delay iron oxidation.* Iron may need to remain in its reduced form (Fe^{2+}) until the precipitation of other metals has occurred and additional alkalinity has been dissolved in the AMD waters. Otherwise, the neutralizing solids may be coated with Fe^{3+} reaction products and rendered ineffective. Anoxic limestone drains are generally relied upon to keep iron in solution and to add alkalinity to the system.
4. *Settling ponds or vat reactors for removing precipitating metals.* The dissolved metals are forced to hydrolyze and precipitate during neutralization. The precipitates initially occur in suspension, and some of them may settle very slowly because of their small particle size (Kairies et al. 2005). Settling of precipitates can be sped up by using flocculants and coagulants (Skousen and Ziemkiewicz 1996; Brown et al. 2002). Such reagents (e.g. inorganic Fe and Al salts, organic polymers) lead to the formation of larger solid aggregates. As a result, voluminous sludge, composed mainly of solid sulfates, hydroxides and carbonates as well as amorphous and poorly crystalline material, is produced which in most cases needs disposal (MEND 1997b; Zinck 1997; Ford et al. 1998; Dempsey and Jeon 2001; Younger et al. 2002; Widerlund et al. 2005). Depending on sludge characteristics, the sludge may have to be protected from

oxidation, leaching, and potential metal mobilization. This is achieved by mixing it with more alkaline reagents prior to its disposal. Alternatively, depending on the mineralogical and chemical characteristics of the AMD sludge, metals can be recovered from the sludges using strong acids. Very pure Fe^{3+} hydroxide sludges may be used as pigments in the production of coloured bricks and concrete.

In many cases, the simple addition of neutralizing agents is not sufficient to reduce metal and metalloid concentrations in mine waters to acceptable levels. These waters need other chemical treatments to lower dissolved metal and metalloid loads. Ponds or wetlands may be required to further improve water quality prior to the discharge to a receiving stream.

3.12.2 Other Chemical Treatments

Waste waters with elevated antimony, arsenic, chromium, iron, manganese, mercury, molybdenum, and selenium contents may require chemical treatment methods other than neutralization. These methods do not employ the traditional addition of alkaline reagents like lime. They use unconventional industrial byproducts such as slags or rely on oxidation, reduction, precipitation, adsorption, or cation exchange processes (Skousen and Ziemkiewicz 1996; Brant et al. 1999; Kuyucak 2000; Younger et al. 2002; Ahn et al. 2003).

Mine waters may have elevated arsenic values at acid to alkaline pH values. A reduction in the dissolved arsenic content of AMD waters is achieved through aeration or the addition of ferric or ferrous salts (Taschereau and Fytas 2000; Seidel et al. 2005). This causes the coprecipitation and adsorption of arsenic with ferric hydroxides, or the precipitation of arsenic as an arsenate phase (AsO_4^{3-}). Some arsenates are considered to be stable (e.g. ferric arsenate) or unstable phases (e.g. calcium arsenate) under neutral pH conditions. Consequently, arsenate formation may result only in the temporary fixation of arsenic.

Other treatment techniques involve the addition of oxidants (e.g. Cl_2, O_2, NaOCl, $CaCl_2$, $FeCl_3$, H_2O_2, $KMnO_4$) which will convert dissolved Fe^{2+} to insoluble Fe^{3+} precipitates, aqueous Mn^{2+} to insoluble Mn^{3+} precipitates, and As^{3+} to less toxic As^{5+}. Also, the addition of barium chloride ($BaCl_2$) or barium sulfide (BaS) forces the precipitation of barium sulfate ($BaSO_4$) and associated lowering of aqueous sulfate and metal concentrations (Maree et al. 2004). The treatment of AMD waters with zerovalent iron causes the formation of secondary metal reaction products and the adsorption of metals onto solids (Herbert 2003).

The removal of metals from solutions may also be achieved through sulfide precipitation. The precipitation process relies on the generation of sulfide activity, either through reagent addition (e.g. compost) or by the microbiological reduction of sulfate to hydrogen sulfide in wetlands or a specially designed reactor. Sulfate reducing bacteria thereby convert the dissolved sulfate to hydrogen sulfide. The sulfide produced reacts with the dissolved metals, which precipitate as insoluble solid sulfides. The presence or even addition of organic-rich materials such as compost to AMD waters does not only promote removal of dissolved metals from AMD solutions but also pH neutralization (Gibert et al. 2005; Johnson and Hallberg 2005a). Moreover, the addition of soluble sulfide reductants (e.g. FeS, BaS, $(NH_4)_2S$, NaS_2) causes the precipitation of mercury and other metals and metalloids as sulfides (Kuyucak 2000). Injection of gaseous sulfide compounds (e.g. H_2S) into AMD waters may also be a successful treatment technique.

The above treatment techniques commonly result in the formation of small suspended particles in the waste water. The addition of coagulants (e.g. $FeSO_4$, $FeCl_3$, $MgCl_2$, $CaCl_2$) and flocculants (e.g. Na_4SiO_4, starch derivatives, bentonite, polysaccharides) improves the settling of these small precipitates (Brown et al. 2002). These added chemicals are also useful in removing an array of metals from solution by adsorption. The addition of zeolites or synthetic polymeric resins to AMD waters can also be successful. The cation exchange capacity of these additives allows the substitution of harmless ions present in the additives for dissolved metals in the AMD waters.

3.12.3
Anoxic Limestone Drains

Calcitic limestone is commonly used to treat AMD as it is highly effective in treating AMD. In contrast, dolomitic limestone is less reactive and hence, ineffective in treating AMD. The calcite dissolution consumes acidity and introduces buffering capacity in the form of bicarbonate ions into AMD waters:

$$CaCO_{3(s)} + H^{+}_{(aq)} \leftrightarrow Ca^{2+}_{(aq)} + HCO^{-}_{3(aq)} \quad (3.43)$$

However, in an oxidizing environment, limestone becomes coated with Fe^{3+} reaction products and rendered ineffective in the production of bicarbonate ions. This disadvantage is overcome by using so-called "anoxic limestone drains" (Fig. 3.11) (Hedin et al. 1994b; Skousen and Ziemkiewicz 1996; Kleinmann 1997; Kleinmann et al. 1998; Cravotta and Trahan 1999; Nuttall and Younger 2000; Brown et al. 2002; Younger et al. 2002; Cravotta 2003). Anoxic limestone drains consist of shallow trenches backfilled with crushed limestone and covered with plastic and impermeable soil or sediment. These backfilled trenches are sealed from the atmosphere in order to maintain iron as dissolved Fe^{2+} species. This prevents oxidation of Fe^{2+} to Fe^{3+} and hydrolysis of Fe^{3+} which would otherwise form Fe^{3+} precipitates coating the carbonates. As a result, the pH can be raised to neutral or even alkaline levels, migration of Fe^{2+}-rich waters is possible, and there is no adsorption of metals onto the precipitating Fe^{3+} phases. The effluent pH of anoxic limestone drains is typically between 6 and 7. Once the pH has been adjusted and the drainage has exited from the channel, controlled aeration permits oxidation of dissolved metals, hydrolysis, and precipitation of metal hydroxides or carbonates.

Alkalinity may also be added in so-called "successive alkalinity producing systems" (SAPS), whereby water passes vertically through successive layers of organic matter and limestone chippings (Skousen and Ziemkiewicz 1996; Demchak et al. 2001; Brown et al. 2002) (Fig. 3.12). These vertical flow systems have a layer of organic substrate which reduces Fe^{3+} to Fe^{2+} and eliminates the oxygen dissolved in the water. The reduced water then enters an alkalinity generating layer of limestone before it is finally discharged.

While anoxic limestone drains provide alkalinity to AMD waters, certain AMD waters are not suitable for anoxic limestone drain treatment. If the mine water con-

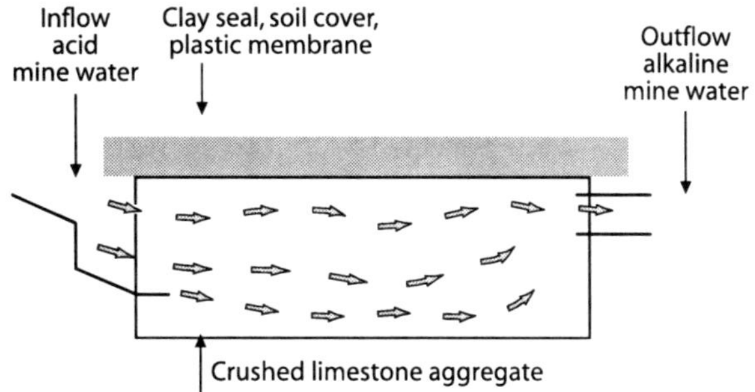

Fig. 3.11. Schematic cross-section of an anoxic limestone drain (after Environment Australia 1997; Younger et al. 2002)

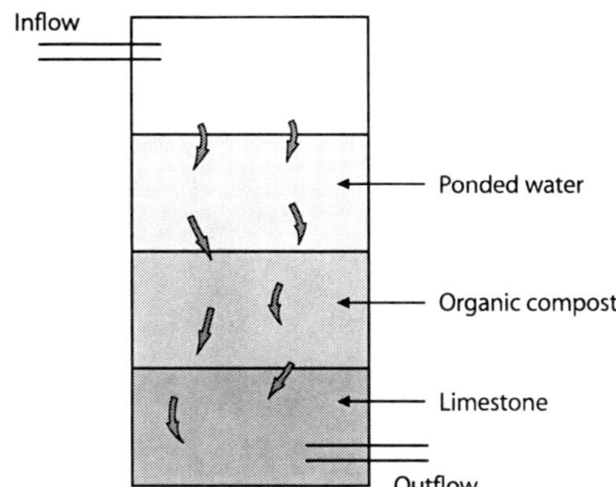

Fig. 3.12. Schematic cross-section of a successive alkalinity producing system (after Environment Australia 1997; Brown et al. 2002)

tains elevated dissolved Fe^{3+} and oxygen (>2 mg l^{-1}) concentrations, armouring of the limestone with iron phases will occur. Such waters require modified limestone beds (Hammarstrom et al. 2003) or anoxic ponds at the inflow to the limestone drain to induce reducing conditions and to convert any Fe^{3+} to Fe^{2+}. Appreciable Al^{3+} contents are also not suitable for anoxic limestone drain treatment. The aluminium will precipitate as hydroxide, causing the clogging of limestone pores, plugging of the drain, and armoring of the carbonates with aluminium precipitates (Demchak et al. 2001). Pre-treament of the drainage water is needed in order to remove aluminium from the waters. This may also be achieved in anoxic abiotic ponds.

While anoxic limestone drains are commonly used in the treatment of AMD waters, their long-term performance remains to be determined. Their effectiveness is based on the dissolution of carbonate over time. The dissolution in a sealed trench will undoubtely lead to cavernous zones, to karst like features and, depending on its structural integrity, to the potential collapse of the subsurface drain.

3.12.4 Wetlands

Wetlands are organic-rich, water-saturated shallow ponds. They are well established treatment options for sewage effluents and other wastewaters, including landfill leachates as well as agricultural and stormwater run-offs. Wetlands have also found their application in the treatment of AMD waters. The treatment is based on a number of physical, chemical and biochemical processes which ameliorate or "polish" AMD waters (Skousen and Ziemkiewicz 1996; Tyrrell 1996; Gazeba et al. 1996; Jones et al. 1998; Mitsch and Wise 1998; Walton-Day 1999; Brown et al. 2002; Younger et al. 2002; Whitehead and Prior 2005; Kalin 2004; Kalin et al. 2006). These processes include sulfide precipitation, oxidation and reduction reactions, cation exchange and adsorption of metals onto the organic substrate, neutralization of proton acidity, adsorption of metals by precipitating Fe^{3+} hydroxides, and metal uptake by plants (Walton-Day 1999) (Fig. 3.13). Other wetland processes are filtering of suspended solids and colloidal matter from the mine water as well as sedimentation and retention of these precipitates by physical entrapment. Consequently, metal-rich sediments and sludges accumulate within wetlands, with the metal loadings increasing over time.

A basic design scheme for constructed wetlands includes an organic substrate and discrete, controlled in- and outflow locations (Skousen and Ziemkiewicz 1996; Walton-Day 1999). In addition, wetlands may grow plants (e.g. reeds, sphagnum moss, cattails) that replenish the organic substrate and support naturally occurring bacteria, vertebrates and invertebrates. Two types of wetlands, aerobic and anaerobic ones, are used for AMD water treatment.

3.12.4.1
Surface Flow or Aerobic Wetland

Aerobic wetlands are generally used for net alkaline waters (i.e. "net alkaline" according to the definition by Hedin et al. (1994a)). These wetlands are with or without vegetation and relatively shallow (~0.3 m). Water flows above the surface of an organic substrate or soil (Fig. 3.13a). The wetlands are designed to encourage the oxidation and precipitation of metals. Most importantly, dissolved iron and manganese ions are oxidized and precipitated as iron and manganese hydroxides and oxyhydroxides. Such reactions are illustrated by the following reactions:

$$4\,Fe^{2+}_{(aq)} + O_{2(g)} + 4\,H^{+}_{(aq)} \rightarrow 4\,Fe^{3+}_{(aq)} + 2\,H_2O_{(l)} \tag{3.44}$$

$$Fe^{3+}_{(aq)} + 3\,H_2O_{(l)} \rightarrow Fe(OH)_{3(s)} + 3\,H^{+}_{(aq)} \tag{3.45}$$

$$Fe^{3+}_{(aq)} + 2\,H_2O_{(l)} \rightarrow Fe(OOH)_{(s)} + 3\,H^{+}_{(aq)} \tag{3.46}$$

Aerobic wetlands use oxidation and hydrolysis reactions to treat mine waters (Iribar et al. 2000). The systems function well in precipitating iron and other metals from mine waters. If the mine waters are metal-bearing and alkaline, only the aerobic wetlands treatment is needed.

The hydrolysis of iron produces acidity (Reactions 3.45, 3.46) and lowers the pH of the mine water in the wetland. The lowered pH reduces the oxidation rate of Fe^{2+} to Fe^{3+} and causes stress to plants growing in the wetland. Hence, the treatment of high iron and low pH AMD waters by aerobic wetlands has not been successful. Surface flow wetlands do not produce enough alkalinity that is required to buffer the acidity pro-

Fig. 3.13. Generalized profiles of **a** aerobic and **b** anaerobic wetlands (after Younger et al. 2002). Physical and chemical processes contributing to metal retention in wetlands are also shown (after Walton-Day 1999)

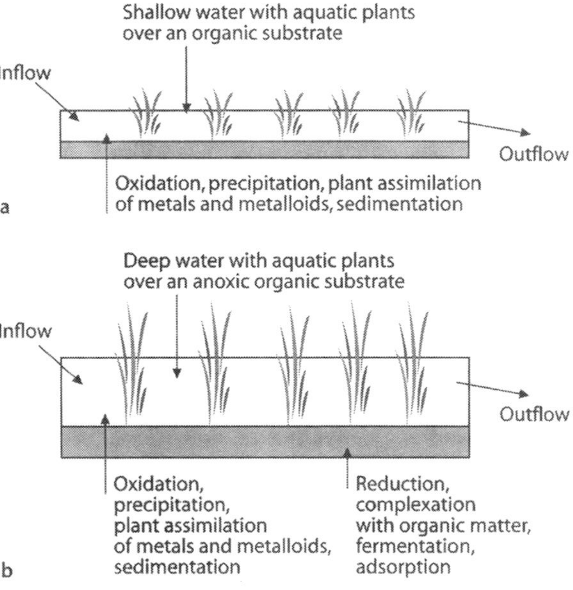

duced from sulfide oxidation at the AMD source and hydrolysis reactions in the wetland. Alkalinity may have to be added to such waters. This can be achieved by growing certain plants in the wetland (e.g. reeds). Reeds are capable of passing oxygen through their root zone and the organic substrate. As a result, oxygen is converted to carbon dioxide. The carbon dioxide gas dissolves in the mine water, consumes hydrogen and adds alkalinity in the form of bicarbonate. Alternatively, anoxic limestone drains may have to be installed at the inflow location of the wetland.

3.12.4.2
Subsurface Flow or Anaerobic Wetland

Anaerobic wetlands are generally used for net acid waters (i.e. "net acid" according to the definition by Hedin et al. (1994a)). Water flows through a relatively deep (~1 m), permeable, and anoxic organic substrate (Fig. 3.13b). Placement of organic waste (e.g. mushroom compost, saw dust, manure) into wetlands helps to establish the reducing conditions. Anoxic conditions favour the proliferation of sulfate reducing bacteria (SRB) (Gould and Kapoor 2003). Bacterial sulfate reduction, or reduction of oxidized acid waters by reactive organic matter (simplified as organic molecule CH_2O), results in a number of chemical reactions (Blowes et al. 1994; Deutsch 1997; Mills 1999; Walton-Day 1999; Mitchell 2000). The most important reaction is the reduction of dissolved sulfate to hydrogen sulfide gas:

$$2\,CH_2O_{(s)} + SO^{2-}_{4(aq)} + 2\,H^+_{(aq)} \rightarrow H_2S_{(g)} + 2\,CO_{2(g)} + H_2O_{(l)} \quad (3.47)$$

$$2\,CH_2O_{(s)} + SO^{2-}_{4(aq)} \rightarrow H_2S_{(g)} + 2\,HCO^-_{3(aq)} \quad (3.48)$$

The hydrogen sulfide gas formed during sulfate reduction may react with dissolved metals. Consequently, solid metal sulfides precipitate:

$$Zn^{2+}_{(aq)} + H_2S_{(aq)} \rightarrow ZnS_{(s)} + 2\,H^+_{(aq)} \tag{3.49}$$

$$Fe^{2+}_{(aq)} + H_2S_{(aq)} \rightarrow FeS_{(s)} + 2\,H^+_{(aq)} \tag{3.50}$$

Precipitation of metal sulfides results in the production of hydrogen ions. However, the sulfate reducing reactions (Reactions 3.47, 3.48) generate more alkalinity; thus, net alkaline conditions prevail. If bacterially mediated sulfate reduction is not achieved, a reduction in dissolved sulfate concentrations occurs through the precipitation of gypsum. Other chemical reactions will consume any dissolved oxygen, cause the precipitation of metal sulfides, convert iron and manganese hydroxides to sulfides, reduce metals, reduce sulfate to sulfur or sulfide, and generate bicarbonate (Tyrrell 1996; Walton-Day 1999). In addition, oxidation/hydrolysis reactions may occur at the wetland's surface.

Alkalinity is produced in the form of bicarbonate in the sulfate reduction reactions (Reactions 3.49 and 3.50). The bicarbonate acts a buffer to neutralize any hydrogen ions. However, the bicarbonate is not necessarily permanent. It will be permanent only if the hydrogen sulfide is removed by degassing. Alternatively, the sulfide ion reacts: (*a*) with an organic compound to form an organic sulfide; or (*b*) with a dissolved metal ion such as Zn^{2+} or Fe^{2+} to form a solid metal sulfide (Reactions 3.49, 3.50) (Walton-Day 1999).

Overall, the chemical reactions result in sulfate reduction, the precipitation of metal sulfides, and an increase in pH and alkalinity. Other wetland processes include sedimentation, physical entrapment of solid particulates and colloids, removal of metals through adsorption onto and coprecipitation with wetland particulates, complexation with organic materials, and plant assimilation (Ledin and Pedersen 1996; Walton-Day 1999). Some aquatic plants growing in wetlands or mine water ponds can take up large amounts of heavy metals and metalloids (Hozhina et al. 2001). Other plant species tolerate high metal concentrations and do not bioaccumulate metals to any significant degree, compared to the overall metal retention by the wetland substrate (Karathanasis and Johnson 2003).

3.12.4.3
Use of Wetlands

Wetland processes aim to decrease acidity and dissolved metal and sulfate concentrations. Effluents produced should be of such quality that they can be discharged to other surface water bodies. Nonetheless, in order to achieve adequate treatment of AMD waters, the aerobic or anaerobic wetlands may require additions. Alkalinity producing systems such as anoxic limestone drains may need to be installed where additional bicarbonate is needed (Barton and Karathanasis 1999). Also, manganese is persistent in mine waters unless the pH has been raised to above 9. Therefore, manganese can be carried for long distances downstream of a source of mine drainage. Aerobic and anaerobic wetlands are incapable of lowering dissolved manganese concentrations to a significant degree. Rock filters, bioreactors or limestone cobble ponds may need to be constructed in order to remove the elevated manganese concentrations (Brant et al.

1999; Johnson and Younger 2005). Such rock filters operate as shallow wetlands and remove dissolved metals and hydrogen sulfide concentrations. Depending on the composition of the AMD waters, they may require a sequential sequence of water treatment stages (Lamb et al. 1998).

Typically, AMD waters pass through a settling pond to remove suspended solids and then through an anaerobic abiotic pond to reduce any Fe^{3+} to Fe^{2+} and to remove aluminium from the waters. The waters then flow through an anoxic limestone drain to induce alkalinity. The discharge passes through an anerobic wetland to induce metal precipitation and sulfate reduction. The treatment is followed by an aerobic wetland – to achieve precipitation of iron – and possibly a rock filter or limestone cobble pond – to remove any dissolved metals such as manganese or hydrogen sulfide.

Generally, the establishment of wetlands is often a preferred option for the complete or partial treatment of AMD waters which have low TDS values. Wetlands are an aesthetically attractive, passive, low-cost, low-maintenance, and sustainable method (SMME 1998; Brown et al. 2002). However, a number of wetlands used to treat AMD waters have failed over time (Woulds and Ngwenya 2004; Kalin et al. 2006) (Fig. 3.14). Furthermore, wetlands are sensitive to pulses of high metal concentrations, and the accumulated metals may be mobilized by microbiochemical processes (Ledin and Pedersen 1996). Furthermore, the accumulation of metals in wetlands creates a metal-rich aquatic environment which may experience changes in its hydrology or climate

Fig. 3.14. Wetland at the Horn Island gold mine, Australia. The wetland has failed because an adjacent damwall was constructed using acid generating waste rocks (foreground), which resulted in significant AMD and caused plant death

in the long or short term. A wetland needs a sufficient year-round supply of water that would ensure that the wetland remains in a permanently saturated condition (Jones and Chapman 1995). Drying out of a wetland will lead to the oxidation of biological materials and sulfides, and the formation of evaporative salts. At the beginning of the next rain period, sulfuric acid, metals, and salts are released. These contaminants are then flushed straight through the wetland and into receiving waters. Wetlands without sufficient water supply become chemical time bombs and sources of metals, metalloids, and sulfate. Therefore, semi-arid areas, polar regions as well as areas with distinct seasonal rainfall and run-off are unsuitable for such a remediation measure.

3.12.5 Adit Plugging

Adits, shafts and tunnels are common point sources of AMD. Some of them were originally installed to drain the underground mine workings, and sealing such mine openings with plugs can reduce the volume of drainage waters (Plumlee and Logsdon 1999; SMME 1998). Adit plugs are concrete and grout hydraulic seals that exert hydraulic control on ground waters emanating from mine openings. The plugs minimize or even prevent ground waters from escaping from underground workings (Banks et al. 1997). The reasoning for this technique is that the plug removes an AMD point source. In addition, ground water will back up in the underground mine workings, precluding atmospheric oxygen from reaching and oxidizing sulfides (Plumlee and Logsdon 1999). Placement of organic material into flooded underground workings may help to induce anoxic conditions and prevent sulfide oxidation and AMD generation (SMME 1998). Adit plugs can prevent the infiltration of oxygen and water into, and the migration of AMD waters out of, underground workings.

A problem may arise if leakage of drainage water occurs around the plug or other hydrologic conduits, and pre-mining springs and water tables can be reactivated carrying now contaminated waters (Plumlee and Logsdon 1999). Moreover, flooding of historic mines with abundant soluble iron salts in the workings may trigger sulfide oxidation. As a consequence, the efficiency and long-term stability of seals are controversial, especially as there have been seal failures and associated massive releases of AMD effluents.

3.12.6 Ground Water Treatment

AMD contaminated ground water requires treatment. Unlike seepage and run-off, *acid ground water* cannot be easily intercepted. Current treatment techniques involve pumping of the water to the surface and treating it there (ex situ), or trying to contain and treat the contaminated ground water in the ground (in situ).

Pump-and-treat methods are well established ex situ techniques used to clean up contaminated waters. Such pump-and-treat systems flush contaminants from the aquifer and treat the pumped ground water at the surface using standard metal removal processes. The major shortcoming of the pump-and-treat approach is that massive amounts of ground water – commonly several times the volume of the contaminated plume – must be pumped to adequately dislodge the metal contaminants. Such aggressive pumping lowers the water table and leaves large pockets of metal-rich subsurface materials. The pump-and-treat technology often fails to achieve ground water clean-up standards in reasonable time frames, and the success rate of of this conventional technology is rather poor. The pump-and-treat technology is recognized, therefore, as being quite inefficient.

In situ treatment techniques aim to take advantage of the natural hydrogeology of a site. They utilize the natural ground water flow. In some cases, a natural reduction in contaminant concentrations can be observed as contaminants migrate from the AMD source into the aquifer. This reduction is primarily due to neutralization reactions. The acid ground water may migrate through a carbonate aquifer, and the natural neutralization reactions can lead to decreased contaminant loads. Other *natural attenuation* mechanisms in aquifers involve dilution, adsorption, precipitation, dispersion and biodegradation processes. In addition, passive chemical or biological treatment systems can be emplaced for the remediation of contaminated ground waters. AMD contaminated ground waters can be remediated using a permeable, reactive zone of organic matter (e.g. sewage sludge, sawdust), calcite, zeolites, phosphates, ferric oxyhydroxides or other materials submerged in the ground water flow path (SMME 1998; Younger et al. 2002; Benner et al. 2002; Amos and Younger 2003; Blowes et al. 2003). These *permeable reactive barriers* are implemented by digging a trench in the flow path of a contaminant plume and backfilling the trench with the reagents (Fig. 3.15). Neutralization, precipitation and adsorption processes in these materials cause the metal and acidity concentrations in the ground water to decrease. The barriers remove the majority of the metals, metalloids and acidity from the polluted waters (Benner et al. 1999; Smyth et al. 2001). Also, the placement of permeable organic matter or liquid organic substances such as methanol in the ground water flow path can favour the abundance of sulfate reducing bacteria (Bilek 2006). The bacteria create a reducing zone and reduce sulfate to sulfide. Any dissolved metal are removed from solution as the metals precipitate as sulfides.

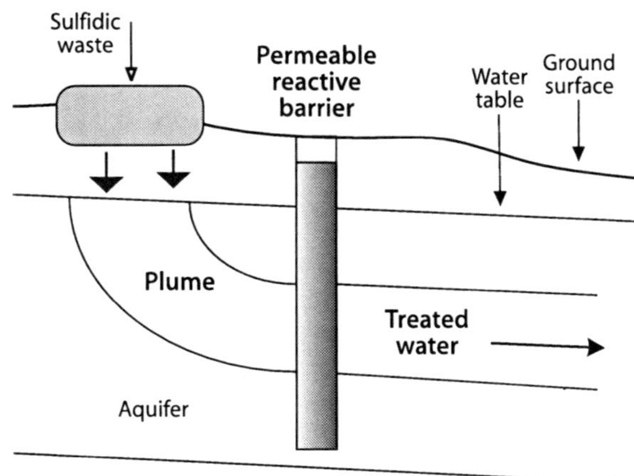

Fig. 3.15. Schematic cross-section showing in situ treatment of acid ground water with a permeable reactive barrier (after Jambor et al. 2000b)

3.13 Summary

The constituents of mine waters are highly variable and include elements and compounds from mineral-rock reactions, process chemicals from mineral beneficiation and hydrometallurgical extraction, and nitrogen compounds from blasting operations. Aqueous solutions in contact with oxidizing sulfides will contain increased acidity, iron, sulfate, metal and metalloid concentrations. While AMD waters are well known for their elevated metal concentrations, neutral to alkaline conditions can also favour the release of metals and metalloids from waste materials. Elevated metal and metalloid concentrations in neutral to alkaline pH, oxidizing mine waters are promoted by: (*a*) the formation of ionic species (e.g. Zn^{2+}), oxyanions (e.g. AsO_4^{3-}) and aqueous metal complexes (e.g. U carbonate complexes, Zn sulfate complexes); and (*b*) the lack of sorption onto and coprecipitation with secondary iron minerals.

Several processes influence the composition of AMD waters. These include biochemical processes, the precipitation and dissolution of secondary minerals, and the sorption and desorption of solutes with particulates. Changes to Eh and pH conditions influence the behaviour, concentrations and bioavailability of metals and metalloids.

The oxidation of Fe^{3+} and hydrolysis of iron in AMD waters produces hydrous ferric oxide (HFO) precipitates (i.e. ochres or yellow boys), which include non-crystalline iron phases as well as iron minerals such as schwertmannite and ferrihydrite. The occurrence of different iron minerals is largely pH dependent. The iron solids occur as colourful bright reddish-yellow to yellowish-brown stains, coatings, suspended particles, colloids, gelatinous flocculants, and precipitates in AMD affected waters. The high specific surface area of hydrous ferric oxide precipitates results in adsorption and coprecipitation of trace metals. Consequently, these solid phases control the mobility, fate and transport of trace metals in AMD waters.

The dissolution of soluble Fe^{2+} sulfate salts can be a significant source of acidity, Fe^{3+} and dissolved metals which were originally adsorbed onto or incorporated in solid phases. Also, the oxidation of Fe^{2+} to Fe^{3+} and subsequent hydrolysis of iron can add significant acidity to mine waters.

Metals are present in AMD waters as simple metal ions or metal complexes. However, significant metal concentrations can also be transported by colloidal materials in ground and surface waters. Colloidal iron precipitates with adsorbed metals can represent important transport modes for metals in mine environments and streams well beyond the mine site.

The monitoring of mine waters is designed: (*a*) to identify the early presence of, or the changes to, dissolved or suspended constituents; and (*b*) to ensure that discharged water meets a specified water quality standard. Sites should measure or estimate flow rates or periodic flow volumes in order to make calculations of contaminant loads possible. Possible tools for the prediction of water chemistry include geological, mathematical and computational modeling. These tools cannot be used, however, to predict the exact chemistry of mine waters.

Sulfidic waste rock dumps are the major sources of AMD because of their sheer volume. The quality and volume of AMD seepages emanating from sulfidic piles are influenced by the properties of the waste materials. Despite their heterogeneity, waste dumps generally exhibit a single continuous water table with a moderate hydraulic gradient. The physical and chemical conditions, and mineralogical composition of waste materials vary on a microscopic scale. Therefore, drainage water from a sulfidic waste dump represents a mixture of fluids from a variety of dynamic micro-environments within the pile. The different rates of the various weathering reactions within the waste can cause temporal changes to the seepage chemistry.

At mine sites, water management strategies aim to protect aquatic environments and to reduce the water volume requiring treatment. Treatment techniques for AMD waters are designed: to reduce volume; to raise pH; to lower dissolved metal and sulfate concentrations; to lower the bioavailability of metals; to oxidize or reduce the solution; and to collect, dispose or isolate any waste waters or metal-rich precipitates. Established AMD treatment options include: neutralization using a range of possible neutralizing materials; construction of aerobic or anaerobic wetlands; installation of anoxic limestone drains; and successive alkalinity producing systems. Acid ground waters are treated using pump-and-treat, natural attenuation, and permeable reactive barrier technologies.

Further information on mine waters can be obtained from web sites shown in Table 3.6.

Table 3.6. Web sites covering aspects of mine waters

Organization	Web address and description
International Mine Water Association (IMWA)	*http://www.imwa.info* Congresses, symposia, meetings
Acid Drainage Technology Initiative (ADTI), US	*http://www.unr.edu/mines/adti* Information dissemination on AMD remediation technologies in the USA
US Environmental Protection Agency (EPA), National Pollutant Discharge Elimination System (NPDES)	*http://cfpub.epa.gov/npdes/indpermitting/mining.cfm?program_id=14* Regulation of water discharge from mine sites
US Geological Survey (USGS) Mine Drainage Interest Group (MDG)	*http://mine-drainage.usgs.gov* Information and reports on AMD monitoring and remediation
Department of Environmental Protection, Pennsylvania, USA	*http://www.dep.state.pa.us/dep/deputate/minres/districts/cmdp/main.htm* Technical reports on AMD and coal mine drainage
West Virginia Water Research Institute, USA	*http://wvwri.nrcce.wvu.edu/index.cfm* Technical reports on AMD prediction, remediation, and prevention
Mine Environment Neutral Drainage Program (MEND), Canada	*http://www.nrcan.gc.ca/mms/canmet-mtb/mmsl-lmsm/mend/* Extensive technical reports and manuals on AMD sampling, analyses, prediction, monitoring, prevention, and treatment
Minesite Drainage Assessment Group (MDAG), Canada	*http://www.mdag.com* Case studies, articles, resources and publications on mine waters
Remediation Technologies Development Forum (RTDF)	*http://www.rtdf.org* Information on reactive barrier technology
Permeable Reactive Barrier Technology Group (PRB-NET)	*http://www.prb-net.org* Information on reactive barrier technology
Partnership for Acid Drainage Remediation in Europe (PADRE)	*http://www.padre.imwa.info* Information on mine water management and treatment
Mining Information Service EnviroMine	*http://technology.infomine.com/enviromine* News and information on AMD for the mining industry
Minewater Information Web Site	*http://www.minewater.net* Information on all aspects of mine water

Chapter 4

Tailings

4.1 Introduction

Mineral processing of hard rock metal ores (e.g. Au, Cu, Pb, Zn, U) and industrial mineral deposits (e.g. phosphate, bauxite) involves size reduction and separation of the individual minerals. In the first stage of mineral processing, blocks of hard rock ore up to a meter across are reduced to only a few millimeters or even microns in diameter. This is achieved by first crushing and then grinding and milling the ore (Fig. 1.2). Crushing is a dry process; grinding involves the abrasion of the particles that are generally suspended in water. The aim of the size reduction is to break down the ore so that the ore minerals are liberated from gangue phases. In the second stage of mineral processing, the ore minerals are separated from the gangue minerals. This stage may include several methods which use the different gravimetric, magnetic, electrical or surface properties of ore and gangue phases (Fig. 1.2). Coal differs from hard rock ore and industrial mineral deposits as it does not pass through a mill. Instead, the coal is washed, and coal washeries produce fine-grained slurries that are discarded as wastes in suitable repositories. Consequently, the end products of ore or industrial mineral processing and coal washing are the same: (*a*) a concentrate of the sought-after commodity; and (*b*) a quantity of residue wastes known as "*tailings*". Tailings typically are produced in the form of a particulate suspension, that is, a fine-grained sediment-water slurry. The tailings dominantly consist of the ground-up gangue from which most of the valuable mineral(s) or coal has been removed. The solids are unwanted minerals such as silicates, oxides, hydroxides, carbonates, and sulfides. Recoveries of valuable minerals are never 100%, and tailings always contain small amounts of the valuable mineral or coal. Tailings impoundments may also receive very high concentrations of valuable minerals or coal during times of inadequate mineral processing or coal washing.

At nearly every metal mine site, some form of mineral processing occurs, and tailings are being produced. In metal mining, the extracted ore minerals represent only a small fraction of the whole ore mass; the vast majority of the mined material ends up as tailings. More than 99% of the original material mined may finally become tailings when low-grade metal ores are utilized. Tailings represent, therefore, the most voluminous waste at metal mine sites.

This chapter documents the characteristics, disposal options, and environmental impacts of tailings. The main focus is on tailings from metal ores. Other aspects important to tailings such as sulfide oxidation and geochemical processes in AMD waters have already been presented (Chaps. 2, 3). Sulfidic tailings and AMD waters of sulfidic tailings may be characterized and treated with the same type of approaches used to characterize and treat sulfidic waste rocks and AMD waters as discussed in the previous chapters.

4.2 Tailings Characteristics

Tailings vary considerably in their chemical and physical characteristics. These characteristics include: mineralogical and geochemical compositions; specific gravity of tailings particles; settling behaviour; permeability vs. density relationships; soil plasticity (i.e. Atterberg limits); consolidation behaviour; rheology/viscosity characteristics; strength characteristics; pore water chemistry; and leaching properties (Environment Australia 1995). Detailed procedures for tailings sampling, preparation and analysis are found in Ficklin and Mosier (1999). Laboratory methods for the mineralogical and geochemical analysis of tailings solids and waters are given by Jambor (1994), Crock et al. (1999), and Petruk (2000).

Tailings consist of solids and liquids. The solids are commonly discharged with spent process water into a tailings storage facility, most commonly a tailings dam. As a result, the waste repository contains liquids in the form of surface and pore waters. These tailings liquids tend to contain high concentrations of process chemicals. The following presentation of tailings is given in terms of process chemicals, tailings liquids, and tailings solids.

4.2.1 Process Chemicals

Many of the mineral beneficiation and hydrometallurgical operations process the crushed ore in water. Ground or surface water is exploited and used in the process circuits as so-called "*process water*". The composition of process water is largely a function of the applied mineral processing and hydrometallurgical techniques. Hydrometallurgical processing requires specific chemicals for different ore characteristics and mineral behaviours. The chemicals can be classified as flotation reagents, modifiers, flocculants/coagulents, hydrometallurgical agents, and oxidants (Table 4.1). Flotation reagents are a group of chemicals used for froth flotation which is a common mineral processing technique to recover sulfides. Froth flotation works on the principle that water and a frothing agent are added to finely crushed mineral particles. The ore minerals cling to air bubbles and form a froth on top of the water. The froth is recovered and dried. The remaining water contains unwanted solids which are pumped to a tailings disposal facility.

Much of the process water accumulates in decant ponds of tailings dams. The tailings water can be decanted for reuse and pumped back to the plant. Recycling of process water and process chemicals makes economic sense and can reduce the load of contaminants contained in ponds and tailings dams. However, a proportion of the discharged process water remains in the tailings disposal facility. Various fractions of the chemical additives ultimately find their way into tailings, and tailings liquids often contain some levels of organic chemicals, cyanide, sulfuric acid, and other reagents used to achieve mineral recovery.

Table 4.1. Examples of common flotation reagents, modifiers, flocculants, coagulants, hydrometallurgical reagents, and oxidants (after Ritcey 1989; Allan 1995; Barbour and Shaw 2000)

Class	Use	Reagent examples
Flotation reagents		
a) Frothers	to act as flotation medium	Surface active organics such as pine oil, propylene glycol, aliphatic alcohols, cresylic acid
b) Collectors	to selectively coat particles with a water repellent surface attractive to air bubbles	Water soluble, surface active organics such as amine, fatty acids, xanthates
Modifiers		
a) pH regulators	to change pH to promote flotation	Lime, hydrated lime, calcite, soda ash, sodium hydroxide, ammonia, sulfuric acid, nitric acid, hydrochloric acid
b) Activators and depressants	to selectively modify flotation response of minerals present	Surface active organics and various inorganics such as copper sulfate, zinc sulfate, sodium sulfide, lead nitrate, lime, sodium silicate
c) Oils	to modify froth and act as collectors	Kerosene, fuel oils, coal-tar oils
Flocculants	to promote larger particle formation and settling efficiency by bridging smaller particles into larger particles	Clays, metal hydroxides, polysaccharides, starch derivatives
Coagulants	to promote larger particle formation and settling efficiency by reducing the net electrical repulsive forces at particle surfaces	Ferric and ferrous sulfate, aluminium sulfate, ferric chloride
Hydrometallurgical reagents	to selectively leach ore minerals	Sulfuric acid, sodium cyanide
Oxidants	to oxidize process water	Hydrogen peroxide, sodium hypochlorite, ferric chloride, potassium permanganate

4.2.2
Tailings Liquids

Water present at the surface of tailings storage facilities and present within the pores of tailings solids is referred to as "*tailings liquid*" or "*tailings water*". It has highly variable compositions depending on the processing technique, and its composition may change over time. Rainfall leads to the dilution of tailings water, and evaporative concentration causes secondary mineral precipitation at and below the tailings surface. In addition, the exploitation of fresh, brackish or saline water for mineral processing will influence the composition of tailings water. In arid regions, the use of saline ground water for the processing of ores can result in extreme saline process waters and tailings. Salt encrustations are common in the waste repositories. Also, tailings solids show poor consolidation behaviour, and the evaporation rate of tailings water is reduced due to the high salinity. Extreme salinities may also originate from chemical reactions in the tailings which are induced by the addition of process chemicals.

Process chemicals, used to extract ore minerals, may influence the behaviour of metals and other elements in tailings. For example, thiosalts ($S_2O_3^{2-}$, $S_3O_6^{2-}$, $S_4O_6^{2-}$) can oxidize to sulfate in waste repositories causing the acidification of waters, and organic chemicals may complex metals present in the tailings. As a consequence, the pore and surface waters of tailings dams may contain strongly elevated concentrations of various elements and compounds, including process chemicals. Some of the process reagents such as organics and cyanides may be destroyed with time by natural bacterial, chemical or photolytic degradation processes (Sec. 5.8). Other compounds may require naturally enhanced or engineered destruction; some elements have to be permanently isolated by the tailings impoundment (e.g. radionuclides, metals, metalloids).

The acidity of tailings waters is influenced by the applied processing technique. For instance, the digestion of bauxite ore and the dissolution of gold using cyanide solutions are conducted under alkaline conditions. In contrast, the hydrometallurgical extraction of copper, nickel, and uranium is based on the use of sulfuric acid under oxidizing conditions. This sulfuric acid or that generated by sulfide oxidation will leach the tailings minerals. For example, the acid leaching of the mineral fluorite (CaF_2) releases fluorine into waters (Petrunic and Al 2005). The fluorine in turn forms strong complexes with aluminium and enhances the dissolution of aluminosilicate minerals. As a result, strongly elevated Al contents can be present in tailings waters. In extreme cases, different processing techniques can create strongly acid or alkaline tailings waters with high concentrations of iron, manganese, aluminium, trace metals, metalloids, fluoride, chloride and sulfate (Bodénan et al. 2004).

4.2.3 Tailings Solids

At metal mines, the amount of the ore mineral(s) extracted from an ore is relatively small, and the vast majority of mined and processed ore ends up as tailings. At modern gold mines, more than 99.999% of the originally mined and processed ore may finally become tailings. The dry weight of tailings produced is nearly equal to the dry weight of the ore mined.

The grain size of tailings is relatively restricted and ranges from clay to sand (i.e. 2 μm to 2 mm). In some cases, the tailings solids have distinct grain sizes, and the solids are then referred to as “*slimes*” and “*sands*”. Dry tailings typically consist of 70 to 80 wt.% sand-sized particles and 20 to 30 wt.% finer clay-sized particles (Fig. 4.1). The grain size depends on the liberation characteristics of the ore and gangue minerals and the applied crushing and grinding process. The grain size influences the resistance of the tailings solids to wind and water erosion and the behaviour and settling characteristics of particles in tailings dams. The mineralogical and geochemical composition of tailings solids is site specific, and such variations provide different challenges. For example, sulfide-rich tailings are potential sources of AMD whereas uranium tailings have elevated levels of radiation.

It is often assumed that tailings contain minerals similar to those of the ore, only in much smaller grain size. Nonetheless, tailings comprise a material which is significantly different to the mined ore in terms of grain size, mineralogy, and chemistry. By nature, mineral processing is designed to change the physical and chemical charac-

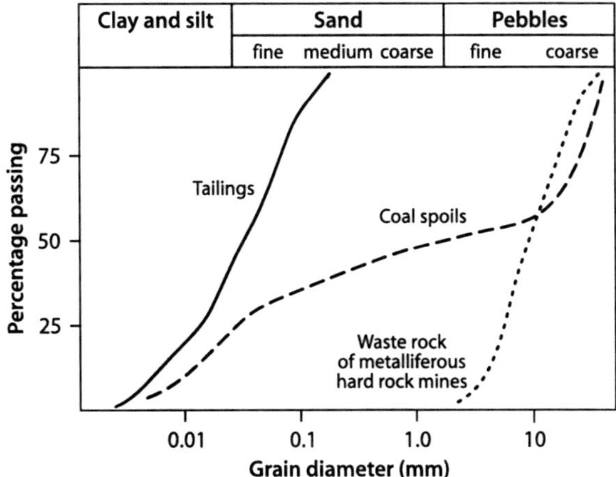

Fig. 4.1. Schematic particle size distribution curves for tailings, coal spoils, and waste rocks (after Robertson 1994; Younger et al. 2002)

teristics of the mined ore. The processing also promotes the dissolution and mobilization of elements present in the ore. The physical and chemical parameters (e.g. pH, Eh) change for individual elements and compounds from the ore deposit to the tailings repository. Consequently, tailings undergo chemical reactions after their deposition in the repository, and their composition changes over time. The tailings solids and the interstitial tailings liquids react and attempt to reach equilibrium. Tailings undergo forms of *diagenesis*. In addition, physical and biological processes occur such as compaction, cementation, recrystallisation as well as mineral dissolution and formation assisted by microorganisms. As a result, tailings contain elements as dissolved species and in potentially soluble and insoluble solid forms (Craw 2003; Sidenko and Sherriff 2005; Bobos et al. 2006).

The diagenetic processes highlight the fact that tailings solids can be the result of different origins. Tailings solids can be: (*a*) primary ore and gangue minerals; (*b*) secondary minerals formed during weathering; (*c*) chemical precipitates formed during and after mineral processing; and (*d*) chemical precipitates formed after disposal in the tailings storage facility. Minerals within tailings can be assigned to several events of mineral formation (Jambor 1994). Primary minerals are ore and gangue minerals of the original ore. Secondary minerals are those minerals which formed during weathering of ore and gangue phases. Chemical precipitates formed during and after mineral processing, including those minerals formed in tailings impoundments, may be labelled as tertiary or quaternary (Jambor 1994).

4.3 Tailings Dams

Most of the tailings mass produced worldwide is pumped into "tailings storage facilities, including large surface impoundments so-called "*tailings dams*". The impoundments are best thought of as purpose-built sedimentation lagoons where fine-grained waste residues and spent process water are captured. There are at least 3500 tailings dams worldwide (Davies and Martin 2000). These dams may range from a few hectares to thousands of hectares in size. Because of their size, they leave the largest "footprint" of any mining activity on the landscape.

Tailings dams are cross valley (i.e. one or two dams constructed across a section of a valley), sidehill (i.e. one dam constructed perpendicular to the slope of a hill), or paddock impoundments (i.e. four sided impoundments constructed on flat land). In some cases, the tailings themselves – in particular the sand-sized fraction – are used to construct the embankments. If the tailings solids and other waste materials are used as a construction material for dam walls and banks, they need chemical and physical characterization prior to such use to ensure that they do not cause tailings dam failure or contaminant release. The storage areas also have to be engineered to optimize the amount of tailings stored and to eliminate any possible environmental impacts. Modern tailings dams and other engineered structures are designed to isolate the processing waste.

Tailings dams are constructed like, and built to standards applicable to, conventional water storage type dams. Generally, they are not constructed initially to completion. They are raised gradually or sequentially as the impoundment fills. The dams are thereby raised upstream towards the tailings, downstream away from the tailings, or in centreline (Vick 1983; Environment Australia 1995; Davies and Martin 2000) (Fig. 4.2). Each construction method has different advantages and disadvantages in terms of construction, use, economics, and seismic stability (Vick 1983). More than 50% of tailings dams worldwide are built using the upstream method although it is well recognized that this construction method produces a structure which is highly

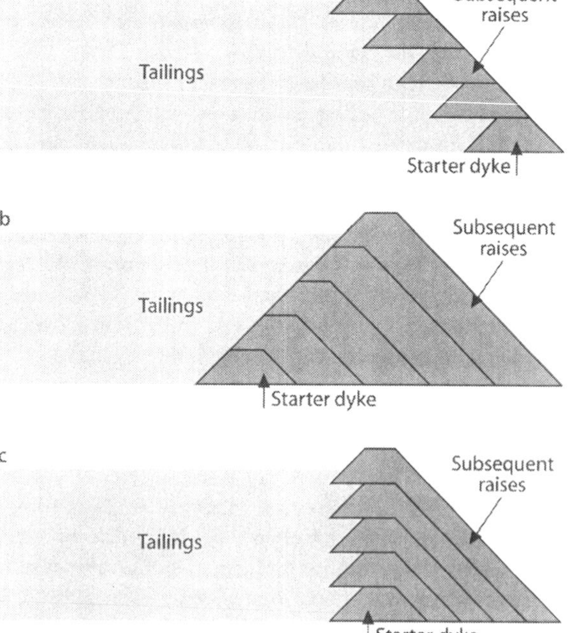

Fig. 4.2. Schematic cross-sections illustrating the construction of embankments for sequentially raised tailings dams; **a** upstream; **b** downstream; **c** centreline method (after Vick 1983)

susceptible to erosion and failure. The failure rate of upstream tailings dams is quite high, and it appears that every twentieth tailings dam fails (Davies and Martin 2000). Major and minor environmental concerns with tailings dams are:

- the structural stability of the dam and the potential release of tailings into the environment through pipeline ruptures, dam spillages and failures;
- air pollution through dust generation;
- release of radiation from tailings;
- seepage from the tailings through the embankment and base into ground and surface waters;
- the visual impact of the large engineered structure; and
- closure and associated capping and vegetation of the tailings dam.

Seepage from a tailings dam is a common environmental concern. The amount of seepage is governed by the permeability of the tailings and the permeability of the liner or ground beneath the impoundment. There are various clay and synthetic liner systems applied to tailings dams so that they reduce leakage into ground water aquifers (Hutchison and Ellison 1992; Environment Australia 1995; Asher and Bell 1999). Plastic geotextiles and clay liners represent effective methods to reduce seepage from a tailings dam. A seepage collection system may need to be put into place, consisting of liners and filter drains placed at the base of the tailings. A toe drain, which will intercept emerging tailings waters, may need to be incorporated into the embankment. Capping tailings dams with dry covers will reduce seepages to even lower rates.

4.3.1 Tailings Hydrogeology

The slurry pumped into tailings dams commonly contains 20 to 40 wt.% solids (Robertson 1994). The tailings may be discharged via one or several so-called "spigotting points" and spread out over large areas. This allows maximum drying and produces a uniform surface.

The solid tailings are transported and deposited in an aqueous environment similar to sediments (Fig. 4.3). Sedimentary textures are common and analogous to those of fluvial and lacustrine environments (Robertson 1994). Stratification, graded and cross bedding as well as lenticular and sinuous textures are typical. Extensive layering is uncommon in tailings. The position of the spigotting points – where tailings are discharged into the dam – usually changes many times. Therefore, depositional environments within the impoundment change over time, and each tailings pile represents a unique sedimentary heterogeneous mass.

Tailings particles have diverse mineralogical compositions and different specific gravities. This gravity difference controls how individual particles segregate and settle within tailings dams. Preferential accumulation and settling of different grain sizes and minerals may occur depending on the discharge rate, the density of the discharged slurry, the method of slurry entry into the impoundment, and whether the disposal surface has a distinct slope (Robertson 1994; Environment Australia 1995). In general, larger, heavier and sand-sized particles settle near the slurry outlet; smaller, lighter and finer-grained particles are located well away from the outlet. If sulfides occur

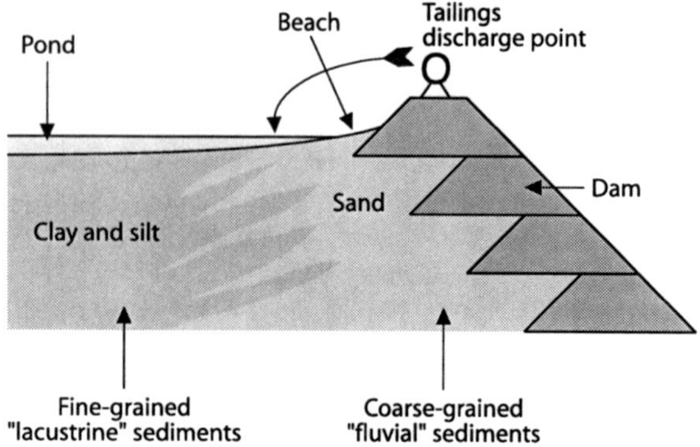

Fig. 4.3. Depositional environments of a tailings dam receiving tailings slurries (after Robertson 1994)

within a particular grain size, this may result in the formation of sulfide-rich tailings sediment. Prolonged surface exposure of these tailings to atmospheric oxygen or subsurface exposure to dissolved oxygen in the vadose zone of tailings may lead to localized AMD generation.

The hydraulic sorting of tailings solids results in coarse-grained size fractions near discharge points and smaller grain size fractions towards the decant pond. This grain size distribution also results in an increased hydraulic conductivity of the coarser grained tailings mass. The coarser tailings drain quicker than the finer material, and there will be a lower ground water level near the dam wall (Environment Australia 1995; Younger et al. 2002).

Ground water storage and flow conditions in tailings dams are site specific and controlled by the grain size and hydraulic conductivity of the tailings, the prevailing climatic conditions as well as the impoundment geometry and thickness (Robertson 1994; Younger et al. 2002). If the tailings have been placed on a permeable base, regional ground water may migrate into the tailings, or tailings seepage may enter the aquifer underlying the tailings dam. Hence, tailings impoundments may represent ground water recharge or discharge areas, depending on whether the water table elevation in the tailings is lower or higher than in the surrounding terrain. Active tailings impoundments have higher water table elevations than inactive ones. Inactive tailings impoundments have unsaturated and saturated zones separated by a ground water table (Fig. 3.6). Such hydrological characteristics are similar to waste rock dumps (Sec. 3.9.1).

4.3.2 AMD Generation

Tailings may have a high sulfide content in the form of rejected pyrite and other sulfides. Sulfidic tailings are a potential source of AMD (Jambor 1994; Blowes et al. 1998; Johnson et al. 2000). If sulfidic tailings are exposed to atmospheric oxygen or to dissolved oxygen in the vadose zone of tailings, the oxygen infiltrating the waste may cause sulfide oxidation and trigger AMD. Acid producing and acid buffering reactions and secondary mineral formation will occur, and low pH pore waters with high concentrations of dissolved constituents will be generated (Coggans et al. 1999; Johnson et al.

2000; Lei and Watkins 2005; Petrunic and Al 2005). Indicators for sulfide oxidation – such as acid, sulfate and metal-rich tailings liquids – are generally observed in the upper part and vadose zone of tailings impoundments (Jambor 1994). Acid, saline tailings liquids may also accumulate in ponds at the surface of the tailings repository (Fig. 4.4).

If sulfidic tailings are present in the unsaturated zone and are exposed to atmospheric oxygen, sulfide oxidation will commence in the upper level of the unsaturated zone. The oxidation front in the tailings will then gradually move downward towards the ground water table where the oxidation of sulfides will almost completely stop. Compared to surface water and pore water in the vadose zone, the ground water in the saturated zone of sulfidic tailings impoundments usually has higher pH and lower Eh values and lower concentrations of dissolved constituents (Fig. 3.6) (Blowes and Ptacek 1994; Coggans et al. 1999). Sulfidic tailings below the ground water table remain protected from oxidation. Unoxidized sulfidic tailings may underlie oxidized sulfidic materials. These unoxidized tailings within the saturated zone have the potential to oxidize as soon as the water table within the impoundment falls to lower levels.

In oxidizing sulfidic tailings, pore waters may exhibit a distinct chemical stratification. For example, oxidation of exposed tailings and subsequent burial of the material by renewed tailings disposal may lead to sulfate-, metal-rich pore waters at deeper levels and saturated zones (Jambor 1994). Also, repeated wetting and drying cycles may result in fluctuating water tables and variable oxidation fronts causing extensive sulfide oxidation, secondary mineral formation and dissolution, and AMD development (McGregor et al. 1998; Boulet and Larocque 1998).

The oxidation and weathering reactions will release metals, metalloids, acid, and salts to tailings pore waters. The released contaminants may: (*a*) be retained within the tailings impoundment; (*b*) reach surface and ground water systems; or (*c*) be precipitated at a particular level of the tailings impoundment. Copper may be precipitated, for instance, as covellite below the zone of sulfide oxidation to form a zone of copper enrichment, akin to the formation of supergene copper ores (Holmström et al. 1999; Ljungberg and Öhlander 2001). Massive precipitation of secondary minerals in-

Fig. 4.4. Ponding of acid, saline water on an abandoned sulfidic tailings dam, Croydon, Australia

cluding oxides (e.g. goethite, ferrihydrite, lepidocrocite, schwertmannite), sulfates (e.g. jarosite, gypsum, melanterite), and sulfides (e.g. covellite) may occur at particular levels within the tailings impoundment (McGregor et al. 1998; Dold and Fontbote 2001). Mineral precipitation can cement sulfidic tailings at or near the depth of sulfide oxidation. Prolonged precipitation causes the formation of a hardpan layer which acts as a diffusion barrier to oxygen and limits the downward migration of low pH, saline pore waters (Blowes et al. 1991; Coggans et al. 1999) (Sec. 2.6.4).

If sulfidic tailings are exposed to the atmosphere for extended periods, evaporation of tailings water commonly leads to the formation of sulfate-bearing mineral salts such as gypsum or jarosite (Johnson et al. 2000; Dold and Fontbote 2001). These secondary mineral blooms occur on or immediately below the impoundment surface, and surficial tailings materials may form a cohesive, rigidly cemented material. Soluble secondary minerals may redissolve upon rainfall events or due to increased water levels within the tailings dam.

While the oxidation of sulfidic tailings is possible, most tailings deposits generally remain water saturated during the operating life. This limits the transfer of air into the tailings and the supply of oxygen to sulfide minerals (Blowes and Ptacek 1994). Moreover, if unoxidized sulfidic tailings are flooded, deposition of organic material may occur, and layers rich in iron and manganese oxyhydroxides may develop at the water-tailings interface. These layers can prevent the release of metals into the overlying water column through adsorption and coprecipitation processes (Holmström and Öhlander 1999, 2001).

Tailings are much finer grained and have a much higher specific surface area (i.e. $m^2 g^{-1}$) available for oxidation and leaching reactions than waste rocks. Thus, the onset of AMD development in tailings can be more rapid than in waste rocks of the same deposit, yet crushing and milling of sulfidic ores do not necessarily increase the oxidation rate of pyrite in tailings dams. The rate of AMD generation in tailings is reduced by:

- *Uniform and fine grain size.* Tailings possess a uniform and fine particle size which leads to a much lower permeability than that in waste rock piles. Compared to coarse-grained waste rocks, tailings exhibit: (*a*) less and slower water and oxygen movement into the waste; (*b*) reduced contact of sulfides with oxygen due to slower oxygen transport into the waste; (*c*) slower water movement and slower replenishment of consumed oxygen; and (*d*) very slow seepage of an AMD plume to outlets of the waste impoundment because the generally low hydraulic conductivity of tailings will delay movement of the AMD plume (Ljungberg and and Öhlander 2001). The behaviours of oxygen and water influence the depth and rate of AMD generation. Consequently, tailings permit a smaller depth of active acid generation than coarse-grained wastes, and sulfidic tailings often generate AMD more slowly than coarser, more permeable waste rock from the same deposit (Mitchell 2000).
- *Addition of alkaline process chemicals.* Some tailings have a high pH due to the addition of alkaline materials during mineral processing (Craw et al. 1999). Any acid generated may immediately be neutralized by residual alkaline processing agents. Tailings may also be stabilized through the addition of neutralizing materials such as lime, crushed limestone or fly ash (Stouraiti et al. 2002). This may prevent highly reactive sulfidic tailings from developing AMD waters.

4.3.3 Tailings Dam Failures

Tailings dams should be constructed to contain waste materials indefinitely. The impoundment should be designed to achieve negligible seepage of tailings liquids into ground and surface waters and to prevent failures of tailings dam structures. Therefore, the dams should be engineered for: (*a*) long-term stability against erosion and mass movement; (*b*) prevention of environmental contamination of ground and surface waters; and (*c*) return of the area for future land use. The overall design objective of tailings dams should be to achieve a safe, stable post-operational tailings impoundment (Environment Australia 1995; Davies and Martin 2000).

The stability of tailings dams is controlled by the embankment height and slope as well as the degree of compaction, nature, and strength of foundation and embankment materials (Environment Australia 1995). Conventional tailings dams can exceed 100 m in height and have to hold back a significant pool of water and up to several hundred million cubic meters of water saturated tailings. The easiest way to maintain dam stability during operation is to keep the decant pond as small as possible and as far as possible from the containing embankment as practical (Environment Australia 1995). This ensures that the phreatic surface remains at low levels. Nonetheless, the mechanical stability of tailings is very poor due to the small grain size and the usually high water content. The level of the water table in the impoundment and embankment falls as tailings discharge ceases. This results in the increased stability of the embankment and tailings mass. Any further accumulation of water on the dam is prevented by capping the impoundment.

In the past 70 years, numerous incidents involving operating tailings dams have occurred (e.g. Wagener et al. 1998). There have been about 100 documented significant upstream tailings dam failures (Davies and Martin 2000) (Table 4.2). The causes for tailings dam failures include:

- *Liquefaction.* Earthquakes are associated with the release of seismic waves which cause increased shear stresses on the embankment and increased pore pressures in saturated tailings. Tailings and the dam may liquefy during seismic events (e.g. Veta de Agua, Chile; 03.03.1985). Liquefaction may also be caused by mine blasting or nearby motion and vibrations of heavy equipment.
- *Rapid increase in dam wall height.* If an upstream dam is raised too quickly, very high internal pore pressures are produced within the tailings. High pore pressures decrease the dam stability and may lead to dam failure (e.g. Tyrone, USA; 13.10.1980).
- *Foundation failure.* If the base below the dam is too weak to support the dam, movement along a failure plane will occur (e.g. Los Frailes, Spain; 25.04.1998).
- *Excessive water levels.* Dam failure can occur if the phreatic surface raises to a critical level; that is, the beach width between the decant pond and the dam crest becomes too small (Fig. 4.3). Flood inflow, high rainfall, rapid melting of snow, and improper water management of the mill operator may cause excessive water levels within the impoundment, which then may lead to overtopping and collapse of the embankment (e.g. Baia Mare, Romania; 30.01.2000; Case Study 5.1). If overtopping of the dam crest occurs, breaching, erosion, and complete failure of the impoundment are possible.

- *Excessive seepage.* Seepage within or beneath the dam causes erosion along the seepage flow path. Excessive seepage may result in failure of the embankment (e.g. Zlevoto, Yugoslavia; 01.03.1976).

If failure occurs, tailings may enter underground workings or more commonly, the wastes spill into waterways and travel downstream. Depending on the dam's location, failures of tailings dams can have catastrophic consequences. Streams can be polluted for a considerable distance downstream, large surface areas can become covered with thick metal-rich mud, the region's sediment and water quality can be reduced, and contaminants may enter ecosystems (e.g. Benito et al. 2001; Hudson-Edwards et al. 2003; Macklin et al. 2003). Tailings dam failures in numerous countries have caused the loss of human life and major economic and environmental costs (Table 4.2) (Mining Journal Research Services 1996).

The prevention of tailings dam failures requires: (*a*) an effective geotechnical characterization of the tailings site; and (*b*) a detailed understanding of the risk of local natural hazards such as earthquakes, landslides, and catastrophic meteorological events. It is important to design tailings dams to such a standard that they can cope with extreme geological and climatic events.

Most tailings dam failures have been in humid, temperate regions. In contrast, there have been very few tailings dam failures in semi-arid and arid regions. However, tailings dams in semi-arid climates suffer from other chronic difficulties including seepage problems, supernatant ponds with high levels of process chemicals, dust generation, and surface crusting. Crusting prevents drying out of tailings, and rehabilitation cannot be undertaken for many years after tailings deposition has ceased.

4.3.4 Monitoring

Monitoring of tailings dam structures is essential in order to prevent environmental pollution and tailings dam failures and spillages. Site specific conditions require tailored monitoring programs. The monitoring program should address the following aspects:

- *Dam performance.* Performance monitoring of tailings dams includes measurements of the filling rate, consolidation, grain size distribution, water balance, and process chemical concentrations such as cyanide.
- *Impoundment stability.* Impoundment stability monitoring establishes the phreatic surface in the embankment and tailings as well as the slope stability and pore pressure within the tailings.
- *Environmental aspects.* Environmental monitoring includes: meteorological observations; measurements of radioactivity levels; investigations of the tailings chemistry and mineralogy; performance of geochemical static and kinetic tests for AMD generation potential; and chemical analyses of ground, surface and seepage waters, downstream stream sediments, and dust particles. Ground water monitoring is an integral part of tailings monitoring and allows an evaluation of tailings seepage into aquifers (Robertson 1994; Environment Australia 1995). Piezometers

Table 4.2. Chronology of tailings dam failures since the 1920s (Genevois and Tecca 1993; Morin and Hutt 1997; Wagener et al. 1998; Lindahl 1998; WISE Uranium Project 2006). (Most of it has been reprinted from *www.wise-uranium.org*, with permission from P. Diehl)

Date	Location	Incident	Release	Environmental impact and fatalities
30.04.2006	Miliang, China	Tailings dam failure during raise	?	17 people missing; cyanide release to local river
30.11.2004	Pinchi Lake, Canada	Tailings dam collapse	6 000–8 000 m^3 of rock and waste water	Spill into lake
20.03.2004	Malvesi, France	Tailings dam failure after heavy rain	30 000 m^3 of liquid and slurry	Nitrate contamination of lokal creek
03.10.2004	Cerro Negro, Chile	Tailings dam failure	50 000 t of tailings	Contamination of local system
27.08.2002	San Marcelino, Phillippines	Overflow spillway failure after heavy rain	?	Villages flooded with waste; contamination of lake and stream system
22.06.2001	Sebastiao das Aguas Claras, Brazil	Tailings dam failure	?	At least 2 mine workers killed
18.10.2000	Nandan, China	Tailings dam failure	?	At least 15 people killed, 100 missing; more than 100 houses destroyed
11.10.2000	Inez, USA	Tailings dam failure	950 000 m^3 of coal waste slurry released into local streams	Contamination of 120 km of rivers and streams; fish kills
09.09.2000	Aitik, Sweden	Tailings dam failure	1 million m^3 of water from the settling pond	?
04.05.2000	Grasberg, Irian Jaya	Waste rock dump failure after heavy rain	Unknown quantity of heavy metal bearing wastes	4 people perished; contamination of streams
10.03.2000	Borsa, Romania	Tailings dam failure after heavy rain	22 000 t of heavy metal bearing tailings	Contamination of streams
30.01.2000	Baia Mare, Romania	Tailings dam crest failure after heavy rain and snow melt	100 000 m^3 of cyanide-bearing contaminated liquid and tailings	Contamination of streams; massive fish kills and contamination of water supplies of >2 million people

Table 4.2. *Continued*

Date	Location	Incident	Release	Environmental impact and fatalities
26.04.1999	Surigao del Norte, Philippines	Tailings spillage from pipe	0.7 Mt of cyanide-bearing tailings	17 homes buried; 51 ha covered with tailings
31.12.1998	Huelva, Spain	Dam failure during storm	50 000 m^3 of phospho-gypsum tailings with pH 1.5	Spillage into local river
25.04.1998	Los Frailes, Aznacollar, Spain	Collapse of dam due to foundation failure	4.5 million m^3 of acid, pyrite-rich tailings	2 616 ha of farmland and river basins flooded with tailings; 40 km of stream contaminated with acid, metals and metalloids
22.10.1997	Pinto Valley, USA	Tailings dam slope failure	230 000 m^3 of tailings and waste rock	16 ha covered with tailings
29.08.1996	El Porco, Bolivia	Dam failure	0.4 Mt	300 km of stream contaminated
March 1996	Marinduque Island, Philippines	Loss of tailings through drainage tunnel	1.5 million t	Siltation of water courses
December 1995	Golden Cross, New Zealand	Dam movement	Nil	Nil
02.12.1995	Suriago del Norte, Philippines	Dam foundation failure after earthquake	50 000 m^3	12 people killed; coastal pollution
19.08.1995	Omai, Guyana	Tailings dam failure	4.2 million m^3 of cyanide-bearing tailings	80 km of local river declared environmental disaster zone
22.02.1994	Merriespruit, South Africa	Dam wall breach after heavy rain	600 000 m^3	17 people killed; extensive damage to town
14.02.1994	Olympic Dam, South Australia	Leakage of uranium tailings dam into aquifer	5 million m^3	?
1993	Marsa, Peru	Tailings dam failure from overtopping	?	6 people killed

Table 4.2. *Continued*

Date	Location	Incident	Release	Environmental impact and fatalities
January 1992	Luzon, Philippines	Collapse of dam due to foundation failure	80 Mt	?
1989	Ok Tedi, Papua New Guinea	Collapse of waste rock dump and tailings dam	170 Mt waste rock and 4 Mt tailings	Flow into local river
30.04.1988	Jinduicheng, China	Breach of dam wall	700 000 m^3	20 people killed
19.01.1988	Grays Creek, USA	Dam failure due to internal erosion	250 000 m^3	?
May 1986	Itabirito, Brazil	Dam wall burst	100 000 m^3	Tailings flow 12 km downstream
1986	Huangmeishan, China	Dam failure from seepage/instability	?	19 people killed
19.07.1985	Stava, Italy	Failure of fluorite tailings dam due to inadequate decant construction	200 000 m^3	269 people killed; two villages buried/wiped out
03.03.1985	Veta de Agua, Chile	Dam wall failure due to liquefaction during earthquake	280 000 m^3	Tailings flow 5 km downstream
03.03.1985	Cerro Negro, Chile	Dam wall failure due to liquefaction during earthquake	500 000 m^3	Tailings flow 8 km downstream
08.11.1982	Sipalay, Philippines	Collapse of dam due to foundation failure	28 Mt	Widespread inundation of agricultural land
18.12.1981	Ages, USA	Dam failure after heavy rain	96 000 m^3 of coal refuse slurry	Slurry flow downstream; 1 person killed; fish kill; homes destroyed
13.10.1980	Tyrone, USA	Dam wall breach due to rapid increase in dam wall height	2 million m^3	Tailings flow 8 km downstream
16.07.1979	Church Rock, USA	Dam wall breach	360 000 m^3 of radioactive tailings water; 1 000 t of tailings	Contamination of river sediments up to 110 km downstream

Table 4.2. *Continued*

Date	Location	Incident	Release	Environmental impact and fatalities
1978	Lincoln, Montana	Dam wall breached by flood water following a small landslide	153 000 m^3 of tailings	Tailings flow into local river
31.01.1978	Arcturus, Zimbabwe	Slurry overflow after heavy rain	30 000 t	1 person killed; extensive siltation
14.01.1978	Mochikoshi, Japan	Wall failure of gold-silver tailings dam due to liquefaction during earthquake	80 000 m^3	1 person killed; tailings flow 7–8 km downstream
01.02.1977	Milan, USA	Dam failure	30 000 m^3	Nil
01.03.1976	Zlevoto, Yugoslavia	Dam failure due to excessive water levels and seepage	300 000 m^3	Tailings flow into river
1975	Mike Horse, USA	Dam failure after heavy rain	150 000 m^3	?
11.11.1974	Bafokeng, Impala, South Africa	Embankment failure of platinum tailings dam due to excessive seepage	3 million m^3	15 people killed; tailings flow 45 km downstream
01.06.1974	Deneen Mica, USA	Dam failure after heavy rain	38 000 m^3	Tailings flow into river
26.02.1972	Buffalo Creek, USA	Failure of coal refuse dam after heavy rain	500 000 m^3	150 people killed; 1 500 homes destroyed
1971	Florida, USA	Tailings dam failure caused by excessive seepage	0.8 Mt	Peace River polluted over a distance of 120 km
1970	Mufulira, Zambia	Tailings move into underground workings	1 Mt	89 miners killed
1967 and 1968	Blackpool and Cholwich, Great Britain	Failure of kaolinite tailings dams	?	?
1966	East Texas	Flow of liquefied tailings from impoundment caused by excessive seepage	80 000–130 000 m^3 of gypsum	?

Table 4.2. *Continued*

Date	Location	Incident	Release	Environmental impact and fatalities
21.10.1966	Aberfan, Great Britain	Liquefaction of coal refuse dam after heavy rain	?	144 people killed
1965	El Cobre, Chile	Liquefaction of eleven tailings dams during earthquake	2 Mt	250 people killed
25.02.1963	Louisville, USA	Failure of calcium carbide tailings dam due to freezing of down stream slope	?	?
1944	Aberfan, Great Britain	Failure of coal refuse dam	?	?
December 1939	Abercyan, Great Britain	Liquefaction of coal refuse dam	?	?
1939	Cilfyndd Common, Great Britain	Failure of coal refuse dam	0.18 Mt	Flow into local river
15.12.1928	Barahona, Chile	Liquefaction of copper tailings dam during earthquake	4 Mt	54 people killed

are essential to monitor water pressure and water level in the impoundment. Piezometers and boreholes also need to be established at background points and along the ground water flowpath that is expected to be affected by tailings leachates. Suitable computational tools (e.g. MODFLOW) are available to model ground water flow and to illustrate potential impacts of leachates on ground water aquifers.

4.3.5 Wet and Dry Covers

Tailings disposal areas can occupy large areas of land and render them useless for future land use unless the tailings repositories are covered upon mine closure. Also, uncovered tailings are susceptible to severe water and wind erosion because of their fine grain size. There are numerous examples where wind-blown particles from uncovered tailings have polluted air, streams and soils and created health problems for local communities (Hossner and Hons 1992). Hence, the objectives of tailings covers are twofold: (*a*) to prevent ingress of water and oxygen into the processing waste; and (*b*) to prevent wind and water erosion.

During operation tailings dams are mostly covered with water, and these wet covers prevent wind erosion. Wet covers over sulfidic tailings also prevent sulfide oxidation and acid production (Sec. 2.10.1). Upon closure and rehabilitation, flooding of reactive tailings and establishment of a shallow water cover on top of tailings will curtail sulfide oxidation (Romano et al. 2003). The placement of tailings into an aqueous environment can only be used in regions where climatic conditions will sustain a permanent water cover. The implementation of water covers over oxidized tailings is inappropriate because metals dissolve into the water cover or are present as water-soluble salts. Such flooded tailings impoundments require a protective layer such as peat at the tailings/water interface to inhibit metal transport (Simms et al. 2001).

A common rehabilitation strategy of tailings involves dry capping. This technique is applied to tailings deposited in tailings dams or backfilled into mined-out open pits. Before dry capping can proceed, the tailings need to settle and consolidate (i.e. "to thicken"). Thickening is the process by which water is removed from the tailings and the volume of the waste is reduced. In some cases, appropriate civil engineering techniques such as wicks, drains or filter beds have to be put into place to ensure consolidation and drying out of tailings in tailings dams and open pits. After drying out and consolidation of the tailings, dry covers are constructed from locally available solid materials. Dry cover designs for tailings are numerous and site specific (Fig. 4.5). They include single layer and multi-layered designs and are largely identical to the dry cover techniques for sulfidic waste rock dumps (Sec. 2.10.2). Dry covers range from the direct establishment of native vegetation on the waste to complex, composite covers (Hutchison and Ellison 1992). The latter type has a number of layers, some compacted to reduce their permeability and others uncompacted to support vegetation. Materials used for dry covers include geotextiles, low sulfide waste rocks, oxide wastes, soils, and clay-rich subsoils. A top surface cover of soil on all external surfaces provides a substrate for a self-sustaining plant cover. Suitable drainage installed prevents erosion of the dam. Additional earthworks may be necessary such as diverting creeks, repro-

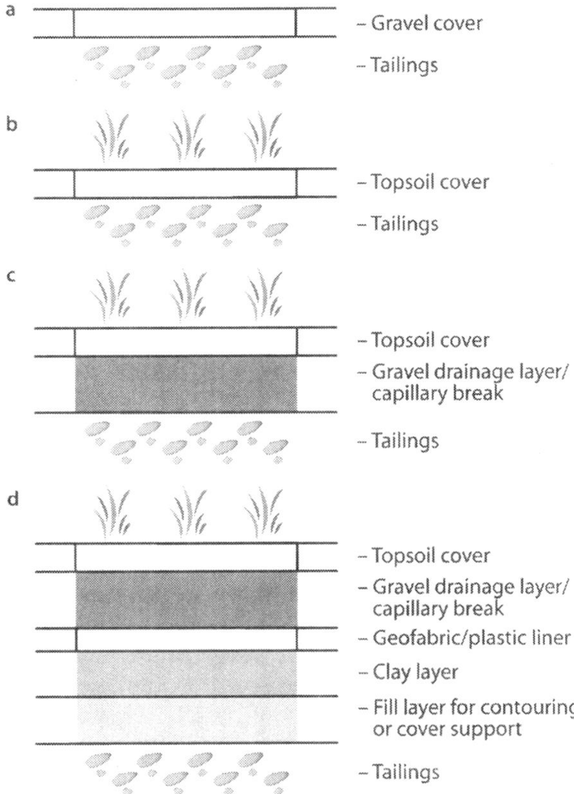

Fig. 4.5. Schematic cross-sections illustrating the principal dry cover designs for tailings storage areas (after Hutchison and Ellison 1992; Environment Australia 1995); **a** gravel layer; **b** single soil cover and revegetation; **c** multilayered soil cover and revegetation; **d** surface recontouring plus a multilayered cover incorporating an infiltration barrier

filing of the slopes of the side walls, and rock armouring of the side walls in order to prevent erosion and slumping of the walls.

4.4 Thickened Discharge and Paste Technologies

The thickened tailings discharge and paste technologies remove a significant proportion of water from the tailings prior to their disposal (Environment Australia 1995; Williams and Seddon 1999; Brzezinski 2001). The water is removed from the tailings at or prior to the point of tailings discharge. Consequently, the waste impoundment does not contain large amounts of water. In an arid to semi-arid environment, solar drying of the tailings completes rapid consolidation. The thickened tailings discharge method is used to construct storage dumps whereby tailings are thickened in a large settler and discharged onto a sloped bed. This technique results in a steep cone of dried tailings. Alternatively, thickened tailings may also be spread out in gently sloped layers for drying. The disposal facility is then constructed progressively as each waste layer is progressively stacked on top of each other. Such stacking of thickened tailings is primarily carried out at alumina processing plants for the disposal of fine tailings (i.e. red mud) from bauxite refining.

The thickened discharge and paste technologies have numerous advantages. Thickened tailings disposal sites do not cover large tracts of lands as conventional tailings dams do. The technologies also reduce water consumption, the risk of water shortage at the mine, and hydraulic sorting of tailings particles (Environment Australia 1995; Williams and Seddon 1999; Brzezinski 2001). Compared to tailings dams, the methods have economic advantages to the operator and reduce the risks of tailings dam failure, water run-off, and leaching. However, thickened tailings operations may also be subject to dust generation or even to failure due to liquefaction of the waste (McMahon et al. 1996).

4.5 Backfilling

Tailings may be pumped into underground workings and mined-out open pits (i.e. in-pit disposal). Such a disposal practice has been used in the mining industry for over 100 years. Tailings are placed directly into the voids from which the ore was won. Backfilling of open pit mines eliminates the formation of an open pit lake. Once backfilling has occurred, ground water will eventually return, approximating the pre-mining ground water table (Fig. 4.6). Within a backfilled open pit, several zones can be identified in terms of oxygen abundance. The saturated anoxic zone is located below the ground water table. Mine waste with the highest acid generating capacity is to be placed into this zone in order to prevent contact of sulfide minerals with oxygen. Wastes with very high acid-generating capacities may require the addition of neutralizing agents prior to or during backfilling.

The advantages of in-pit disposal compared to other tailings disposal methods include:

- the placement of tailings below the ground water table, ensuring limited interaction with the hydrosphere and biosphere;
- no spillages, failures or erosion of tailings dams;
- the backfilling of a large open void with mine waste to such an extent that landscaping and revegetation of the area is possible; and
- a greater depth of cover, ensuring suppression of oxidation of sulfidic wastes or radiological safety of uranium tailings.

While backfilling may appear at first sight a suitable technique to fill and remediate unwanted mine workings, it can, nevertheless, bear several problems:

- The placement of tailings liquids and solids into open pits does not mean that the tailings are as secure chemically and physically as the original mined ore. Chemical and mineralogical changes occur within the processed ore during and after mineral processing. Tailings are not fine-grained ores. They contain very fine-grained, modified and in many cases, reactive ore particles with a large surface area. In addition, tailings contain interstitial pore waters with reactive process chemicals and other dissolved elements and compounds.
- Most pits are deep with a relatively small surface area. In such cases, the evaporation rate of water is slow, and the consolidation of tailings takes considerable time (Environment Australia 1995).

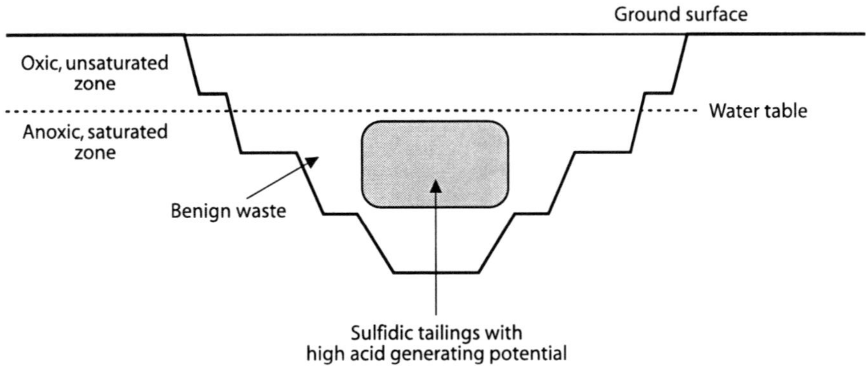

Fig. 4.6. Backfilled open pit showing return of pre-mining ground water table and different hydrological zones (after Abdelouas et al. 1999)

- Any backfill material placed below the water table will become part of the aquifer (Lewis-Russ 1997; Siegel 1997). In particular, if the open pit is not lined with clay or other impermeable liners and if the tailings are disposed of below the post-mining ground water table, the tailings will become part of the local aquifer. Water-rock reactions may lead to the mobilization of contaminants into ground waters. The impact depends on the mineralogical and geochemical characteristics of the tailings and their permeability. The permeability of backfilled material can be very low if significant compaction has occurred. However, if the water is flowing through reactive, permeable and soluble materials, the increased surface area of the waste coupled with its reactivity will result in ground waters enriched in various components. For example, if oxidized sulfidic waste is disposed of in an open pit or a flooded pit, soluble secondary minerals may dissolve in the pore water. This may lead to the release of metals, metalloids, and salts to the ground water or to the overlying water column. Surface or ground water may need to be treated to prevent environmental contamination.

Sulfidic tailings should be disposed of into open pits or underground workings below the post-mining water table once mining ceases. Placement of sulfidic tailings below the post-mining ground water table will preclude any ready access of atmospheric oxygen to sulfidic tailings. The backfill of tailings into underground workings as tailings-cement paste mixtures is also possible. Such a technique is primarily used to provide ground support (Bertrand et al. 2000). The disposal technique also provides limited acid buffering capacity to the waste as cement contains minerals that are able to neutralize acid generated by the oxidation of sulfides.

4.6 Riverine and Lacustrine Disposal

Tailings may also be disposed into rivers and lakes. Once discharged into the streams, the solids and liquids of tailings may be transported considerable distance downstream.

In historic mining areas, the riverine disposal practice was commonly applied (Fig. 4.7) (Black et al. 2004). At these locations, sulfidic tailings may be exposed on the banks of impacted rivers and represent potential AMD sources. Rehabilitation of streams contaminated with tailings may involve dredging, removal, and disposal of the tailings in suitable impoundments, or revegetation of the stream banks with suitable metal-tolerant local plant species. If the deposited tailings are sulfidic, oxidation of sulfides is possible during dredging, draining or erosion. Therefore, precautions must be taken to prevent any change to these potentially reactive, acid materials. While sulfide minerals in discharged mine tailings generally oxidize in oxygenated river waters, sulfide grains can be transported considerable distances downstream from their source zone without weathering (Leblanc et al. 2000).

Today, riverine disposal of tailings and erodible waste dumps are used at copper mine sites in Indonesia (Grasberg-Ertsberg) and Papua New Guinea (Porgera, Ok Tedi, Bougainville) (Jeffery et al. 1988; Salomons and Eagle 1990; Apte et al. 1996; Hettler et al. 1997) (Fig. 4.8). Tailings are discharged directly into the streams or after neutralization. In these earthquake- and landslide-prone, rugged, high rainfall areas, the construction of tailings dams has been proven to be geotechnically impossible, so riverine disposal of tailings is preferred. In these environments, natural erosion rates are very high, and landslides are common. As a result, high natural sediment loads occur in local streams and rivers which can dilute tailings discharges. This disposal practice comes at a price. It causes increased sedimentation of the river system, increased turbidity, associated flooding of lowlands, and contamination of stream and floodplain

Fig. 4.7. Sulfidic tailings layer in ephemeral creek bed downstream from the derelict lead-zinc Webbs Consols mine, Australia. The tailings contain wt.% levels of arsenic, lead and zinc

Fig. 4.8. Earthworks in the Ajkwa River, Irian Jaya, Indonesia. Tailings are discharged (~8.7 Mt per year) from the giant Grasberg-Ertsberg mine into the Ajkwa River, causing severe aggradation and flooding of vegetated lowland downstream of the mine. Areas inundated with tailings are to be turned into productive and sustainable agricultural land

sediments with metals (Case Study 4.1). Diebacks of rainforests and mangrove swamps occur while the impact of elevated copper concentrations in sediments and waters on aquatic ecosystems will become clearer in the long term.

4.7 Marine Disposal

The discharge of tailings into oceans is a preferred option for handling tailings in coastal settings (Hesse and Ellis 1995; Jones and Ellis 1995; Rankin et al. 1997; Berkun 2005) (Case Study 4.2, Fig. 4.9). The aims of this disposal method are: (*a*) to place the tailings into a deep marine environment which has minimal oxygen concentrations; and (*b*) to prevent tailings from entering the shallow, biologically productive, oxygenated zone. Prior to their discharge, tailings are mixed with seawater to increase the density of the slurry. Upon their discharge from the outfall at usually more than 50 m below the surface, the tailings are transported down the seafloor to their ultimate resting place. The disposal technique achieves a permanent water cover and inhibition of sulfide oxidation, yet the physical dispersion of tailings must be assessed, and the extent of metal uptake from the tailings solids by fish and bottom dwelling organisms has to be evaluated (Rankin et al. 1997; Johnson et al. 1998; Blanchette et al. 2001; Powell and Powell 2001). Some tailings may contain bioavailable contaminants in the form

Case Study 4.1. Riverine Tailings Disposal at Ok Tedi, Papua New Guinea

The Ok Tedi Mine

The Ok Tedi mine is one of the largest copper mines in the world. The mine is also a major contributor to the economy of Papua New Guinea (PNG), accounting for much of PNG's total annual export income (Murray et al. 2000a). Ok Tedi is located in highly mountainous terrane at an altitude of approximately 1 600 m. Annual rainfall is at 8 to 10 m per year, and the mine is located in an earthquake- and landslide-prone area. The construction of a tailings dam was attempted; however, the waste repository could not be completed due to landslides. Since 1984, the large-scale open cut operation produces a copper-gold-silver concentrate as well as significant quantities of mining and processing wastes.

Discharge of Waste into Ok Tedi River

Since 1986, tailings have been discharged and waste rock dumps have been left to erode into the headwaters of the Ok Tedi and Fly River system. The discharge rate amounts to about 160 000 t of waste per day. Mine-derived wastes are transported from the mine site into the Ok Tedi River which flows into the Fly River, 200 km downstream of the mine site. The Fly River joins the even larger Strickland River. The combined river system flows via a large estuary and discharges about 120 Mt of sediment a year into the Gulf of Papua. The massive input of tailings and waste rock at the Ok Tedi mine has distinct hydrological, sedimentological and geochemical impacts on the Fly River system throughout the entire length of the river, a distance of over 1 000 km (Salomons and Eagle 1990; Hettler and Lehman 1995; Hettler et al. 1997):

- *Increased turbidity.* The small grain size (<100 µm in diameter) and large quantity of wastes add to the suspended sediment load of the Ok Tedi River system. As a consequence, the suspended matter content in the Middle Fly River is 5 to 10 times higher than the natural background. The increase in suspended sediment load possibly impacts on aquatic organisms.
- *Increased sedimentation.* The tailings and mine waste solids are transported significant distances downstream from the mine. The wastes are then deposited as sediments on the floodplains of the Middle and Lower Fly River, in the delta of the Fly River, and in the Gulf of Papua. Very high deposition rates of mine-derived sediment are found in floodplain environments. Floodplain lakes, creeks, channels, and cut-off meanders receive the greatest quantities of mine-derived sediment. The increased sedimentation on flood plains has buried large areas of tropical lowland rainforests and mangroves with a thin veneer of waste and has caused dieback of the vegetation on a large scale.
- *Metal contamination of sediments.* The near surface sediments – deposited along river channels, in lakes and swamps, and on floodplains – are enriched in copper and gold. Mobilization of copper occurs from the contaminated sediments into surface waters. Chronic copper toxicity to aquatic communities in the affected streams and floodplains are expected, and negative ecological impacts on the fish population may occur in the long term.

Over the entire mine life from 1984 to 2007, there will be a release of 1 400 Mt of waste into the tropical river system. Riverine disposal is allowed under and occurs in compliance with PNG laws and regulations. Increasing concern over environmental impacts has led to dredging trials in the lower Ok Tedi (Murray et al. 2000a). Sand and gravel particles are being dredged from the river bed and stored on the river bank. The dredging efforts may create suitable sediment traps which prevent the burial of lowland rainforests and mangroves with a thin veneer of mine waste.

of soluble metal hydroxides and sulfates. In addition, the mobility and bioavailability of metals and metalloids may increase once the tailings have been injected into the sea (Blanchette et al. 2001). Therefore, the submarine tailings discharge method requires waste characterization as well as modeling and monitoring of the disposal site.

Case Study 4.2. Submarine Tailings Disposal at the Black Angel Mine, Greenland

The Black Angel Lead-Zinc Mine

The Black Angel lead-zinc mine site is located at 71 degrees north on the west coast of Greenland, some 500 km north of the Arctic Circle. The mine site is situated in an alpine landscape, where the 4 km long Affarlikassaa Fjord joins the 8 km long Qaamarujuk Fjord. The name Black Angel derives from the angelic appearance of an exposed contorted band of pelite, clearly visible on the face of the 700 m cliff face that overlooks the Affarlikassaa Fjord (Fig. 4.9). The orebodies were mined from 1973 to 1990 using underground methods, with a total production of 11 million tonnes of ore grading 4.1% Pb, 12.5% Zn and 30 ppm Ag (Asmund et al. 1994). The mineralisation consists of pyrite, sphalerite and galena and the sulfide orebodies are enclosed by thick marble units and clastic metasediments.

Tailings Discharge

The entrance to the mine was 600 m up the cliff facing Affarlikassaa Fjord. The mined ore was transported by an aerial cable car across the Fjord to an industrial area for processing. Mineral recovery was by conventional selective flotation and the tailings were discharged into the Affarlikassaa Fjord. The total amount of discharged tailings was about 8 million tonnes, containing elevated arsenic, cadmium, copper, lead and zinc values (Poling and Ellis 1995).

Environmental monitoring of the fjord system prior to mining showed elevated metal concentrations in local seaweed and mussels (Poling and Ellis 1995). These metal distributions were due to the natural exposure of the sulfide orebodies to weathering and erosion. Within a year from starting the submarine tailings disposal, distinctly elevated lead and zinc values were found in waters and biota of the entire fjord system. This led to extensive investigations into the origin and dispersal mechanisms of metal contaminants. Prior to marine discharge, it was assumed that all residual metals would be present in the tailings only as insoluble sulfides. Yet, the tailings contained lead and zinc phases that were soluble in seawater. In addition, oceanographic investigations contradicted an earlier assumption that the discharged tailings are permanently protected by stagnant bottom waters of the Affarlikassaa Fjord. The studies demonstrated that the disposal site did not have a permanently stratified water column and that complete mixing of the water took place during winter (Poling and Ellis 1995).

Consequently, several changes were made to the mineral processing and tailings discharge which significantly reduced the levels of metal release to the environment. Firstly, improvements in mineral processing led to tailings with lower lead values released to the ocean (1973, 0.4% Pb; 1989, 0.18% Pb). Also, the tailings were treated with chemicals, deaerated prior to discharge and mixed with seawater. These initiatives led to the coagulation and flocculation of small tailings solids, prevented the dispersion of metal-rich particles via mineral-laden air bubbles, and increased the tailings density which led to the deposition of tailings in bottom waters. As a result, the lead and zinc concentrations in water and biota of the Affarlikassaa Fjord decreased (Asmund et al. 1994).

Transfer of Metals into the Fjord

Despite the improvements to the submarine tailings disposal system, metal contamination of the area could not be prevented. The tailings discharge resulted in the metal enrichment of water, suspended particulate matter, sediment and biota in the Affarlikassaa and Qaamarujuk Fjords up to 70 km away from the tailings outfall (Loring and Asmund 1989; Elberling et al. 2002). During mining, analyses of seals and fish species largely revealed no metal contamination, while deep sea prawns and capelins as well as the livers of certain fish species and seabirds contained lead concentrations above the safe consumption limit (Asmund et al. 1994). Mussels and seaweed obtained increasingly elevated cadmium, lead and zinc concentrations, depending on their location relative to the tailings outfall.

Since mine closure, metal concentrations have declined in fjord waters as well as animal and plant life (Asmund et al. 1994). Yet, dispersion and release of metals from the tailings continues (Elberling et al. 2002). In hindsight, detailed mineralogical, geochemical and oceanographic studies prior to tailings discharge would have allowed an informed decision on whether submarine tailings disposal was appropriate at this particular site (Poling and Ellis 1995).

Fig. 4.9. View of the 700 m cliff face that overlooks the Affarlikassaa Fjord at the Black Angel mine, central West Greenland. Over 8 million tonnes of tailings and a large shoreline waste rock pile of 0.4 million tonnes were dumped into the Affarlikassaa Fjord. The two small black spots beneath the left wing of the angel-like figure are the cable car entrances to the mine. The ship *MS Disko II* is 50 m long

4.8 Recycling

An alternative to the disposal of tailings is to put the waste to good use. For example, manganese tailings may be used in agro-forestry and as coatings, resin cast products, glass, ceramics, glazes as well as building and construction materials (Verlaan and Wiltshire 2000). Base metal tailings and low grade metal ores can also be planted with suitable plant species which extract the metals from the substrate (Scientific Issue 4.1). Phytomining not only recovers metals from wastes, but the technique may also turn hazardous materials into benign wastes with much lower metal concentrations.

4.9 Summary

Mineral processing methods of metalliferous hard rock deposits involve crushing, grinding and milling the ore, and concentrating the ore minerals. Chemical additives are often employed to help separate or leach the ore minerals from the gangue phases. The additives include flotation reagents, modifiers, flocculants/coagulents, oxidants,

Scientific Issue 4.1. Phytoremediation and Phytomining of Metalliferous Wastes

Introduction

Plants and microorganisms (fungi, yeast cells, algae, bacteria) can influence the behaviour of metals and metalloids in surface environments. While some organisms tolerate elevated concentrations of dissolved precious and heavy metals and metalloids, others can dissolve, contain or immobilize such elements. Organisms – which contain or immobilize metals and metalloids – extract the dissolved elements by adsorption, intracellular uptake, and chemical transformations. The metals and metalloids, accumulated by these living plants and microorganisms or their dead biomass, may amount up to several weight percent of the cell dry weight.

Phytoremediation

The ability of particular plants and microorganisms to influence the cycling of metals can be used in the clean-up of metal contaminated soils, sediments, waters, and wastes. Plants have been used for some time to remove or immobilize environmental contaminants, a process which is commonly referred to as "phytoremediation" (Brooks 1998). Phytoremediation was initially applied in the 19th century when authorities began to treat municipal wastewaters in constructed wetlands and spray irrigation systems. Today, phytoremediation is an emerging technology for the rehabilitation of contaminated sites (Johansson et al. 2005). The techniques of phytoremediation can be grouped into the following strategies:

- *Phytostabilization.* Plants transform toxic forms of metals into non-toxic forms. This technique uses metal-tolerant plants to immobilize heavy metals in the root zone. Metals are absorbed into and accumulated by roots, adsorbed onto roots, or precipitated in the root zone. The processes reduce mobility and bioavailability of the metals.
- *Rhizofiltration.* Dissolved heavy metals – surrounding the root zone – are adsorbed or precipitated onto plant roots, or the metals are absorbed into the roots. The technique may be applied to decontaminate ground water rather than soils and wastes.
- *Phytoextraction.* This strategy involves the uptake of precious and heavy metals by roots into the above ground portions of plants (i.e. metal accumulators). A phytoextraction operation entails planting a hyperaccumulator crop over the contaminated site. This is followed by harvesting and incineration of the biomass to produce a metal concentrate for subsequent disposal. The ideal plant to use for phytoextraction has a large biomass, grows well in metal-rich environments, and accumulates metals to high concentrations. Plants that accumulate 1000 mg kg^{-1} metal in their dry biomass are termed "hyperaccumulators" (Brooks 1998). There are natural hyperaccumulators for a range of metals (Cd, Co, Cu, Mn, Ni, Pb, Se, Tl, U, Zn). However, many hyperaccumulators tend to grow slowly and produce little biomass. An alternative approach is to genetically engineer fast-growing species to improve their metal tolerance and metal accumulating capacity (Pilon-Smits et al. 2000). The key limitation to metal uptake by any plant species is the solubility of the metal in the root zone since metals must be dissolved in the soil solution for uptake to occur. The natural solubility of metals can be artificially induced by adding suitable chemicals to the substrate in which the plants grow. The complexing agents dissolve additional metals which become bioavailable and are subsequently taken up by the plants. Induced phytoextraction is of particular interest to companies as the application of chelates can increase the natural uptake of metals and induce the uptake of other metals by plants (Anderson et al. 1998).

Phytomining

An alternative strategy to phytoremediation is phytomining (Robinson et al. 1997; Anderson et al. 1999). In this technology, hyperaccumulators are not only used to remove metals from substrates, they are used to yield a metal-rich biological ore for smelting. The plants accumulate metals in their harvestable tissue, followed by the removal of plant tissue from the site, ashing of the plants, and extraction of the metals. In most cases, a phytomining operation would be conducted on low grade ores or wastes such as tailings, containing metal concentrations too low for conventional extraction. Phytomining represents an emerging technology and "cleans" ores and wastes to shallow depths. It has many advantages and unique features which could make it part of future metal mining operations.

and hydrometallurgical agents. The desired fraction containing the ore minerals is recovered, and other materials are disposed of as wastes. Such wastes are referred to as tailings. Tailings consist of solids and liquids. The liquids contain process chemicals which influence the chemical behaviour and mobility of elements within the waste repository.

The pH of tailings waters is influenced by the applied processing technique. Any sulfuric acid added as process chemical or generated by sulfide oxidation will leach ore and gangue minerals. In extreme cases, the applied processing techniques may create highly acid or alkaline tailings with high concentrations of dissolved and soluble salts, metals and metalloids.

Tailings contain minerals similar to those of waste rocks from the same deposit, only in much smaller grain size. Primary minerals are the ore and gangue minerals of the original ore. Secondary minerals crystallize during weathering of the ore. Tertiary and quaternary minerals form before, during, and after the deposition of the tailings in their impoundment as a result of chemical reactions. Evaporation of tailings water commonly leads to the formation of mineral salts such as gypsum at and below the tailings surface.

The accumulation of sulfide-rich tailings sediments may occur in an impoundment, and subsequent sulfide oxidation and AMD generation may be possible. In particular, if sulfidic tailings are present within the unsaturated zone of an inactive impoundment, sulfide oxidation will occur. However, tailings remain water saturated during operation and are deposited as fine-grained sediments that have a relatively low permeability. Consequently, tailings often generate AMD more slowly than coarser, yet more permeable waste rocks from the same deposit.

Disposal of tailings commonly occurs into engineered tailings dams. Tailings dams are constructed like conventional water storage type dams whereby the tailings are pumped as a water-rich slurry into the impoundment. The safety record of tailings dams is poor. On average, there has been one major tailings dam failure or spillage every year in the last thirty years. Tailings dam failures are due to liquefaction, rapid rise in dam wall height, foundation failure, excessive water levels, or excessive seepage. Tailings dam failures and spillages have caused environmental impacts on ecosystems, loss of life, and damage to property. Other concerns with tailings dams include air pollution through dust generation, release of radiation from tailings, and seepage from the tailings through the embankment into ground and surface waters. Monitoring of tailings dams is essential and includes investigations on dam performance, impoundment stability, and environmental impacts. Dry capping techniques of tailings are highly variable and site-specific, and consist of single layer and multi-layered designs using compacted and uncompacted materials.

Tailings disposal, using the thickened discharge or paste technology, results in a cone or stack of dried tailings. The methods are based on the discharge of dewatered tailings into the impoundment. Containment of tailings in open pits and underground workings is also possible; nonetheless, consolidation of the wastes may take considerable periods of time. If the tailings are placed below the post-mining ground water table, the restricted access of atmospheric oxygen will restrict any sulfide oxidation. An interaction of the tailings with ground water occurs if permeable, reactive, soluble tailings materials are present.

In historic mining areas, tailings were commonly discharged into nearby rivers. Today, such a disposal is applied in earthquake- and landslide-prone, rugged, high rainfall areas where the construction of engineered tailings disposal facilities has proven to be impossible. This disposal practice comes at a price. It can result in considerable metal contamination and increased sedimentation downstream. Similarly, submarine tailings disposal is applied in areas where high rainfall, increased seismicity, and limited land make the construction of tailings dams impossible. Tailings are released into the ocean at significant water depth well below wave base and commercial fishing grounds. While the oxidation of sulfides in deep sea environments is insignificant, the long term uptake of metals and metalloids from the tailings by bottom dwelling organisms remains to be investigated.

The ever increasing volume of tailings has stimulated research into their recycling potential. The alternative uses of tailings are dependent on the chemical and mineralogical characteristics of the wastes. Phytoremediation is one emerging technology whereby plants remove or immobilize metals in tailings. The technique has the potential to convert hazardous materials into benign wastes.

Further information on tailings can be obtained from web sites shown in Table 4.3.

Table 4.3. Web sites covering aspects of tailings

Organization	Web address and description
United Nations Environmental Programme (UNEP) Mineral Resources Forum	*http://www.mineralresourcesforum.org* Information on tailings dams
Tailings.Info	*http://www.tailings.info* Detailed information on tailings
Mining Information Service TailingsMine	*http://technology/infomine.com/tailingsmine* Reports on tailings
World Information Service on Energy (WISE) – Uranium Project	*http://www.wise-uranium.org* Case studies, information, and record on tailings dam safety and failures

Chapter 5

Cyanidation Wastes of Gold-Silver Ores

5.1 Introduction

Cyanide is a general term which refers to a group of chemicals whereby carbon and nitrogen combine to form compounds (CN^-). Cyanide leaching is currently the dominant process used by the minerals industry to extract gold (and silver) from geological ores. Gold extraction is accomplished through the selective dissolution of the gold by cyanide solutions, that is, the so-called "process of cyanidation". This hydrometallurgical technique is so efficient that the mining of low-grade precious metal ores has become profitable, and modern gold mining operations extract a few grams of gold from a tonne of rock. As a result, exceptionally large quantities of cyanide-bearing wastes are produced for a very small quantity of gold. The wastes of the cyanidation process are referred to as *cyanidation wastes*. At modern gold mining operations, cyanidation wastes occur in the form of heap leach residues, tailings, and spent process waters. While the bulk of the cyanide used in the mining industry is applied to gold ores, cyanide is also added as a modifier in the flotation of base metal sulfide ores in order to separate base metal sulfides from pyrite. Therefore, some process circuits of base metal ores may generate cyanide-bearing tailings and spent process waters.

This chapter documents the fundamental characteristics of cyanidation wastes. Since an understanding of such wastes requires a knowledge of the chemistry of cyanide, the chemistry of cyanide is introduced prior to an overview on the use of cyanide in precious metal extraction. This is followed by a documentation of cyanidation wastes and the monitoring and analysis of cyanide. The release of cyanide to the environment can cause contamination of surface waters, and cyanide can be toxic to humans and animals. The chapter concludes with the different treatment strategies used to reduce cyanide levels in mine waters.

5.2 Occurrences and Uses of Cyanide

Cyanide occurs naturally. It is found in minute quantities in several foods including cassava, maize, bamboo, sweet potatoes, sugarcane, peas, and beans, as well as in the seeds and kernels of almond, lemon, lime, apple, pear, cherry, apricot, prune, and plume (Hynes et al. 1998; Logsdon et al. 1999; Mudder and Botz 2001). In addition, cyanide is an important component of vitamin B-12 where it is securely bound and not harmful. There are about 2 650 cyanide producing plant species, and levels of cyanide in cyanogenic plants can reach hundreds of parts per million (Logsdon et al. 1999; Mudder and Botz 2001). Furthermore, cyanide is produced and used by certain soil bacteria, algae, fungi, insects, and arthropods (Logsdon et al. 1999). In nature, the presence of cyanide in plants and organisms provides protection against predators.

Apart from these natural occurrences of cyanide, there are man-made emissions of cyanide. Cyanide is released during coal combustion, refining of petroleum, iron and steel production, burning of plastics, incineration of household waste, combustion of fossil fuels, and cigarette smoking (Environment Australia 1998). Significant amounts of cyanide may enter surface waters from municipal sewage treatment works and the dissolution of road salt (Mudder and Botz 2001). Cyanide is also synthesized and used for numerous industrial processes such as steel hardening and plastics production. The mining industry consumes only a minor proportion (13%) of the total world hydrogen cyanide production (Mudder and Botz 2001). In the mining industry, cyanide is used in hydrometallurgical processes to extract precious metals from geological ores and to separate sulfides in the flotation of base metal sulfide ores. Cyanide has been chosen for the extraction of gold and silver because: (*a*) it forms soluble complexes with gold and silver; and (*b*) its use is extremely efficient and cost effective. The potential toxicity of cyanide has spurred much research into alternative leaching reagents. Ammonia (NH_3), thiocyanate (SCN^-), thiosulfate ($S_2O_3^{2-}$), and aqua regia (HCl-HNO_3) are also able to dissolve gold (McNulty 2001). These alternative leaching agents, however, can be toxic and cause significant environmental harm, so the chemicals are not applied in the mining industry.

5.3 Cyanide Chemistry

Cyanide chemistry is complex since there are different cyanide compounds present within process waters as well as solid and liquid mine wastes. The different cyanide compounds have been classified into five general groups: free cyanide; simple cyanide compounds; weakly complexed cyanide; moderately strong complexed cyanide; and strongly complexed cyanide (Smith and Mudder 1991, 1999) (Table 5.1). "*Weak acid dissociable*" or "*WAD*" cyanide consists of two cyanide forms: (*a*) cyanide which is already present as free cyanide; and (*b*) cyanide which is released as free cyanide from weak and moderately strong complexes when the pH is lowered in a sample to a pH of approximately 4.5. The five cyanide groups constitute the "total cyanide" content or the sum of the different forms of cyanide present (Table 5.1). The term "cyanide" is loosely used in this book to refer to all five cyanide forms.

5.3.1 Free Cyanide

The term "free cyanide" refers to two species; the cyanide anion (CN^-) dissolved in water and the hydrogen cyanide or hydrocyanic acid (HCN) formed in solution. For example, when solid sodium cyanide (NaCN) dissolves in water, it forms a sodium cation (Na^+) and a cyanide anion (CN^-). The cyanide anion then undergoes hydrolysis and combines with hydrogen ions to form the molecule hydrogen cyanide:

$$CN^-_{(aq)} + H_2O_{(l)} \leftrightarrow HCN_{(aq)} + OH^-_{(aq)} \quad (5.1)$$

Table 5.1. Nomenclature of cyanide compounds in cyanidation solutions (Smith and Mudder 1991, 1999; Environment Australia 1998; Botz 2001)

Common name	Common name	Cyanide group	Examples of cyanide compounds and their chemical formula
Total cyanide	WAD cyanide	Free cyanide	$CN^-_{(aq)}$
			$HCN_{(aq)}$
		Simple compounds	$NaCN_{(s)}$
			$KCN_{(s)}$
			$Ca(CN)_{2(s)}$
			$Hg(CN)_{2(s)}$
			$Zn(CN)_{2(s)}$
			$Cd(CN)_{2(s)}$
			$CuCN_{(s)}$
			$Ni(CN)_{2(s)}$
			$AgCN_{(s)}$
		Weak complexes	$Zn(CN)_4^{2-}$
			$Cd(CN)_3^-$
			$Cd(CN)_4^{2-}$
		Moderately strong complexes	$Cu(CN)_2^-$
			$Cu(CN)_3^{2-}$
			$Cu(CN)_4^{3-}$
			$Ni(CN)_4^{2-}$
			$Ag(CN)_2^-$
		Strong complexes	$Au(CN)_2^-$
			$Co(CN)_6^{4-}$
			$Fe(CN)_6^{4-}$
			$Fe(CN)_6^{3-}$
		Thiocyanate	SCN^-
		Cyanate	CNO^-

The amount of cyanide converted to hydrogen cyanide depends on the salinity and especially the pH of the solution (Fig. 5.1). At alkaline pH conditions greater than approximately 10.5, most of the free cyanide is present as cyanide anion. Equal concentrations of cyanide and hydrogen cyanide are present at a pH value of approximately 9.3 (Fig. 5.1).

At near neutral to acid pH conditions (i.e. pH < ~8.3), all free cyanide is present as hydrogen cyanide which is reasonably volatile and can be dispersed into the atmosphere:

$$HCN_{(aq)} \leftrightarrow HCN_{(g)} \quad (5.2)$$

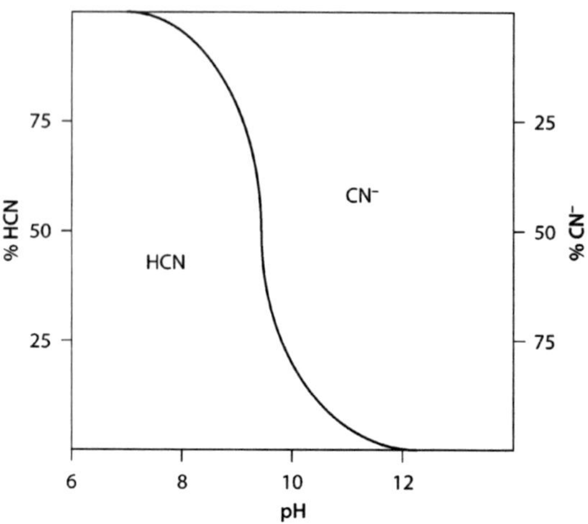

Fig. 5.1. Diagram illustrating the pH dependence of the cyanide–hydrogen cyanide equilibrium (after Logsdon et al. 1999; Smith and Mudder 1999). At alkaline pH values greater than approximately 10.5, nearly all of the free cyanide is present as cyanide anion

The formation of hydrogen cyanide is also influenced by the salinity of the solution and is promoted at high salinities. The form of free cyanide in process waters is of fundamental importance to the gold extraction process. Volatilization of hydrogen cyanide reduces the concentration of dissolved free cyanide ($HCN_{(aq)}$) in the process solution, causes elevated cyanide loadings in the atmosphere, and may expose workers to cyanide gas ($HCN_{(g)}$). Thus, it is important that the solution pH of hydrometallurgical gold extraction is maintained at a pH value of approximately 10.3. This is achieved through the addition of quicklime (CaO) or hydrated lime ($Ca(OH)_2$) to the process water. Higher pH values than 10.3 decrease the efficiency of the gold extraction.

5.3.2 Simple Cyanide Compounds

Simple cyanides are the salts of hydrocyanic acid. The salts contain only one type of cation, typically an alkaline or alkaline earth ion. Sodium cyanide used in the gold extraction process is one of these simple cyanide compounds. Simple cyanides exist as solid cyanides, some of which are also water soluble to form free cyanide and dissolved cations (Smith and Mudder 1991, 1999):

$$NaCN_{(s)} \rightarrow Na^{+}_{(aq)} + CN^{-}_{(aq)} \tag{5.3}$$

$$KCN_{(s)} \rightarrow K^{+}_{(aq)} + CN^{-}_{(aq)} \tag{5.4}$$

$$Ca(CN)_{2(s)} \rightarrow Ca^{2+}_{(aq)} + 2\,CN^{-}_{(aq)} \tag{5.5}$$

5.3.3 Complexed Cyanide

The presence of metals other than gold in the ore results in the formation of soluble cyanide metal complexes. In fact, most of the cyanide in solution combines with metals and metalloids and forms dissolved complexes. The formation of the iron complex ($Fe(CN)_6^{4-}$) is particularly common:

$$Fe^{2+}_{(aq)} + 6\,CN^-_{(aq)} \rightarrow Fe(CN)^{4-}_{6\,(aq)} \tag{5.6}$$

The stability of metal complexes varies. Some complexes are exceptionally stable (i.e. strong complexes), others are moderately strong (i.e. moderately strong complexes), whereas few readily degrade (i.e. weak complexes) (Smith and Mudder 1991, 1999). Weak and moderately strong complexes are formed by zinc, cadmium, copper, nickel, and silver (Table 5.1). Strong cyanide complexes include complexes of gold, iron and cobalt, and their destruction is slow under natural conditions. The strength of the gold complex ($Au(CN)_2^-$) forms the basis of the use of cyanide in gold extraction. A change in processing or environmental conditions such as pH, water temperature, salinity, complex concentration, oxidant concentration, and intensity of sunlight or UV radiation reduces the stability of strong cyanide complexes. The complexes decompose at varying rates, some quickly, others quite slowly.

5.4 Gold Extraction

The production of gold from geological ores is achieved by hydrometallurgy. Gold extraction and recovery is a two-stage process. In the extraction stage, gold is dissolved using cyanide. In the recovery stage, the dissolved gold is recovered from the cyanide solution generally using cementation with zinc or adsorption onto carbon. The extraction stage starts with the preparation of the cyanide solution at the mine site. Solid sodium cyanide (NaCN) and lime are dissolved in water producing a "barren" cyanide solution. The solution is alkaline and typically contains 100 to 500 ppm sodium cyanide (Logsdon et al. 1999). There are two types of cyanide extraction processes used by the modern gold mining industry: (*a*) heap leach; and (*b*) vat/tank leach processes.

5.4.1 Heap Leach Process

In the heap leach process, oxide gold ore is extracted, sometimes crushed, and then piled onto plastic lined pads (Figs. 5.2, 5.3). The heap leach pad needs to have an impermeable, engineered base to prevent rupture of the plastic liner as well as ground and surface water contamination. The channels and bunds surrounding the pad have to be designed to contain excessive rainfall events. During operation, a dilute solution of sodium cyanide is sprayed onto the heap and allowed to percolate down through the ore. Over time, the cyanide solution trickles through the heap and dissolves gold and silver along its path. The solution is captured by the plastic or rubber pad underlying the heap. The "pregnant" leachate is then channelled into a holding pond for further processing and gold extraction. Once gold and silver are stripped, the barren cyanide solution is recycled, recharged with additional cyanide, returned to the heap, and sprayed over the pile again.

Fig. 5.2. Cyanide heap leach pile and plastic lined collection ponds, Wirralee gold mine, Australia

Fig. 5.3. Large heap leach piles at the Yanacocha gold mine, Peru. The siliceous gold ore is so porous that it can be heap leached without crushing. (Photo courtesy of P. Williams)

5.4.2 Vat/Tank Leach Process

In the vat or tank leach process, the crushed and ground ore is leached in large enclosed tanks where gold is dissolved and then adheres to pieces of carbon. The carbon with the gold is stripped of the solution, and the barren solution together with the leached ore is disposed of. Both the heap and the vat/tank leaching processes use the ability of cyanide to form highly soluble gold complexes. The other necessary components are oxygen and water. The cyanide dissolves gold according to the following reaction:

$$4\,Au_{(s)} + 8\,NaCN_{(aq)} + O_{2(g)} + 2\,H_2O_{(l)} \rightarrow 4\,NaAu(CN)_{2(aq)} + 4\,NaOH_{(aq)} \quad (5.7)$$

The equation highlights the fact that free cyanide is needed for the dissolution of gold and that the pH has to be maintained at alkaline levels of approximately 10.3. The reaction with silver is analogous to Reaction 5.7.

Gold ores commonly not only have elevated silver but also elevated metal and metalloid concentrations which exceed the gold content by several orders of magnitude. The cyanide anion not only dissolves gold and silver, it may also dissolve other metals and metalloids (e.g. As, Cd, Co, Cu, Fe, Hg, Mo, Ni, Pb, Zn). The dissolution of these metals and metalloids depends on how these elements are present within the ore. In completely weathered and oxidized ores, metals are present as secondary minerals; most of the metal-bearing minerals and the metals themselves are insoluble in the alkaline process waters. Process waters of the oxidized ores will have low metal concentrations, yet arsenic and antimony can be present in strongly elevated concentrations in cyanide process waters as these metalloids are soluble at alkaline pH conditions. Hence, the extraction of gold from oxidized ores is extremely efficient because the process waters do not react with other metals. All of the cyanide is used to form free cyanide as well as gold and silver cyanide complexes. In contrast, the treatment of sulfide-rich gold ores may lead to sulfide oxidation, acid generation, and a reduction in solution pH. In addition, the sulfide minerals will react with the cyanide process waters, and soluble thiocyanate (SCN^-) and metal cyanide complexes are formed. The sulfide content of some gold ores can be so high that these ores require roasting prior to cyanide leaching. Consequently, the dissolution of sulfide minerals in cyanide solutions leads to: (*a*) higher cyanide consumption rates during mineral processing; and (*b*) a solution containing a range of cyanide species and complexes as well as significant metal and metalloid concentrations.

5.5 Hydrometallurgical Wastes

Precious metal mining and the hydrometallurgical recovery of gold and silver not only result in the production of gold and silver bars, but also in the generation of waste rock dumps, spent process waters, heap leach piles, and tailings. Spent process waters, heap leach residues, and tailings are cyanidation wastes:

- *Process water.* Once the ore has been leached with cyanide and gold has been recovered from the pregnant solution, the process water contains aqueous free cyanide (HCN, CN^-), cyanide metal complexes, cyanates (CNO^-), thiocyanates (SCN^-), and numerous other dissolved and colloidal chemical species. Much of this barren process water is recovered and recycled. However, a solution containing excessive amounts of thiocyanates and metal cyanide complexes is discarded from the process circuit because the extraction capacity of such process water is exhausted. Thus, a portion of the barren solution has to be discharged. The water may be treated to reduce cyanide levels. It is then pumped to tailings impoundments and ponds, or it is discharged into rivers and oceans.

- *Heap leach residues.* Cyanide heap leach operations result in piles of spent ore after operations have ceased. Sulfide oxidation and AMD generation are not commonly found in heap leach piles because gold ores containing major amounts of sulfides are usually not amenable to cyanide leaching. If sulfidic ores are leached, the process is conducted with a high pH solution (pH 10.3), which introduces a significant neutralizing factor. The leached ore may contain carbonate minerals and therefore, the piles may be strongly alkaline once heap leaching has ceased. A strong alkalinity does not allow direct planting of vegetation on the wastes. Alkaline substrates are characterized by reduced nutrient availability because many elements are poorly soluble at high pH and deficiencies of trace elements such as copper, manganese, iron and zinc may occur. Many plants will not be able to tolerate the alkaline pH values greater than 8.5. In addition, the high sodium content of such wastes due to the application of sodium cyanide will have adverse impacts on the soil structure since the sodium will promote the dispersion of clay particles (i.e. sodicity). These strongly alkaline cyanidation wastes not only require attenuation of cyanide but also pH reduction to appropriate levels, using native sulfur as a soil amendment. Alternatively, the wastes may need to be capped with benign materials.
- *Tailings.* Cyanide leached ore accumulates from the vat/tank leaching process. The leached ore may be pumped as a slurry to a tailings impoundment, which commonly is a tailings dam. In the waste repository, the pores of the tailings are filled with dilute cyanide solutions, and part of the solution will be collected in a decant pond on top of the tailings. This decanted tailings water is recycled to the processing plant to extract more gold. Alternatively, the leached ore is used as backfill in mined-out open pit and underground workings, or it is dumped into rivers and oceans (Jones and Ellis 1995).

5.6 Cyanide Analysis and Monitoring

Cyanide analysis is performed on liquid and solid sample types. For cyanide monitoring, water samples need to be obtained from background areas, waste discharge points, process water ponds, tailings dams, heap leach pads, seepage points, and saturated and unsaturated ground water zones (Environment Australia 1998). Solid sample types include heap leach residues and tailings. Samples should be refrigerated, and cyanide determinations should be conducted immediately after sampling. Details on sampling, sample preservation, sample treatment, measurement, and analytical methods are found in Environment Australia (1998) and Smith and Mudder (1991, 1999).

Free cyanide is considered the most toxic cyanide form as it causes toxicity at relatively low cyanide concentrations. Higher concentrations of other cyanide species are needed to induce toxicity. For example, hydrogen cyanate and cyanate ions are less toxic than hydrogen cyanide, and thiocyanate is relatively non-toxic compared to free cyanide. The toxicity of the different cyanide complexes is influenced by their stability. The more stable the complex, the less toxic it is, especially to aquatic life. WAD cyanide is considered to be an appropriate measure for assessing potential toxicity of a cyanide-bearing solution to humans and animals.

The concentrations of free cyanide in pore and seepage waters of sulfidic heap leach piles and tailings may increase over time due to decreasing pH conditions. Such increases in free cyanide concentrations are due to the destruction of stable cyanide complexes. For instance, cyanide complexes like the ubiquitous iron complex ($Fe(CN)_6^{4-}$) may become unstable under decreasing pH conditions. Sulfide oxidation and AMD generation could result in the dissociation of the iron complex and the release of cyanide and iron. As a consequence, once acid conditions develop in heap leach piles or tailings dams, increasing free cyanide levels may be generated.

Cyanide analysis should involve the characterization of all cyanide species present including total cyanide, free cyanide, WAD cyanide, strong metal cyanide complexes, cyanate, and thiocyanate. Analysis for only WAD cyanide or only total cyanide is inappropriate as it will not provide important information on the abundance of individual cyanide forms. Alternatively, the analysis of both WAD cyanide and total cyanide provides limited but useful information. Subtraction of the WAD cyanide value from the total cyanide value provides a measure of the essentially non-toxic and stable complexed cyanide forms (Mudder and Botz 2001). In addition, water samples need to be characterized for their pH, heavy metal concentrations, and contents of the various nitrogen species (i.e. total nitrogen N, nitrite NO_2^-, nitrate NO_3^-, ammonia NH_3). The nitrogen compounds should be determined as they are the result of cyanide degradation reactions (Mudder and Botz 2001). These degradation products can increase the solubility of metals and are also toxic to organisms at sufficiently high concentrations.

5.7 Environmental Impacts

The use of cyanide is well established in the mining industry, yet it remains controversial to some. Much of this concern relates to the much publicized use of cyanide in genocidal crimes, suicides, murders, and judicial executions. On the other hand, human cyanide poisoning from a natural source is less publicized and involves the ingestion of cyanogenic plants, in particular the consumption of cassava in developing nations. Several thousand cases of natural cyanide poisoning have been documented whereas the overall safety record of the mining industry has been exemplary (Mudder and Botz 2001). The concern about cyanide has been so significant that new mining operations using cyanide heap leach extraction methods have been banned in certain parts of the world (e.g. Montana). The bans have occurred in spite of the natural occurrence of cyanide, the chronic exposure of humans to cyanide from sources other than mining, and numerous cases of natural cyanide poisoning. The controversy has also been fuelled by cyanide spillages, transportation accidents, bursting of pipes containing cyanide solutions, overtopping of dams, and tailings dam failures (Case Study 5.1, Table 5.2).

If released at sufficiently high concentrations, cyanide can pollute surface waters. It is a potentially toxic substance and can be lethal if sufficient amounts are taken up by fish, animals, and humans (Table 5.2). Cyanide is not toxic to plants, and a major cyanide spill in the Kyrgyz Republic had no impact on plant life (Table 5.2) (Hynes et al. 1998). The non-toxic effects on plants are probably due to the fact that soils tightly adsorb cyanide. Arid and semi-arid conditions and an abundance of clay may lead to

Case Study 5.1. Cyanide Spill at Baia Mare, Romania

Mining History

The Baia Mare region in northwest Romania has been a center of mining, metallurgical and smelting activities for some time. There are numerous gold, silver, lead, zinc, copper and manganese mine sites, approximately 215 tailings dams and ponds, several metallurgical plants including copper and lead smelters as well as a sulfuric acid plant (UNEP/OCHA 2000). Over the years and especially during communist rule, the industrial activity was pursued with improper waste management, treatment and disposal. This caused contamination of soil, water, and air in the region. Historic mining activities resulted in the presence of numerous tailings dams and ponds filled with solid and liquid wastes (UNEP/OCHA 2000). Contaminated seepages leaked from these repositories, and particles were eroded from the open waste dumps, leading to high dust loads. The local population living close to the waste repository sites is exposed to a range of contaminants. Residents have elevated lead levels in their blood, and the contamination is detrimental to their health, particularly among children. Therefore, the reprocessing, removal and safe disposal of mining, processing and metallurgical wastes were encouraged to ensure an environmental improvement of the area. Recycling operations of historical tailings began by various companies in order to extract gold and silver. One of the operations was conducted by the Aurul mining company. After cyanide treatment and extraction of the metals, the wastes were redeposited by the company into a newly constructed dam, well away from residential areas.

Tailings Dam Failure

On 30.01.2000, the dam crest of the Aurul operation was breached and washed away. The breach in the retention dam was probably caused by a combination of factors including tailings dam design deficiencies, unforeseen operating conditions, and excessive rainfall and snow melt (UNEP/OCHA 2000). About 100 000 m^3 of cyanide-, metal and metalloid-rich tailings spilled and escaped into the river system (Macklin et al. 2003). Between 50 to 100 t of cyanide and unknown amounts of heavy metals and metalloids were released. The contaminants travelled into local and regional rivers and finally into the Danube before reaching the Black Sea. The spill crossed several national boundaries and impacted on rivers in Romania, Hungary, and Yugoslavia. Some 2 000 km of the Danube catchment area were affected. As the contaminant plume progressively flowed down the river system, it caused contamination of drinking water supplies, partial or total loss of phyto- and zooplankton, and massive fish kills (UNEP/OCHA 2000).

Recovery

Biological recovery of the river system in Hungary and Yugoslavia was relatively quick due to the inflow of uncontaminated water from upstream. Recovery of the local ecosystem, immediately downstream from the spill, did not eventuate. The cyanide spill occurred in an area already stressed, deteriorated and contaminated with heavy metals from historic mining and mineral processing operations. Similar cyanide spills and tailings dam failures are likely to occur in the future (UNEP/OCHA 2000).

a distinct enrichment of cyanide in soils (Shehong et al. 2005). As a result, cyanide is immobilized and prevented from being transferred into plants.

Plants are capable of tolerating elevated cyanide concentrations as indicated by cyanogenic plants. In contrast, cyanide poisoning of fish, animals, and humans can occur through inhalation, ingestion or skin adsorption. Loss to wildlife and stock has occurred at operating and rehabilitated mine sites where dissolved free cyanide levels in waters exceeded safe levels. Individual countries have, therefore, implemented national guidelines and criteria for free cyanide and total cyanide levels in food products, drinking water, air, soil, and mining related effluents. For instance, a level of 50 mg l^{-1} WAD cyanide in tailings slurry entering an impoundment has been recom-

Table 5.2. Chronology of mining-related cyanide accidents and spillages since 1990 (Hynes et al. 1998; UNEP/OCHA 2000; Mudder and Botz 2001)

Year	Location	Incident	Release	Impact
2000	Papua New Guinea	Transportation accident	150 kg of solid sodium cyanide	Spillage into local river
2000	Romania	Tailings dam crest failure after heavy rain and snow melt	100 000 m^3 of cyanide-bearing waste waters and tailings	Contamination of streams, massive fish kills; contamination of water supplies of >2 million people
1999	Philippines	Tailings spillage from pipe	0.7 Mt of cyanide-bearing tailings	17 homes buried; 51 ha covered with tailings
1998	Kyrgyz Republic	Transportation accident	1.7 t of solid sodium cyanide	Spillage into local river
1998	United States	Pipe failure	Several tonnes of cyanide-bearing waters	?
1995	Australia	Dam overtopping	5 000 m^3 of cyanide-bearing waters	?
1995	Australia	Dam failure	40 000 m^3 of cyanide-bearing wastes	?
1995	Omai, Guyana	Tailings dam failure	4.2 million m^3 of cyanide-bearing tailings	80 km of local river declared an environmental disaster zone
1991	United States	Dam overtopping	39 000 m^3 of cyanide-bearing wastes	?

mended or is used as a regulatory guideline by United States and Australian agencies and the World Bank (Mudder and Botz 2001). This level is to provide protection for animals coming in contact with the tailings.

The safe use of cyanide at an operating mine requires: (*a*) appropriate transport, storage, handling, and mixing of cyanide chemicals; (*b*) monitoring of operations, discharges, and the environment for cyanide losses; and (*c*) treatment of cyanidation wastes (Hynes et al. 1998; Environment Australia 1998). In most cases, the discharge of waste waters from the processing plant and the mine site has to comply with discharge limits set by statutory authorities. Compliance below allowed limits can be achieved through cyanide treatment. Cyanide can be persistent in waste repositories; consequently, safe levels of cyanide in seepage and pore waters of rehabilitated or abandoned heap leach piles and in waters of tailings dams need to be achieved through cyanide destruction.

5.8 Cyanide Destruction

Low levels of cyanide in waters can be achieved through the recovery of cyanide from process waters prior to their discharge. Once gold and silver are stripped from the process water, the barren cyanide solution is recycled and recharged with additional cyanide. This recovery based process aims to remove cyanide from solution and to reuse it in the metallurgical circuit (Botz 2001).

Low levels of cyanide in waters can also be achieved by converting cyanide into less toxic compounds. A large number of processes exist for the destruction of cyanide in wastes (Environment Australia 1998). The degradation of cyanide takes time, and the rates of destruction of the different cyanide species are affected by numerous factors including the intensity of light, water temperature, pH, salinity, oxidant concentration, and complex concentration (Smith and Mudder 1991, 1999). Degradation of cyanide occurs naturally in contaminated soils, process waters, tailings, and heap leach piles. The natural reduction of dissolved cyanide in mine wastes can be sped up by applying enhanced natural or engineered treatment processes.

5.8.1 Natural Attenuation

Natural attenuation rates depend on local climatic and environmental conditions and are, therefore, mine site specific. At some mine sites, these natural attenuation processes are sufficient to achieve cyanide levels which meet regulatory requirements for the release of cyanide-bearing waters into local drainage systems. Natural attentuation processes include: volatilization of hydrogen cyanide gas; precipitation of insoluble cyanide compounds; adsorption; degradation by bacteria; oxidation to cyanate; degradation of cyanide through UV radiation; and formation of thiocyanate:

- *Volatilization.* Volatilization of cyanide is the prominent natural attenuation process. Cyanide is lost from waste waters, tailings, heap leach piles, and contaminated soils to the atmosphere as a result of the volatilization of gaseous hydrogen cyanide ($HCN_{(g)}$) (Reaction 5.2). This can be triggered by adding water with a lower pH to the cyanide solutions and wastes. Volatilized cyanide will disperse in the atmosphere to background concentrations, or it will undergo chemical and biochemical reactions to degrade to ammonium and carbon dioxide (Logsdon et al. 1999).
- *Precipitation.* The precipitation of complexed cyanide can be achieved through the addition of a complexing metal. For example, ferrocyanide ($Fe(CN)_6^{4-}$) and ferricyanide ($Fe(CN)_6^{3-}$) ions are present in solution. If metals (Fe, Cu, Pb, Zn, Ni, Mn, Cd, Sn, Ag) are added to the solution, the ferrocyanide and ferricyanide ions react with the dissolved metals to form insoluble metal cyanide solids (Smith and Mudder 1991, 1999). The solids precipitate and settle, removing cyanide from solution.
- *Adsorption.* Free cyanide ions and complexed cyanide can be adsorbed from solution onto mineral surfaces. Clay-rich ores are particularly effective in adsorbing cyanide. Unless the adsorbed cyanide is oxidized by organic matter, the removal of cyanide from solution is only temporarily, and the different cyanide forms can be desorbed from the mineral surfaces back into solution.
- *Biological oxidation.* Certain bacteria degrade cyanide in order to generate nutrients (i.e. carbon and nitrogen) for their growth. The microflora is indigenous to the waste. The naturally present bacteria (e.g. *Pseudomonas*) degrade reactive cyanide into harmless products such as dissolved formate ($HCOO^-$), nitrate, ammonia, bicarbonate, and sulfate (Mosher and Figueroa 1996; Oudjehani et al. 2002;

Zagury et al. 2004). The following equation represents the simplified oxidation of cyanide by bacteria to bicarbonate and ammonia:

$$CN^-_{(aq)} + 1/2\, O_{2(g)} + 2\, H_2O_{(l)} \rightarrow HCO^-_{3(aq)} + NH_{3(aq)} \tag{5.8}$$

- *Oxidation to cyanate.* Dissolved free cyanide can be oxidized to cyanate in the presence of a strong oxidant. The resulting cyanate is substantially less toxic to aquatic life than cyanide. Thus, some oxidizing chemicals (e.g. ozone, gaseous chlorine, hypochlorite, hydrogen peroxide) are used in the engineered attenuation process to decompose cyanide in wastes. The oxidation of free cyanide occurs according to the following reaction:

$$2\, CN^-_{(aq)} + O_{2(g)} \rightarrow 2\, CNO^-_{(aq)} \tag{5.9}$$

This degradation process can also be achieved through UV light and catalysts (e.g. titanium dioxide, cadmium sulfide, zinc oxide) (Smith and Mudder 1991, 1999). Overall, the process causes a reduction in dissolved cyanide and elevated cyanate levels (Scharer et al. 1999). The cyanate is unstable in aquatic environments and, depending on the pH conditions, cyanate slowly decomposes to nitrate and carbon dioxide or to ammonia and bicarbonate:

$$CNO^-_{(aq)} + 2\, O_{2(g)} \rightarrow NO^-_{3(aq)} + CO_{2(aq)} \tag{5.10}$$

$$CNO^-_{(aq)} + H^+_{(aq)} + 2\, H_2O_{(l)} \rightarrow NH^+_{4(aq)} + HCO^-_{3(aq)} \tag{5.11}$$

- *Photolytic degradation.* In the presence of UV radiation, even strong cyanide complexes such as the stable Fe^{2+} and Fe^{3+} cyanide complexes ferrocyanide ($Fe(CN)_6^{4-}$) and ferricyanide ($Fe(CN)_6^{3-}$) can be broken down to yield free cyanide (Johnson et al. 2002). The free cyanide is further oxidized with the help of UV light to yield the cyanate ion.
- *Formation of thiocyanate.* Oxidation of sulfide minerals occurs in sulfidic ores, and this yields sulfur reaction products such as polysulfides (S_x) or thiosulfates (S_2O_3). The free cyanide may react with these forms of sulfur to form thiocyanate (SCN^-):

$$S^{2-}_{x(aq)} + CN^-_{(aq)} \rightarrow [S_{(x-1)}]^{2-}_{(aq)} + SCN^-_{(aq)} \tag{5.12}$$

$$S_2O^{3-}_{3(aq)} + CN^-_{(aq)} \rightarrow SO^{2-}_{3(aq)} + SCN^-_{(aq)} \tag{5.13}$$

Natural attenuation processes occur in tailings dams and heap leach piles. Volatilization is the major mechanism of natural cyanide attenuation in surface waters of tailings dams. In these environments, the discharged slurries are alkaline and have a pH value of approximately 10. Over time, the pH of the surface waters is reduced by rainfall and uptake of carbon dioxide from the atmosphere. The carbon dioxide dissolves in water as carbonic acid and lowers the pH of the cyanide-bearing water. As the pH is lowered, the hydrogen cyanide concentration in the waters increases at the expense of the dissolved cyanide anion. Volatilization of hydrogen cyanide gas occurs

($HCN_{(g)}$) (Fig. 5.4); as a consequence, cyanide levels in tailings dams naturally decrease due to volatilization. Any seepage of cyanide from the tailings dam into the ground water likely occurs in the form of stable iron complexes ($Fe(CN)_6^{4-}$) (Murray et al. 2000a) (Fig. 5.4).

Volatilization and biological oxidation are the major mechanisms of cyanide degradation in heap leach piles (Smith 1994; Smith and Mudder 1991, 1999). Heap leach piles commonly have a distinct geochemical zonation with a reduced, water saturated lower zone which is overlain by an oxidized, unsaturated upper horizon. The upper unsaturated zone has a low pore water content and abundant atmospheric oxygen – conditions which are amenable for cyanide volatilization and biological oxidation (Smith 1994; Smith and Mudder 1991, 1999). Hence, the rate of cyanide degradation is most pronounced in the unsaturated zone.

5.8.2 Enhanced Natural Attenuation

Enhanced natural degradation techniques speed up the rates of natural cyanide destruction. For instance, increased exposure of cyanide-bearing waters to UV radiation may result in the degradation of cyanide complexes (Fig. 5.5). Also, ponds and tailings dams tend to have large surface areas. Enlarging the surface areas even further provides greater contact of the surface waters with atmospheric carbon dioxide. This leads to the formation of more carbonic acid, decreasing pH values, and increasing volatilization of hydrogen cyanide. Therefore, increasing the surface area of ponds and tailings dams is one option to naturally enhance the degradation of cyanide in tailings. Similarly, the stirring of ponds, associated aeration and mixing of the water will increase the volatilization rate of cyanide.

At the end of heap leach operations, cyanide still remains in the heaps, either in the pore water or adsorbed onto minerals. The remediation of cyanide in inactive heap leach piles can be achieved by repeated rinsing of the piles with water. This dilutes the cyanide remaining in the heap leach pores and lowers the pH to less than 9.3, at which much of the free cyanide volatilizes (Smith 1994; Smith and Mudder 1991, 1999). Heaps are regarded successfully rinsed when monitoring shows that the WAD cyanide content is below a particular concentration acceptable to statutory agencies. However, repeated rinsing and large volumes of water are needed for this remediation process until the dissolved WAD cyanide content has reached acceptable levels. If the cyanide

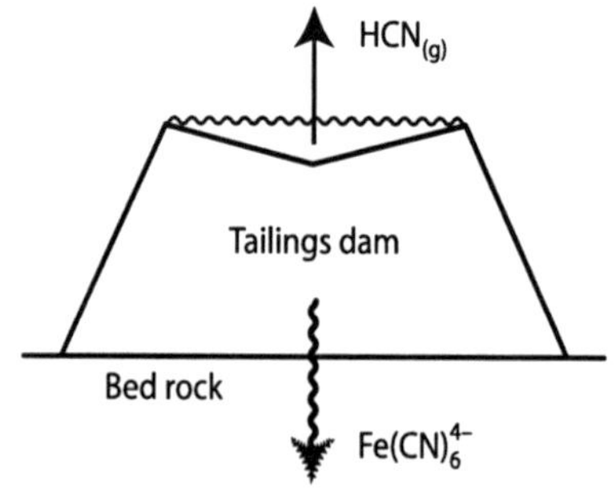

Fig. 5.4. Simplified model of the fate of cyanide in tailings dams (after Murray et al. 2000a). Cyanide levels in tailings dams decrease naturally due to volatilization of hydrogen cyanide gas and seepage of stable iron cyanide complexes into the ground water

Fig. 5.5. Collection of cyanide-bearing seepage waters in shallow ponds at the base of the tailings dam, Red Dome gold mine, Australia. UV radiation causes the destruction of dissolved Cu cyanide complexes, precipitation of cyanate salts and attenuation of total cyanide from 300 mg l^{-1} to <1 mg l^{-1} in the successive ponds

concentration cannot be reduced to the compliance limit, the cyanide must be removed using other attenuation processes. Enhanced biological oxidation represents one of these attenuation processes. When cyanide degrading bacteria and nutrients are added in a solution to the piles, the bacteria reduce the cyanide concentrations via biological oxidation. The addition of only nutrients to inactive heap leach piles can provide the necessary stimulation of naturally present cyanide degrading bacteria.

5.8.3 Engineered Attentuation

Many mine wastes contain cyanide levels which require engineered attenuation (Wang et al. 1997). Water treatment methods are frequently used to reduce the concentrations of cyanide, metals, metalloids, and suspended solids in process waters and slurries. These engineered methods rely on chemical, physical or biological processes to reduce cyanide levels.

- *Chemical oxidation.* Chemical oxidation processes are well established treatment techniques. In order to destroy cyanide and to convert cyanide to harmless substances, oxidants are added (e.g. chlorine gas; sodium or calcium hypochlorite; ozone;

hydrogen peroxide; sulfuric acid + hydrogen peroxide = H_2SO_5, i.e. Caro's acid or peroxymonosulfuric acid; sulfur dioxide + air = SO_2-INCO process; activated carbon). The reaction products include cyanate, carbon dioxide, bicarbonate, nitrogen, nitrate, ammonia, and sulfate. The hypochlorite ion (OCl^-) oxidizes, for instance, cyanide to cyanate:

$$OCl^-_{(aq)} + CN^-_{(aq)} \rightarrow CNO^-_{(aq)} + Cl^-_{(aq)} \tag{5.14}$$

- *Physical precipitation.* The precipitation process uses the addition of iron which causes the precipitation of insoluble metal cyanides. Metals can also be precipitated as solid metal hydroxides and carbonates through the addition of ferric chloride or sulfate and lime. The iron reagent causes coagulation of suspended solids, and adsorption of metals onto coagulated solids occurs. Many gold deposits, especially epithermal gold ores, have elevated mercury contents, and the mercury dissolves during cyanidation and reports into the waste products. The addition of sodium sulfide (Na_2S) to the waste stream can force the precipitation of mercury as mercury sulfide (i.e. cinnabar). The physical precipitates accumulate as sludges in the treatment plant. Sludges then require disposal in suitable waste repositories such as tailings storage facilities.
- *Biological oxidation.* In most cases, engineered degradation of cyanide relies on chemical oxidation or physical precipitation processes. Both processes result in a reduction of dissolved cyanide. Increasingly, biological methods are used to treat cyanide wastes (Botz 2001). The natural capability of microorganisms to decompose cyanide is used on spent heap leach piles and in artificial bioreactor tanks and wetlands (Gessner et al. 2005). Bacteria, however, require the correct nutrient balance for their growth, and nutrients may have to be added to facilitate biological oxidation. Both passive and active biological treatment methods remove dissolved cyanide from solution using either aerobic or anaerobic microorganisms (Mosher and Figueroa 1996; Fischer et al. 1996; Botz 2001).

5.9 Summary

Cyanide, a compound made up of carbon and nitrogen, is ubiquitous in nature. It is naturally produced by plants and lower organisms. It is also released into the atmosphere during resource processing, manufacturing and consumption. Furthermore, cyanide is synthesized for numerous industrial applications including the extraction of gold, silver and base metals from geological ores.

Cyanide combines with hydrogen to form hydrogen cyanide and with metals and metalloids to form complexes. The total cyanide content of a process water consists of five different cyanide forms: free cyanide (i.e. cyanide anion + hydrogen cyanide); simple cyanide compounds; weak complexes; moderately strong complexes; and strong complexes. Free cyanide plus weak cyanide and moderately strong complexes are termed "weak acid dissociable" or "WAD" cyanide. In addition, process waters contain dissolved thiocyanate (SCN^-) and cyanate (CNO^-) species. These different cyanide compounds and derivatives vary in their stability and toxicity. Iron cyanide complexes are very stable and essentially non-toxic whereas WAD cyanide complexes can break down at pH 4.5 to form highly toxic free cyanide forms. An analysis of effluents for WAD cyanide is considered to be an appropriate measure for assessing potential toxicity of the solutions.

The extraction and recovery of gold from geological ores using cyanide solutions is a two-stage process. In the extraction stage, gold is dissolved using cyanide whereby solid sodium cyanide and lime are dissolved in water. This barren cyanide solution is used either in a heap leach or as part of a vat leach process to dissolve the gold. Once the ore has been leached with cyanide, and gold has been recovered from the pregnant solution, the process water contains aqueous free cyanide, cyanide metal complexes, cyanates, thiocyanates, and numerous other dissolved and colloidal chemical species. Solutions containing excessive amounts of thiocyanates and metal cyanide complexes are discarded from the process circuit as the extraction capacity of such process waters is exhausted. These waters may be treated to reduce cyanide levels. After treatment, the waters are pumped to a tailings impoundment or are discharged into rivers and oceans. Cyanide-leached ore from the tank leaching process is either pumped as a slurry to the tailings storage facility, backfilled into open pits and underground workings, or dumped into rivers and oceans.

Cyanide leaching of gold ores is usually undertaken at a pH of approximately 10.3. The pH is maintained at such a level in order to maximize the amount of dissolved cyanide anions. A reduction to lower pH conditions leads to the loss of hydrogen cyanide to the atmosphere via volatilization. This can pose a safety problem during operation, but it may be highly desirable in the remediation of cyanidation wastes.

Monitoring of cyanide at a mine site involves the complete characterization of all cyanide species present including total cyanide, free cyanide, WAD cyanide, strong metal cyanide complexes, and thiocyanate. In most cases, cyanide concentrations in discharges from the processing plant and the mine site have to comply with discharge limits set by statutory authorities. Thus, safe levels of cyanide in seepages of rehabilitated heap leach piles and tailings dams need to be achieved through attenuation processes. There are numerous natural, enhanced natural, and engineered attenuation processes which can be very effective in reducing dissolved cyanide concentrations. Volatilization of hydrogen cyanide is the dominant natural process responsible for a reduction of cyanide levels in mine wastes.

Further information on cyanidation wastes can be obtained from web sites shown in Table 5.3.

Table 5.3. Web sites covering aspects of cyanidation wastes

Organization	Web address and description
United Nations Environmental Programme (UNEP) Mineral Resources Forum	*http://www.mineralresourcesforum.org* Cyanide fact sheet; cyanide spill incidents
International Cyanide Management Code for the Gold Mining Industry	*http://www.cyanidecode.org* Information on the use of cyanide in the gold mining industry
Department of Primary Industries, New South Wales, Australia	*http://www.minerals.nsw.gov.au/prodServices/minfacts/minfact_63* Gold mining and cyanide fact sheet

Chapter 6

Radioactive Wastes of Uranium Ores

6.1
Introduction

Uranium ores have the specific issue of radioactivity, and uranium mine wastes are invariably radioactive. This property differentiates uranium mine wastes from other mine waste types. For example, gold mine tailings contain cyanide, and the cyanide can be destroyed using natural, naturally enhanced or engineering techniques. Sulfidic wastes have the potential to oxidize, and oxidation of sulfidic wastes can be curtailed using covers. In contrast, the decay of radioactive isotopes and the associated release of radioactivity cannot be destroyed by chemical reactions, physical barriers or sophisticated engineering methods. Therefore, appropriate disposal and rehabilitation strategies of radioactive uranium mine wastes have to ensure that these wastes do not release radioactive substances into the environment and cause significant environmental harm.

The fundamental characteristics of uranium mine wastes are given in this chapter with a particular focus on radioactivity, uranium tailings, and the behaviour of uranium in mine waters. Some uranium ores contain abundant sulfides. The presence of these minerals in pit faces, underground workings, tailings, and waste rock piles may lead to sulfide oxidation and AMD. The principles of sulfide oxidation and AMD have already been presented in Chaps. 2 and 3. Sulfidic uranium mine wastes and their acidic waters may be characterized and treated with the same type of approaches used to characterize and treat sulfidic wastes and AMD waters.

6.2
Mineralogy and Geochemistry of Uranium

6.2.1
Uranium Ores

Uranium ore minerals can be classified as reduced and oxidized species. Reduced uranium minerals incorporate uranium as U^{4+} whereas oxidized species have uranium as U^{6+}. Uraninite (UO_{2+x}) is the most common reduced U^{4+} mineral species, and it is the main ore mineral in many uranium deposits (Burns 1999). Other important uranium ore minerals are: brannerite ($(U,Ca,Y,Ce)(Ti,Fe)_2O_6$); coffinite ($USiO_4 \cdot n\,H_2O$); and pitchblende (i.e. amorphous or poorly crystalline uranium oxide) (Finch and Murakami 1999). Uranium ore minerals commonly contain minor amounts of thorium, rare earth elements, lead, calcium, and other elements as cation substitutions (Lottermoser 1995). Uranium may also be present as minor substitution in rock-forming minerals or in accessory phases such as allanite, xenotime, monazite, zircon, apatite, and sphene. Uranium ore minerals weather easily, and numerous secondary uranium (U^{6+}) minerals are known from weathered uranium ores (Burns 1999). The gangue mineralogy of uranium ores is highly variable and deposit specific. Gangue minerals include framework, ortho, ring, chain and sheet silicates as well as sulfides, oxides, hydroxides, sulfates, phosphates, and carbonates.

Uranium ore grades are expressed in terms of triuranium octoxide (U_3O_8) which is 85% elemental uranium. Uranium ores exploited have ore grades as low as 0.01% to >0.5% U_3O_8. In addition, uranium ores have variably elevated metal and metalloid values (As, Cu, Mo, Ni, Pb, Ra, Re, Sc, Se, Th, V, Y, Zr) (e.g. Langmuir et al. 1999; Pichler et al. 2001). Arsenic, copper, lead, molybdenum, nickel, and selenium are commonly hosted by sulfide minerals. Sulfide minerals are ubiquituous in some uranium ores (e.g. roll front type uranium deposits). Uranium ores which contain abundant pyrite or marcasite may oxidize upon exposure, and AMD may develop in mine workings or mine wastes (Fernandes et al. 1998).

6.2.2 Placer and Beach Sands

Radioactive wastes not only accumulate as a result of uranium and phosphate mining but also during mining and processing of placer and beach sands. Placer and beach sands are sedimentary deposits which contain minerals with a high density, chemical resistance to weathering, and mechanical durability. Ore minerals include silicates, oxides, phosphates, and native elements (Table 6.1). The deposits are principally mined for their gold, diamond, sapphire, ruby, titanium (ilmenite, rutile), tin (cassiterite) or

Table 6.1. Examples of heavy minerals found in placer deposits and beach sands

Mineral name	Chemical formula
Cassiterite	SnO_2
Chromite	$FeCr_2O_4$
Columbite	$(Fe,Mn)Nb_2O_6$
Diamond	C
Gahnite	$ZnAl_2O_4$
Garnet group	$(Mg,Fe,Mn,Ca)_3(Al,Fe,Ti,Cr)_2Si_3O_{12}$
Gold	Au
Ilmenite	$FeTiO_3$
Magnetite	Fe_3O_4
Monazite	$(Ce,La)PO_4$
Platinum	Pt
Ruby	Al_2O_3
Rutile	TiO_2
Sapphire	Al_2O_3
Spinel	$MgAl_2O_4$
Tantalite	$(Fe,Mn)Ta_2O_6$
Xenotime	YPO_4
Zircon	$Zr(SiO_4)$

rare earth element (monazite, xenotime) contents. Some deposits may possess elevated uranium and thorium concentrations. In these deposits, the radioactive elements are present within chemically and physically stable, weathering resistant gangue minerals (e.g. monazite, xenotime, zircon, tantalite, columbite). The minerals effectively lock up the uranium and thorium in the crystal lattice and prevent mobilization of these elements and their radioactive decay products into the environment (Yusof et al. 2001). However, the minerals still emit radiation. Thus, exposure to radiation needs to be controlled, dust control measures need to be implemented, and appropriate handling methods of the radioactive wastes or ores need to be applied.

While the mining and mineral processing of alluvial tin deposits and heavy mineral sands do not result in the mobilization of radioactive elements into the environment, the metallurgical extraction of ore elements from such ores may. For example, the processing of titanium ores obtained from heavy mineral sands is based on sulfuric acid leaching. Acid leaching of titanium ores not only liberates titanium from the crystal lattices of rutile and ilmenite but also releases radionuclides and other metals from gangue phases. Consequently, spent process waters of titanium processing plants can exhibit very low pH values and high TDS concentrations (Schuiling and van Gaans 1997a,b).

6.3 Aqueous Chemistry of Uranium

Uranium is the typical constituent of surface and ground waters at uranium mine sites. It originates from the leaching of uranium ores as well as from the leaching of solid mining, processing and metallurgical wastes. An understanding of the leaching processes requires a knowledge of the aqueous chemistry of uranium. The aqueous behaviour of uranium can be described in terms of oxidative dissolution of uranium minerals, uranium solubility, and uranium precipitation.

6.3.1 Oxidative Dissolution of Uranium Minerals

The most reduced form of uranium (i.e. U^{4+}) is found in uraninite, coffinite, brannerite, and pitchblende. This oxidation state is the least soluble and least mobile form of uranium. In contrast, the most oxidized state of uranium (i.e. U^{6+}) is the most soluble and most mobile (Wanty et al. 1999). As a consequence, if oxygen is available as dissolved or gaseous species, U^{4+} can be oxidized to U^{6+}, and uranium dissolves in water as the uranyl oxyanion (UO_2^{2+}). The oxidative dissolution of uranium minerals is achieved progressively through a sequence of chemical reactions (Ragnarsdottir and Charlet 2000). A simplified overall reaction for the oxidative dissolution of uraninite is as follows:

$$2\,UO_{2(s)} + 4\,H^+_{(aq)} + O_{2(g)} \rightarrow 2\,UO^{2+}_{2\,(aq)} + 2\,H_2O_{(l)} \quad (6.1)$$

The oxidation of U^{4+} minerals (i.e. uraninite, coffinite, brannerite, pitchblende) can also be achieved by the oxidant Fe^{3+} (Abdelouas et al. 1999). The production of Fe^{3+} is possible through the indirect oxidation of iron sulfide minerals within the waste:

$$4\,FeS_{2(s)} + 14\,O_{2(g)} + 4\,H_2O_{(l)} \rightarrow 4\,Fe^{2+}_{(aq)} + 8\,H^+_{(aq)} + 8\,SO^{2-}_{4(aq)} + \text{energy} \quad (6.2)$$

$$4\,Fe^{2+}_{(aq)} + 4\,H^{+}_{(aq)} + O_{2(g)} \rightarrow 4\,Fe^{3+}_{(aq)} + 2\,H_2O_{(l)} + \text{energy} \tag{6.3}$$

$$FeS_{2(s)} + 14\,Fe^{3+}_{(aq)} + 8\,H_2O_{(l)} \rightarrow 15\,Fe^{2+}_{(aq)} + 16\,H^{+}_{(aq)} + 2\,SO^{2-}_{4(aq)} + \text{energy} \tag{6.4}$$

In the initial step (Reaction 6.2), pyrite reacts with oxygen and water to produce dissolved Fe^{2+}, sulfate and hydrogen ions. The release of hydrogen with the sulfate anion results in an acidic solution unless other reactions occur to neutralize the hydrogen ions. The second step (Reaction 6.3) involves the oxidation of Fe^{2+} to Fe^{3+} and occurs at a low pH. The Fe^{3+} turn oxidizes pyrite via Reaction 6.4, which in turn produces more Fe^{2+}, and so on. The Reactions 6.2, 6.3 and 6.4 represent the indirect oxidation of pyrite (Sec. 2.3.1). However, the Fe^{3+} produced in Reaction 6.3 may not only oxidize reduced minerals such as pyrite but also reduced U^{4+} minerals like uraninite:

$$UO_{2(s)} + 2\,Fe^{3+}_{(aq)} \rightarrow UO^{2+}_{2\,(aq)} + 2\,Fe^{2+}_{(aq)} \tag{6.5}$$

The oxidation of uraninite (Reaction 6.5) generates dissolved uranyl oxyanions and Fe^{2+}. This Fe^{2+} can then be oxidized to Fe^{3+} via Reaction 6.3, which in turn oxidizes uraninite via Reaction 6.5, which in turn produces more Fe^{2+}, and so on. Reactions 6.3 and 6.5 form a continuing cycle of Fe^{2+} conversion to Fe^{3+} and subsequent oxidation of uraninite by Fe^{3+} to produce Fe^{2+} etc. This cycle continues until the supply of uraninite or Fe^{3+} to the reaction system is exhausted. While oxygen is not required for the Reaction 6.5 to occur, it is still needed to convert Fe^{2+} to Fe^{3+} (Reaction 6.3). Thus, pyrite oxidation produces acid and Fe^{3+}, and these two products enhance the leaching and destruction of uranium minerals. The resulting low pH conditions favour the dissolution of uranium, metals, and metalloids within solid uranium mine wastes.

Once uraninite, coffinite, brannerite, and pitchblende are exposed to atmospheric oxygen and water, they readily undergo oxidation and dissolution. Other reduced minerals such as sulfides in the wastes will also be oxidized. If the waste materials are depleted in acid neutralizing minerals, acid pore waters and leachates will develop. The AMD conditions favour the enhanced dissolution of uranium minerals and the transport of dissolved uranium as sulfate complexes. Consequently, the release of uranium, metals, and metalloids from solid uranium mine wastes is controlled by the availability of oxygen and water.

The oxidative dissolution of uranium minerals appears to be very similar to that of sulfides (Sec. 2.3): (*a*) there are certain bacteria which act as catalysts in the destruction of sulfides and uranium minerals (Ragnarsdottir and Charlet 2000); (*b*) the destruction of both mineral types can occur through biotic or abiotic and direct or indirect oxidation processes; (*c*) the destruction of the minerals is achieved progressively through a sequence of chemical reactions; and (*d*) various factors including solution pH, oxygen abundance, microbiological activity, and mineral surface area influence the rate of mineral dissolution.

6.3.2 Uranium Solubility

Uranium occurs in natural waters as U^{4+} and U^{6+} ions. The abundance of these uranium ions in water is influenced by the presence and abundance of complexing agents and the prevailing pH and Eh conditions.

Under reducing conditions and acid pH, uranium is present in water as the U^{4+} species. In the absence of ligands, the concentrations of dissolved U^{4+} never exceed μg l^{-1} levels (Ragnarsdottir and Charlet 2000). Under oxidizing conditions, conversion of U^{4+} to the higher state U^{6+} results in the formation of the highly soluble, stable and mobile uranyl ion (i.e. UO_2^{2+}). Solubility is enhanced by complexation with sulfate, carbonate, hydroxide, fluoride, chloride, nitrate, phosphate or organic ligands such as humic or fulvic acids (Langmuir 1997; Ragnarsdottir and Charlet 2000). Uranium is then transported as an anion or complexed anion in oxidized ground and surface waters. The concentrations of U^{6+} species in oxidized ground and surface waters reach tens or even thousands of mg l^{-1}, depending on pH and ligand concentrations (Ragnarsdottir and Charlet 2000). Alkaline pH values and elevated bicarbonate concentrations in oxidized ground and surface waters favour the stabilization and mobilization of uranium as uranyl carbonate complex (Abdelouas et al. 1998a). In contrast, the predominant species in acid, oxygenated waters are the uranyl ion and uranyl-sulfate complex (Wanty et al. 1999; Abdelouas et al. 1999). Thus, acid or alkaline oxygenated mine waters may carry significant concentrations of dissolved uranium.

Natural ground and surface waters contain several μg l^{-1} uranium, and much higher uranium concentrations are found in mineralized areas (Ragnarsdottir and Charlet 2000). At uranium mine sites, oxidative dissolution of uranium minerals can be extreme, causing strongly elevated uranium concentrations in mine waters. Leachates of uranium mine wastes may contain tens of mg l^{-1} uranium. The dissolved uranium can migrate many kilometers from its source until changes in solution chemistry lead to the precipitation of uranium minerals.

Leachates of uranium mine wastes may also have elevated thorium and radium-226 levels. Thorium is less soluble than uranium under all conditions. However, thorium may be mobilized in low pH mine waters (Langmuir 1997). Once in solution, the element readily coprecipitates with iron and manganese oxyhydroxides, or it precipitates if the pH increases to values above 5 (Wanty et al. 1999).

Thorium-230 is the long-lived parent (half-life: 80 000 years) of the radium isotope Ra-226. This radium isotope is commonly present in waters of uranium mines since it is soluble and mobile under acid conditions; yet high sulfate concentrations in acid mine waters limit the solubility of Ra-226. Sulfate minerals such as barite ($BaSO_4$), celestite ($SrSO_4$), anglesite ($PbSO_4$), gypsum ($CaSO_4 \cdot 2\,H_2O$), or jarosite may precipitate from the waters. These minerals incorporate radium in their crystal lattices, limiting Ra-226 mobility.

6.3.3
Uranium Precipitation

Precipitation of dissolved uranium is primarily achieved through the destruction of uranyl complexes. The stability of dissolved uranium is greatly influenced by: coprecipitation; adsorption onto clays or oxyhydroxides; decreasing Eh; microbiological activity; and composition of the solution. Hence, uranium can be removed from solution through the following processes (e.g. Erskine et al. 1997; Abdelouas et al. 1998a,b; Wanty et al. 1999; Ragnarsdottir and Charlet 2000):

- *Coprecipitation.* Coprecipitation of uranium with carbonates and iron or aluminium leads to the incorporation of uranium into carbonate minerals and iron or aluminium oxyhydroxides. The coprecipitation of uranium is reversible. For example, reduction and dissolution of solid uranium-rich iron oxyhydroxides release uranium back into the water column (cf. Sec. 3.5.9).

- *Adsorption.* Positively charged uranium may be adsorbed onto negatively charged surfaces of sulfides, clays and organic matter as well as iron, manganese and aluminium oxyhydroxide particles. Uranium adsorption is also reversible (cf. Sec. 3.5.4). Desorption of uranium back into solution may occur if the water chemistry allows redissolution of uranium (e.g. increasing abundance of carbonate complexes at an alkaline pH).
- *Decreasing Eh.* Uranium is a redox-sensitive element, and reaction of U^{4+}-bearing waters with organic matter or hydrogen sulfide leads to uranium precipitation. For example, reduction due to a reaction of uranyl complexes with organic matter (simplified as molecule CH_2O) results in the precipitation of U^{6+}. The precipitated uranium may form an insoluble uranium oxide (UO_2) or one of its hydrates (Reaction 6.6) that is stable under reducing conditions. Also, the reduction and precipitation of uranium can be due to reactions of uranyl complexes with hydrogen sulfide, which has been generated by bacteria living in an anaerobic environment (Reaction 6.7). Such reduction reactions can occur in stream sediments of humid temperate or tropical environments where organic material is abundantly available to accumulate uranium.

$$4\,UO_{2\,(aq)}^{2+} + CH_2O_{(s)} + H_2O_{(l)} \rightarrow 4\,UO_{2(s)} + CO_{2(aq)} + 4\,H_{(aq)}^{+} \quad (6.6)$$

$$4\,UO_{2\,(aq)}^{2+} + HS_{(aq)}^{-} + 4\,H_2O_{(l)} \rightarrow 4\,UO_{2(s)} + SO_{4(aq)}^{2-} + 9\,H_{(aq)}^{+} \quad (6.7)$$

- *Microbiological activity.* Microorganisms are capable of removing dissolved uranium from solution and incorporating it into their cell structure. Uranium may also be adsorpted onto the cell surface. In addition, microorganisms can directly convert dissolved U^{6+} to solid U^{4+} compounds. As a consequence, the amount of uranium dissolved in mine waters is reduced.
- *Decreasing pH and removal of the complexing agent from the water.* Uranium may precipitate with phosphate or carbonate ions to form secondary phosphate- or carbonate-uranium minerals. However, the precipitation of secondary uranium minerals is unlikely since mine waters are commonly undersaturated with respect to secondary uranium minerals (Wanty et al. 1999). Significant evaporation of waters has to occur to cause the formation of secondary uranium minerals.

6.4 Radioactivity

6.4.1 Principles of Radioactivity

An element is characterized by its defined number of protons whereas the number of neutrons for a particular element may vary, giving rise to several isotopes (Attendorn and Bowen 1997). For instance, uranium has 92 protons and either 146, 143 or 142 neutrons, giving rise to three different isotopes (i.e. U-238, U-235, U-234). Most lighter elements possess stable isotopes whereas the heavier elements tend to have unstable isotopes, which eventually disintegrate into new isotopes. These unstable or radioactive isotopes have an excess energy that is released during the decay in order to achieve a lower and more stable energy state. Unstable, radioactive atomic nuclei change their structure by absorbing or, more commonly, releasing protons, neutrons, electrons or electromagnetic waves to become nuclei of new isotopes. Such radioactive atomic nuclei are also referred to as "*radionuclides*" in order to stress the radioactive nature of these nuclides. Naturally occurring "*parent nuclides*" break up yielding "*daughter nuclides*". If the daughter nuclides are radioactive, they will decay further. Any radioactive decay will end with the formation of a stable isotope. The decay of radionuclides in a series of steps is a so-called "*decay series*". The various decay products of a decay series are referred to as "*progeny*". In this chapter, the term "radionuclides" is used to refer to the parent and daughter radionuclides of the uranium-238 (U-238), uranium-235 (U-235), and thorium-232 (Th-232) series.

The process of radioactive decay is associated with the release radiant energy (i.e. "*radiation*"), which takes the form of particles and electromagnetic waves. Radiation levels are assessed using the activity of the radioactive material. "*Radioactivity*" is defined as the number of disintegrations of an atomic nucleus per unit time. The decay of a radionuclide over time can be measured. The term "*half-life*" is used to describe the period of time during which half of the atoms of a particular radionuclide decay. Measured half-lifes vary from a fraction of a second to several thousand million years.

Radioactive nuclides differ from stable nuclei in that radioactive nuclides emit "*ionizing radiation*" in the form of: (*a*) alpha particles (i.e. a helium nucleus of atomic mass four consisting of two neutrons and two protons); (*b*) beta particles (i.e. highly energetic electrons); and (*c*) gamma-rays (i.e. high energy electromagnetic waves). Alpha particles do not penetrate matter deeply because of their large size and double positive charge. However, they cause an enormous amount of ionization along their short path of penetration. They travel only a few centimeters in air and can be stopped by a piece of paper or the outer layer of the skin. Hence, alpha particles present an insignificant hazard outside the body, yet they are a potential hazard when ingested or inhaled. Beta particles are more penetrating than alpha particles because of their smaller size and negative charge, but beta particles produce much less ionization than alpha particles. Beta particles travel a few meters in air and require a couple of centimeters of plastic to stop them. Gamma-rays have neither mass nor charge and are much more penetrating than particulate radiation. Gamma-rays have an infinite range in air and require several centimeters of lead to absorb them.

The three kinds of radiation have very different properties but all are energetic enough to break chemical bonds. The primary effects of alpha particles, beta particles, and gamma-rays are the removal of electrons from atoms and the production of ions in the materials they strike; therefore, the term "ionizing radiation" is used. Ionizing radiation has the potential to cause biological damage by two principal mechanisms. Firstly, the ionization radiation induces direct damage to atoms and molecules of living cells and tissue. Secondly, the ionization radiation causes indirect damage to water molecules in the organism, resulting in highly reactive free radicals and molecules (e.g. hydrogen peroxide) that are chemically toxic to the organism.

6.4.2 Radioactive Decay of Uranium and Thorium

Uranium is a naturally occurring radioactive element in the Earth's crust, and uranium ores are concentrations of such natural radioactivity. The radioactivity of uranium ores and wastes is caused by the decay of radioactive isotopes. Uranium has three natural isotopes: uranium-238 (U-238); uranium-235 (U-235); and uranium-234 (U-234); all of which are radioactive. The relative abundances of U-238, U-235, and U-234 are 99.28%, 0.71%, and 0.006%, respectively (Attendorn and Bowen 1997; Ragnarsdottir and Charlet 2000). These parent nuclides are unstable; they decay to daughter nuclides and release radiation in the form of alpha and beta particles as well as gamma-rays. In fact, U-238 and U-235 are the parent isotopes for various intermediate radionuclides, and U-234 is a decay product of U-238 (Ragnarsdottir and Charlet 2000). These intermediate nuclides decay further – via alpha or beta emission – to new nuclides which in turn decay to other radionuclides. Uranium-238 and U-235 are the head of two separate decay series, which ultimately yield the stable, non-radioactive daughter nuclides Pb-206 and Pb-207, respectively (Table 6.2). Uranium-235 is not as prevalent in nature as U-238 because of its shorter half-life and lower abundance than U-238. Consequently, radiation originating from the U-235 decay series is significantly less than that of the U-238 series.

Uranium occurs together with thorium in uranium ore deposits. Thorium has several isotopes including Th-234, Th-232, Th-231, Th-230, Th-228, and Th-227 (Attendorn and Bowen 1997). Thorium-232 is the most abundant radioactive isotope and head of the Th-232 series. It eventually decays via intermediate daughter isotopes to a stable lead isotope (Pb-208) (Table 6.2). Thus, the concentrations of uranium and thorium are naturally decreasing. The uranium and thorium isotopes have long half-lifes comparable to the age of our planet and hence, these isotopes are still present.

6.4.2.1 *Radium*

The radiogenic isotope radium-226 (Ra-226) is a daughter product of the U-238 series and a direct descendent of Th-230 (half-life: 80 000 years) (Table 6.2). While Th-230, Pb-210, and radon isotopes are of concern, Ra-226 is of most concern in uranium mining and processing operations because (Landa and Gray 1995; Kathren 1998; Landa 1999; Ewing 1999):

1. Radium-226 has a half-life of 1 622 years and therefore, persists in uranium mine wastes.
2. Radium-226 has similar geochemical and biogeochemical properties to its fellow group II elements (Ca, Ba, Sr) and forms compounds that can be taken up by humans, plants, and animals.
3. Radium-226 has a high radiotoxicity and affinity for accumulating in bones.
4. Compared to uranium and thorium, Ra-226 is readily liberated from the uranium ore minerals during natural weathering and mineral processing; it is more soluble than uranium and thorium; it may leach from soils, rocks, ores and mine wastes; and it is readily mobilized into ground and surface water.

Table 6.2. Simplified decay pathways and half-lifes of the U-238, U-235 and Th-232 series (after Brownlow 1996; Kathren 1998; Langmuir 1997)

Series	Isotope	Particle emitted	Half-life	
U-238	U-238	α	4.5×10^9	yr
	Th-234	β	24	d
	Pa-234	β	1.17	min
	U-234	α	247 500	yr
	Th-230	α	80 000	yr
	Ra-226	α	1 622	yr
	Rn-222	α	3.8	d
	Po-218	α	3	min
	Pb-214	β	26.8	min
	Bi-214	α, β	19.7	min
	Po-214	α	164	μs
	Tl-210	β	1.32	min
	Pb-210	β	22.5	yr
	Bi-210	β	19.7	min
	Po-210	α	140	d
	Pb-206	Stable	–	
U-235	U-235	α	713×10^6	yr
	Th-231	β	25.6	h
	Pa-231	α	34 300	yr
	Ac-227	β	22	yr
	Th-227	α	18.6	hr
	Ra-223	α	11.2	d
	Rn-219	α	3.9	s
	Po-215	α	1.83	μs
	Pb-211	α	36.1	min
	Bi-211	α, β	2.16	min
	Po-211	α	0.52	s
	Tl-207	β	4.79	min
	Pb-207	Stable	–	
Th-232	Th-232	α	1.4×10^{10}	yr
	Ra-228	β	6.7	yr
	Ac-228	β	6.13	hr
	Th-228	α	1.9	yr
	Ra-224	α	3.64	d
	Rn-220	α	54.5	s
	Po-216	α	0.15	s
	Pb-212	β	10.6	hr
	Bi-212	α, β	60.4	min
	Po-212	α	0.29	μs
	Tl-208	β	3.1	min
	Pb-208	Stable	–	

5. Radium-226 decays by alpha emission to the important radon isotope radon-222 (Rn-222). Consequently, Ra-226 is the head and source of the important subseries that includes radon (Table 6.2).

6.4.2.2
Radon

Radon is a colourless, odourless and tasteless noble gas with three naturally occurring radioactive isotopes (i.e. Rn-219, Rn-220, Rn-222). The term "radon" commonly refers only to Rn-222 (Sharma 1997). Radon-219 is the daughter product of the U-235 series, Rn-220 is a member of the Th-232 series, and Rn-222 is a daughter product of the U-238 series and the direct descendent of Ra-226. The most abundant isotope is Rn-222 which is due to the abundance of its parent isotope U-238. The other two isotopes Rn-219 and Rn-220 are significantly less abundant, have shorter half-lifes, and are therefore of little environmental concern (Sharma 1997). In contrast, Rn-222 has a half-life of 3.8 days. Radon-222 is of concern in uranium mining and processing because:

1. Radon-222 is a descendant of parent radionuclides with long half-lifes (Th-230: half-life 80 000 years; Ra-226: half-life 1 622 years). As a result, Rn-222 represents a long-term hazard despite its short half-life of 3.8 days.
2. Radon-222 is a noble gas and soluble in water. Such properties allow radon to move freely in ground and surface waters.
3. Radon-222 itself decays by the emission of an alpha particle. The daughter products are polonium-218 (Po-218), lead-214 (Pb-214), and bismuth-214 (Bi-214), that is, the so-called "*radon progeny*". These solid daughter products are highly radioactive and emit alpha and beta particles as well as gamma-rays (Table 6.2).
4. Once Rn-222 is inhaled by humans, its radioactive solid decay products are deposited directly within the lungs. The lodged radon progeny will cause ionizing radiation and induce lung cancer.

In uranium mine environments, radon gas particularly emanates from uranium ores, mineralized waste rocks, and uranium mill tailings. Radon – released to pore spaces of mine wastes – migrates to the surface of the material or is dissolved in pore waters. Radon is also emitted from undergound workings and open pits. After the termination of mining, radon will continue to emanate from the walls and floor of mine workings. Since most of the ore is generally removed from mine workings, the primary sources of radon are the waste rock piles and any ore stockpiles. Local meteorological effects can control radon emanation from the ground, and radon emanation depends on air pressure, soil moisture, soil structure, ground cover, wind, and temperature (Nielson et al. 1991).

6.4.3 Units and Measurements of Radioactivity and Radiation Dose

6.4.3.1 *Units*

There is a confusing array of units to measure radioactivity. Fortunately, radiation units can be classified in two major categories (Nielson et al. 1991; Wilson 1994) (Table 6.3).

1. *Radioactivity.* The radioactivity of a material is measured by the number of nuclear disintegrations per unit of time. Thus, the units to describe radioactivity are based on nuclear disintegrations which are counted in units of time for a specific volume of radioactive material. The basic and fixed unit of radiation is the Curie (Ci) that is defined as the number of nuclear disintegrations per second. One Curie is equivalent to 3.7×10^{10} disintegrations per second which is the measured activity of 1 g ra-

Table 6.3. Radioactivity and radiation dose measurement units and their conversion (Nielson et al. 1991; Sharma 1997; Wilson 1994)

Unit	Definition	Conversion			
Radioactivity = Nuclear disintegration per unit time per volume of radioactive material					
Bq (Becquerel)	One disintegration per second of a radioactive isotope	1 Bq	=	27	pCi
Bq kg^{-1}	Specific activity of a radioactive isotope per unit mass				
Ci (Curie)	Activity of 1 g radium: 3.7×10^{10} disintegrations per second	1 Ci	=	3.7×10^{10}	Bq
Ci kg^{-1}	Specific activity of a radioactive isotope per unit mass				
pCi l^{-1}	Specific activity of a radioactive isotope measured in water	1 pCi l^{-1}	=	37	Bq m^{-3}
R (Roentgen)	Radiation required to produce one electrostatic unit of charge in one cubic centimeter of dry air	1 R min^{-1}	=	1	Ci
Radiation dose = Biological effects of nuclear disintegration					
Gy (Gray)	Absorbed radiation dose; corresponding to one joule of radiation absorbed per kilogram of tissue	1 Gy 1 Gy	= =	100 1	rad J kg^{-1}
Sv (Sievert)	Absorbed radiation dose measured in gray times the quality factor for the type of radiation and a weighting factor for the tissue irradiated; unit to describe the damage to tissue	1 Sv	=	100	rem
REM	Roentgen equivalent in man; ionizing radiation equal to damage to humans of one roentgen of high voltage X-rays	1 rem	=	0.01	Sv
rad	Radiation absorbed dose per unit mass	1 rad	=	0.01	J kg^{-1}

dium. Radioactivity is also measured in Becquerel (Bq) which is defined as one disintegration per second of a radioactive isotope. The *specific radioactivity* of a nuclide in solid materials refers to the number of nuclear disintegrations per second per unit of mass (e.g. a tailings sample has 100 Bq kg^{-1} of Ra-226). In radon measurements, picoCuries per liter (pCi l^{-1}) are commonly used, and potable water standards and water analyses are given in this unit (e.g. a water sample has 10 pCi l^{-1} of Ra-226).

2. *Radiation dose.* The effects of radiation are assessed using the amount of radiation received by biological material. The amount of radiation received by a person is of prime interest in human health studies. The amount of radiation is measured by the "*radiation dose*" which refers to the amount of energy imparted per unit of biological mass. Rems, rads, Grays (Gy), and Sieverts (Sv) are all related to the radiation effects on humans. These units are not based on pure physical measurements but on radiation research, statistics, and multiplication factors. The term "rem" is often used to describe the dose that can be imposed upon humans. Rem is an acronym for "roent-

gens equivalent in man". Rad is the "radiation absorbed dose". Rem and rad relate to the Standard International (SI) radiation units Gray (Gy) and Sievert (Sv). The dose of energy that is absorbed in the body tissue is measured in Grays (Gy). Equal exposure to different types of radiation does not necessarily produce equal biological effects. One Gray of alpha radiation has a greater effect than one Gray of beta radiation. Radiation effects, regardless of the type of radiation, are measured in Sieverts (Sv). One Sievert of radiation deposits 0.01 J of energy per kilogram of tissue. Radiation exposure levels for the general public and maximum permissible radiation doses are given in milliSievert per year (mSv $year^{-1}$).

6.4.3.2
Measurements

Curies and Becquerels are laboratory measurements whereas counts per minute (cpm), milliSievert per hour (mSv $hour^{-1}$), and microRoentgens per hour (μR $hour^{-1}$) can be established in the field (Wilson 1994). Field gamma-ray spectrometers are used to measure the latter three units. Alpha and beta particles are best measured in counts per minute (cpm), using Geiger counters or scintillometers whereby the measured counts per second (cps) are roughly equivalent to disintegrations per second (Bq) (Wilson 1994). The levels of radionuclides found in water (U-238, U-234, Th-230, Th-232, Ra-226, Pb-210, Po-210) are typically measured in units of milliBecquerel per liter (mBq l^{-1}) or picoCuries per liter (pCi l^{-1}). The radionuclide of most concern in water is Ra-226.

The detection and measurement of radon is based on sniffer-type radon detectors. The soil gas or degassed water is thereby analyzed for its alpha activity since Rn-222 decays by alpha emission to Po-218 (Table 6.2). Alternatively, alpha-sensitive films buried in soil allow the determination of radon gas concentrations (Nielson et al. 1991; Sharma 1997). Details on field procedures and operational considerations of gamma-ray spectrometry and radon emanometry are given by Sharma (1997) and Nielson et al. (1991).

6.4.4
Radioactive Equilibrium and Disequilibrium

Uranium ores have a radioactive equilibrium or disequilibrium in the U-238, U-235, and Th-232 decay chains. In an equilibrium state, radioactive parent and daughter isotopes decay, and the decay products remain isolated with the parent radionuclide. An equilibrium is reached between the parent and daughter nuclides, and the specific activity (ie. the number of radioactive disintegrations per unit time per unit mass) of all decay products will be approximately equal. For example, an equilibrium is in place if the activity (pCi g^{-1}) in the U-238 decay chain will be the same for the parent nuclide U-238 and for the daughter nuclides (e.g. Th-230, Ra-226, Rn-222 and so forth). Disequilibrium is found in naturally disturbed deposits where mobilization of nuclides has occurred during or after radioactive decay, for instance, as a result of weathering. In such deposits, the specific activity of all decay products will be different. The difference is caused by the behaviour of individual nuclides according to their own chemistry and the preferential mobilization of individual nuclides. For example, the loss of parent nuclides from the ore reduces the activity of daughter nuclides which follow the parent in the decay series.

Mining and beneficiation do not significantly disrupt the radioactive equilibrium or disequilibrium of uranium ores whereas chemical processing does. Chemical processing of uranium ores aims to concentrate uranium and to reject all other elements, including the daughter nuclides of uranium isotopes (Table 6.2). The daughter nuclides become separated from the parent uranium nuclides and a radioactive equilibrium is no longer established. The concentrations of the daughter nuclides are no longer controlled by the parent nuclides. Uranium mill tailings are in pronounced radioactive *disequilibrium* since the parent nuclides (U-238, U-235) have been removed.

The disequilibrium of radionuclides in tailings materials or seepage precipitates can be used to evaluate radionuclide mobility in and from a tailings repository (e.g. Naamoun et al. 2000). Also, dissimilar specific activities of parent and daughter nuclides in waste rock dumps are an indication of the mobility of individual radionuclides. For example, if the waste material has Ra-226 activities (Bq kg^{-1}) different to those of the immediate parent isotope Th-230 (Bq kg^{-1}), the parent-daughter nuclide ratio (i.e. Th-230/Ra-226) is not unity, disequilibrium is present, and preferential leaching and mobilization of individual radionuclides have occurred.

6.5 Uranium Mining and Extraction

Traditionally, uranium is extracted from open pits and underground mines (Figs. 6.1, 6.2), and the mined ore is treated in hydrometallurgical plants on and near the mine site to yield uranium (i.e. hydrometallurgy). In addition to this conventional mining and extraction process, an alternative technique, the so-called "in situ leach" operation, is applied to some uranium ores.

6.5.1 Conventional Mining and Extraction

In conventional uranium mining and extraction, ores are obtained from underground or open pit mines. The ore is first crushed and/or powdered for uranium extraction and then leached. The process chemicals include acid (e.g. sulfuric acid, nitric acid) or alkaline solvents (e.g. sodium carbonate-bicarbonate, soda) as well as oxidants (e.g. sodium chlorate, ferric ion, hydrogen peroxide). Oxidizing conditions are essential to allow formation of the highly soluble U^{6+} complexes. The acid or alkaline leaching process is applied to heap leach piles, or more commonly, under controlled conditions in a hydrometallurgical plant. The acid or alkaline process of uranium extraction oxidizes the uranium (U^{4+}) present in the uranium minerals (e.g. uraninite) and dissolves the oxidized uranium (U^{6+}) as a sulfate or carbonate complex:

$$UO_{2(s)} + 4\,Na^+_{(aq)} + 2\,CO^{2-}_{3(aq)} \rightarrow UO_2(CO_3)^{2-}_{2(aq)} + 4\,Na^+_{(aq)} \tag{6.8}$$

$$UO_{2(s)} + 2\,H^+_{(aq)} + SO^{2-}_{4(aq)} \rightarrow UO_2(SO_4)^0_{(aq)} + 2\,H^+_{(aq)} \tag{6.9}$$

The process chemicals dissolve the uranium minerals, and an uranium enriched liquor is formed. This liquor is commonly referred to as "pregnant liquor". It contains apart from uranium a range of other elements such as rare earth elements which were

Fig. 6.1. Open pit of the Mary Kathleen uranium mine, Australia. Elevated gamma-ray readings in the open pit correspond to exposed ore lenses, the former haul road, and abandoned ore stockpiles (up to 16 mSv per year). Wallrock oxidation of reactive sulfides produces acidic solutions, yet buffering reactions prevent low pH conditions from developing. The open pit lake contains saline surface waters at a pH of 6.11 (salinity 0.15%; elevated Ca, SO_4, Cu, Fe, Mn, Ni, U and Zn values)

Fig. 6.2. Headframe of the Ronneburg underground uranium mine, Germany. Waste rock pile of uraniferous, carbonaceous shale is present behind the headframe

originally present as cation substitutions within the uranium minerals (Lottermoser 1995). After the solids have been removed by filtration or other methods, the solution is concentrated and purified using ion exchange or solvent extraction techniques. The impurities (e.g. metals, metalloids) are collected and disposed of with the tailings.

Ammonia is added to the liquor, and the uranium precipitates as yellow coloured ammonium diuranate which is referred to as "*yellowcake*". The yellowcake is then converted and refined to uranium trioxide concentrate (UO_3). Uranium in yellowcake is mostly U-238 (>99%), containing a small proportion of U-235 and trace amounts of other elements such as thorium and radium.

6.5.2 In Situ Leach (ISL) Operations

"*In situ leach mining*" (ISL) or "*solution mining*" leaves the buried ore in place. The technique extracts uranium by reversing the natural process which originally deposited the uranium. The uranium was naturally emplaced by uranium laden (U^{6+}), oxidized ground waters which moved through permeable aquifers. Once the uranium-rich waters encountered reducing conditions in the aquifers, the uranium precipitated to form reduced (U^{4+}) uranium ore minerals.

The ISL process uses an acid or alkaline leaching solution that is introduced to the ore by means of injection wells (OECD 1999) (Figs. 6.3, 6.4). The leaching solution consists of ground water and process chemicals. The solution attacks the (U^{4+}) uranium minerals and oxidizes the uranium which forms soluble uranium (U^{6+}) complexes. The uranium-bearing solution is then pumped to the surface through production wells, and the uranium is recovered from the solution. The barren solution is regenerated with process chemicals and circulated back into the aquifer for continued leaching.

An acid or an alkaline leaching process is used to extract uranium. In the acid ISL operation, the ground water will be reduced to a pH of approximately 2. The process uses sulfuric acid – in some cases hydrofluoric acid – and oxidants such as nitric acid, nitrate, hydrogen peroxide or dissolved oxygen (Reaction 6.9). The alkaline ISL scheme applies alkaline reagents including ammonia, ammonium bicarbonate, and sodium carbonate/bicarbonate (Reaction 6.8). The alkaline ISL scheme is considered to be more "environmentally acceptable" than acid ISL mining because ISL affected ground waters are technically easier to restore once mining ceases (OECD 1999). The addition of bacteria to the leaching solution has also been trialled. *Acidithiobacillus ferrooxidans* has been employed as a leaching agent to enhance the dissolution of uranium ore minerals and to extract uranium from low grade ores in Canada and the United States.

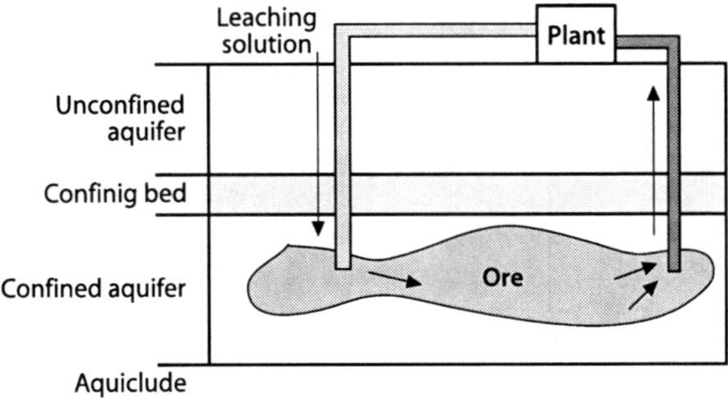

Fig. 6.3. Schematic cross-section of an ISL uranium mining operation. The ISL process uses an acid or alkaline leaching solution that is introduced to the ore by injection wells. The leaching solution dissolves uranium minerals; the uranium-bearing solution is then pumped to the surface through production wells; and finally, the uranium is recovered from the solution in the processing plant

Fig. 6.4. Borefield of the Honeymoon ISL uranium mine, Australia. An acid leaching process is applied to a confined aquifer

The ISL mining technique is well established and has been used in various countries including former East Germany, Bulgaria, Ukraine, China, Kazakhstan, Uzbekistan, Czech Republic, the United States, and Australia (Mudd 2001a,b). Uranium deposits suitable for ISL mining occur in near surface permeable sand or sandstones, are located below the ground water table, and are confined by impermeable strata in the foot and hanging walls (Fig. 6.3).

Proponents of the ISL technique argue that the ISL method is applied to an unusable ground water resource and that it is controllable, safe, and environmentally benign compared to conventional open pit or underground mining techniques (Uranium Information Centre 2001a). In many cases, mining is conducted on a confined aquifer. The aquifer is naturally saline and contains naturally elevated levels of radionuclides and sulfate, which are present at concentrations well in excess of those considered safe for stock or domestic use. Furthermore, an ISL mining operation does not produce tailings, ore stockpiles, and waste rock dumps. Surface disturbances are minimized and environmental impacts are much less than for conventional mining. Finally, most of the radionuclides remain in the ore horizon, and there is a minimal increase in radon release and dust contamination from mined and processed ore. Mine waters contain radionuclides, and small quantities of radioactive sludges accumulate at ISL processing plants. The sludges need to be disposed of in lined waste repositories. The mine waters are either reinjected back into the aquifer, or they are applied as irrigation waters to designated land.

Opponents of the ISL technique argue that it is not an acceptable mining method. The acid ISL scheme requires monitoring and controlling of ground water conditions to ensure isolation of the affected aquifer from other ground water resources. Post-mining concentrations of some constituents (U, Ra-226) have been found to be elevated in the ground water when compared to pre-mining conditions (Caldwell and Johnson

1997). Once mining ceases, ground water restoration of the leached aquifer to pre-mining conditions is difficult if not impossible to achieve. Over time, mobilized and precipitated mineral particles such as clays, gypsum or jarosite will become trapped in the pore spaces (Mudd 2001a). The permeability of the leached ore horizon will be reduced, and the hydrology of the horizon is irreversibly changed. Acid ISL mining was not successful in the United States because of high acid consumption in carbonate-bearing aquifers and problems with ground water restoration after acid leaching (Mudd 2001a). In contrast, acid ISL operations were common in Eastern Europe and the former Soviet Union (Mudd 2001a). Where used in inappropriate locations, acid ISL mining resulted in significant ground water contamination in Bulgaria, the Czech Republic, and East Germany (Case Study 6.1).

Ground water restoration programs at former ISL sites rely on various techniques (e.g. reverse osmosis, volume reduction by evaporation, pump-and-treat). Contaminants are extracted from the ground waters, and some components may even be recycled including sulfuric acid, aluminium oxide, ammonia, and gypsum. Moreover, natural attenuation processes within aquifers can restore contaminated ground waters in ISL environments. The key geochemical processes which naturally remove dissolved radionuclides and heavy metals from ground waters are: (*a*) anaerobic conditions resulting in precipitation of reduced compounds such as uranium minerals and sulfides; (*b*) coprecipitation of uranium with iron oxyhydroxides, organic matter, and carbonates; and (*c*) adsorption of uranium on clay minerals, silicates, and aluminosilicates (Erskine et al. 1997). Such natural restoration mechanisms will of course only operate if the aquifer is anaerobic or contains elevated organic carbon, sulfide, clay as well as iron and manganese oxyhydroxide concentrations.

6.6 Mining, Processing and Hydrometallurgical Wastes

Wastes of uranium mining, processing and hydrometallurgical extraction are invariably radioactive. Most uranium mine wastes can be classified as non-radioactive (e.g. overburden, waste rocks) or low-level radioactive wastes (e.g. mine water, heap leach piles, waste rocks, tailings). Low-level radioactive wastes contain low levels of nuclides, emitting alpha and beta particles and gamma-rays. The greatest volume of non-radioactive and low-level radioactive waste products is generated at uranium mines that use conventional mining and extraction methods. At these locations, mining, processing and hydrometallurgical uranium recovery generate mine waters, sludges, waste rock dumps, heap leach piles, and tailings. In comparison, uranium mines using ISL techniques produce significant less waste and generate smaller volumes of radioactive mine waters and sludges. Wastes of uranium mining and extraction can be grouped into the following four categories:

- *Tailings.* Uranium extraction at many operating uranium mines is based on: (*a*) crushing of the ore to a fine powder; and (*b*) leaching the powdered ore with process chemicals in a hydrometallurgical plant. The solids are removed from the process circuit and pumped with spent process waters to a tailings repository. These wastes are referred to as "*uranium tailings*" or more commonly as "*uranium mill tailings*".

Case Study 6.1. Waste Production and Environmental Impacts of the Wismut Uranium Mines, Germany

Mining

During the Cold War, much of the Soviet Union's uranium for nuclear energy and weapons was obtained from mines in former East Germany. The operation was known as "Wismut" and subject of utmost secrecy (Mager and Vels 1992). The mining activities were extraordinary large. Underground workings include 280 mine shafts and several thousand kilometers of drifts, ramps, and drives. Between 1946 and 1990, 220 000 t of uranium were produced. The uranium production – in terms of historical uranium production from 1946 to 1989 – was the third largest in the world after the United States (330 000 t) and Canada (230 000 t). The production exceeded that of Australia (45 000 t), South Africa (138 000 t), and even the Soviet Union (190 000 t). However, the production costs of Wismut's uranium were probably ten times the actual market value of uranium. Resource extraction was among the world's most inefficient (Mager and Vels 1992).

Environmental Impacts and Rehabilitation

From 1946 to 1953, mining was practically carried out without considering damage to the environment or effects on humans. During that time, almost no radiation protection existed for the German miners. Mean annual radiation exposures were at 40 times higher than current radiation protection standards, and thousands of miners suffered from lung and liver cancer as well as leukaemia. There was no restoration concept or funds set aside for rehabilitation. Only minor rehabilitation work was carried out during exploitation. Mining continued from 1954 to 1990 and ceased with the reunification of Germany in 1990. Following mine closure, the total area of mining and milling sites held by Wismut was 37 km^2, and the total area affected by mining was 240 km^2. In 1991, the Soviet Union was released from its obligations for the decommissioning and rehabilitation, and ownership was transferred to the German government. The German government initiated a gigantic program to clean-up the former underground and open pit mines, mill sites, dump sites, buildings, ore stockpiles, soils, river beds, tailings ponds, and contaminated aquifers. The former uranium mining and processing sites were scattered over an area of 100 km by 60 km, and a total of 120 km^2 were assessed for the environmental database. A total surface area of 37 km^2 had to be remediated, including 22.5 km^2 of mine waste rock dumps and tailings. The environmental problems were serious because of the high population density in the region.

Waste Production

A total volume of 240 Mt of ore and 760 Mt of waste rock were produced from numerous underground and open pit mines. Open pit mining produced 600 million m^3 of ore and waste rock (Mager and Vels 1992; OECD 1999; Mudd 2001b). Most of this material has been backfilled into open pits. Rehabilitation efforts also focus on the stabilization, reshaping, cover, and revegetation of waste rock piles. The waste materials contain considerable amounts of pyrite, resulting in significant acid generating potential. In total, there were 70 waste dumps with a combined volume of 317 million m^3, covering an area of about 15 km^2. Mineral processing activities used numerous milling plants, ponds, sedimentation plants, and tailings dams. Tailings were produced using alkaline and acid leaching techniques, and the tailings were disposed of in very large ponds. At Crossen, the pond was 2 km^2 in size and contained approximately 45 million m^3 of tailings and 6 million m^3 of water. At Seelingstädt, the tailings were disposed of in two ponds, covering 3.4 km^2 with a volume of 107 million m^3. Some of the tailings dams were constructed on geological faults and are located in an seismically active region. The repositories require careful rehabilitation considering seismic stability, liquid and solid contamination potential, and radiation exposure.

- *Mine water.* Water contaminated with radionuclides, heavy metals, metalloids, and other constituents is commonly generated at an uranium mine site. Such water includes: mill water; process water; mining water from the dewatering of underground

Case Study 6.1. ***Continued***

ISL Operations

The Wismut operation also used in situ leach (ISL) techniques to recover uranium from subsurface ores (Mager and Vels 1992; OECD 1999; Mudd 2001b). At Königstein, mining activities and ISL operations severly disturbed a layered ground water system. Dilute sulfuric acid was injected into the orebody after blasting. The acid dissolved the uranium minerals, and the uranium-bearing solutions were pumped to the surface for processing and uranium recovery. This extraction technique left 2 million m^3 of dilute sulfuric acid with dissolved heavy metals in local aquifers. The leaching process chemically affected more than 55 million m^3 of rock and aquifer. Approximately 2650 million liters of mine water – containing sulfuric acid and uranium – circulated or were trapped in the pore space of the rocks. The trapped waters have arsenic, cadmium, nickel and uranium levels up one to three orders of magnitude higher than German water quality guidelines. While the mine was in operation, there was little ground water contamination because the water flowing into the mine was pumped away from the acidified orebody. However, with the cessation of mining, pumping of working areas ceased, and the ground water may become contaminated because of the mixing of inflowing ground water with the remaining acid. Ground water restoration is essential because one aquifer of the layered ground water system is used by residents of the region for their water supply.

Conclusions

Intense uranium mining was conducted by the Soviet Union in former occupied East Germany over 45 years. The mining activities left a massive environmental disaster in a densely populated area. Like all centrally planned economies, at the time of mine closure there were no financial reserves to fund these activities. The total cost for decommissioning and rehabilitation has been estimated at 7000 million EURO, and this will take at least until 2007 to complete.

and open pit mines; and seepage water emanating from waste rock dumps, heap leach piles and ore stockpiles. Contaminated water is also produced in association with ISL operations and during ground water restoration of former ISL activities. Contaminated waters require treatment prior to disposal or even discharge of the mine lease.

- *Waste rocks.* Country rocks enclosing uranium ores possess no or subeconomic amounts of uranium minerals, and the mined waste rocks display variable uranium concentrations as well as radioactivity and radon levels.
- *Heap leach residues.* At some mine sites, uranium is removed from low-grade ore using heap leaching. The leaching solution is sprayed on top of the pile. The solution percolates downwards until it reaches a liner below the pile where it is caught and pumped to a processing plant. Together with waste rock piles, heap leach piles represent a potential hazard during operation because of the possible release of dust, radon, and seepage waters. If waste rock dumps or heap leach piles are sulfidic, there may be a potential for AMD development in the long term.

6.7 Tailings

Uranium mill tailings are generally disposed of as a slurry. The tailings are best described in terms of solids and liquids. The solids can be further subdivided based on their particle size into slimes and sands. Each of these three tailings components (i.e. sands, slimes, tailings liquids) have distinct chemical, mineralogical and radiochemical properties (Table 6.4).

Following placement into a tailings repository, the slimes and sands will settle, and the tailings liquids can be decanted for reuse in the uranium extraction circuit. The pH of sulfuric acid tailings is generally less than 4 whereas the pH of the alkaline tailings is generally greater than 9 (Table 6.4). In both leaching processes, most of the uranium minerals are dissolved, and the gangue minerals present in the ore are attacked by the solutions. This creates tailings with high concentrations of dissolved and soluble salts, metals and radionuclides.

6.7.1 Tailings Radioactivity

The hydrometallurgical processing of powdered uranium ore is very selective for uranium and removes most of the uranium from the ore. The extracted uranium is only weakly radioactive because of the long half-life of U-238, and the uranium oxide concentrate contains approximately 15% of the initial radioactivity of the uranium ore (OECD 1999).

Table 6.4. The components of uranium mill tailings and their characteristics (US DEO 1999)

Tailings component	Particle size (µm)	Chemical composition	Mineralogical composition	Uranium content and radioactivity
Sands	>75	Major SiO_2 with <1 wt.% Al, Fe, Mg, Ca, Na, K, Se, Mn, Ni, Mo, Zn, V, U	Mostly gangue minerals of the original ore	0.004–0.01 wt.% U_3O_8 *Acid leaching:*[a,b] 26–100 pCi Ra-226 g^{-1}; 70–600 pCi Th-230 g^{-1}
Slimes	<75	Major SiO_2 with <1 wt.% Al, Fe, Mg, Ca, Na, K, Se, Mn, Ni, Mo, Zn, V, U	Mostly fine-grained gangue minerals of the original ore and clay minerals, oxides, fluorides, sulfates and amorphous substances	U_3O_8 and Ra-226 concentrations are almost twice the concentrations present in the sands *Acid leaching:*[b] 150–400 pCi Ra-226 g^{-1}; 70–600 pCi Th-230 g^{-1}
Liquids	–	*Acid leaching:* pH 1.2–2.0; Na^+, NH_4^+, SO_4^{2-}, Cl^-, and PO_4^{3-}; total dissolved solids up to 1 wt.%	–	*Acid leaching:* 0.001–0.01% U; 20–7 500 pCi Ra-226 l^{-1}; 2 000–22 000 pCi Th-230 l^{-1}
		Alkaline leaching: pH 10–10.5; CO_3^{2-} and HCO_3^-; total dissolved solids up to 10 wt.%	–	*Alkaline leaching:* 200 pCi Ra-226 l^{-1}; essentially no Th-230 (insoluble)

[a] U_3O_8 content is higher for acid leaching than for alkaline leaching.

[b] Separate analyses of sands and slimes from the alkaline leaching process are not available. However, total Ra-226 and Th-230 contents of up to 600 pCi g^{-1} of each isotope have been reported for the combined sands and slimes.

Hydrometallurgical processing does not achieve complete extraction of uranium from the ores, and discarded tailings always contain small amounts of uranium. Moreover, most of the undesired daughter radionuclides of the U-238 and U-235 series end up in the tailings, and the tailings carry the remaining 85% of the initial radioactivity of the uranium ore (Landa 1999; OECD 1999; Abdelouas et al. 1999). As a consequence, uranium mill tailings constitute a high volume, low-level radioactive waste. The radioactivity is caused by: (*a*) highly radioactive nuclides with relatively short half-lifes (e.g. Po-218, Po-214, Po-210, Pb-210); and (*b*) less radioactive nuclides with long half-lifes (e.g. Th-230, Ra-226) (Table 6.2). If the uranium ore is enriched in thorium, the tailings will possess abundant daughter nuclides of the Th-232 series. Hence, radiation levels and radon emissions from tailings are potentially significant, and the radioactivity of uranium tailings is greater than that of the yellowcake or uranium oxide concentrate.

6.7.2 Tailings Solids

In general, tailings solids can be: (*a*) primary ore and gangue minerals; (*b*) secondary minerals formed during weathering; (*c*) chemical precipitates formed during and after mineral processing; and (*d*) chemical precipitates formed after disposal in the tailings storage facility (Sec. 4.2.3). In the case of uranium mill tailings, the chemical precipitates form: (*a*) during and after hydrometallurgical extraction; (*b*) upon neutralization of acid tailings with lime prior to their disposal; and (*c*) after disposal of the tailings in the storage facility.

The tailings solids can be crystalline, poorly crystalline, and/or amorphous in nature, and they contain radionuclides, heavy metals, and metalloids (e.g. Langmuir et al. 1999; Pichler et al. 2001). The solids are insoluble or potentially soluble (Willett et al. 1994; Landa 1999; Donahue et al. 2000). Detailed analyses have revealed that radionuclides (i.e. Th-230, Ra-226, U-235, U-238), heavy metals, and metalloids are present in: (*a*) ion exchangeable, carbonate and readily acid soluble forms; (*b*) iron and manganese hydrous oxides; (*c*) fluorides; (*d*) alkaline earth sulfates (e.g. $BaSO_4$, $SrSO_4$); (*e*) organic matter; (*f*) sulfides; and (*g*) arsenates (Willett et al. 1994; Landa and Gray 1995; Somot et al. 2000; Donahue et al. 2000; Pichler et al. 2001; Donahue and Hendry 2003; Martin et al. 2003). The stability of these solid phases determines the potential mobilization of radionuclides, heavy metals, and metalloids into tailings pore waters.

Radium-226 is a radionuclide of significant concern in uranium mining. It tends to be concentrated in the fine fraction of uranium tailings (Landa and Gray 1995) (Table 6.4). Barium and radium have similar geochemical properties, including the low solubility of their sulfate salts and the coprecipitation of radium with barium as solid sulfates ($Ba(Ra)SO_4$). In addition, adsorption and coprecipitation processes lead to the fixation of Ra-226 in iron oxyhydroxides, feldspars, clays, amorphous silica, organic matter, sulfides, barite, and other sulfate grains. Iron oxyhydroxides and alkaline earth and lead sulfates play an important role in the fixation of Ra-226 and other radionuclides in acid tailings materials (Landa and Gray 1995; Goulden et al. 1998; Landa 1999; Somot et al. 2000; Martin et al. 2003).

Uranium tailings like all tailings undergo chemical reactions in a tailings repository (cf. Sec. 4.2.3). Over time, the tailings mineralogy and pore water composition may change. Dissolved radionuclides, metals, and metalloids may: (*a*) persist in solution;

(*b*) precipitate or coprecipitate by interacting with other components in the tailings; or (*c*) be adsorbed by tailings solids such as quartz, kaolinite, clays or amorphous substances (Landa and Gray 1995; Landa 1999). The potential release of radionuclides, metals and metalloids from tailings solids into pore waters and the presence of these elements in solution are undesirable. Tailings liquids laden with contaminants can escape from the tailings storage area and may impact on aquifers or surface waters. Consequently, the mineralogical siting and chemical reactions in uranium tailings need to be understood in order to determine the long-term behaviour of radionuclides, metals and metalloids in the waste repositories. The mobilization of radionuclides, heavy metals, and metalloids from tailings solids into tailings liquids can be induced through various factors (Landa and Gray 1995):

- *AMD development.* Tailings may contain abundant sulfide minerals that can oxidize in the tailings impoundment. The acid producing reactions may not be sufficiently buffered by acid neutralizing reactions, and this leads to the formation of AMD. The development of AMD in uranium tailings, heap leach piles and waste rock dumps will enhance the dissolution of uranium minerals and the mobility of radionuclides.
- *Presence of process chemicals.* Hydrometallurgical extraction schemes add significant amounts of sulfuric acid, alkaline materials, nitrate, chloride and/or organic solvents to the processed ore. The process reagents can leach host phases and act as complexing agents for radionuclides and heavy metals. The contaminants may be mobilized from their host minerals and dissolved in tailings waters.
- *Acid leaching or reduction of iron and manganese oxyhydroxides.* Iron and manganese oxyhydroxides represent important host phases to radionuclides, metals, and metalloids, in particular arsenic. These hosts can become unstable under acid or reducing conditions (e.g. Pichler et al. 2001). The establishment of acid or reducing conditions in uranium tailings may lead to dissolution of oxyhydroxides and to the mobilization of the previously fixed radionuclides, metals, and metalloids.
- *Bacterial reduction.* Sulfate and iron reducing bacteria within tailings can favour enhanced dissolved Ra-226 concentrations. The bacteria reduce solid sulfates and iron oxyhydroxides to sulfides and allow the release of radium from sulfate salts (Martin et al. 2003). The bacteria also convert dissolved sulfate to sulfide and keep the dissolved sulfate concentrations at relatively low levels. Low sulfate levels prevent the formation of insoluble sulfates and the fixation of Ra-226 as $Ba(Ra)SO_4$. The undesired release of Ra-226 into solution can occur through engineering measures or natural processes that cause low dissolved oxygen concentrations in tailings liquids.
- *Presence of clay minerals.* Clay minerals act as sinks for barium and strontium cations in tailings dams. The cations are incorporated into the clay structure which prevents the formation of insoluble alkaline earth sulfates and the coprecipitation of Ra-226 as $Ba(Ra)SO_4$. Abundant clay minerals within tailings may greatly enhance the concentration of Ra-226 in solution.

6.7.3 Tailings Liquids

Tailings liquids represent the surface and pore waters of tailings storage facilities. The tailings liquids tend to contain high concentrations of radionuclides, heavy metals, and metalloids. If neutralization has not been applied to the liquids of an acid leach plant, the tailings liquids are typically acid, saline and enriched in heavy metals, metalloids, radionuclides, iron, manganese, aluminium, alkalis (Na, K, Ca, Mg), sulfate, chloride, fluorine, and process chemicals (Landa 1999). Acid tailings liquids are commonly in equilibrium with a range of secondary salts such as gypsum and aluminium sulfates, which invariably precipitate and redissolve in the waste repository.

At some mine sites, tailings of an acid leach uranium processing plant are partly or completely neutralized with lime prior to discharge to a tailings storage facility. The tailings become saturated with calcium sulfate. The neutralization may result in the precipitation of gypsum and of numerous elements that were in solution in the acid process water (e.g. Fe, Cu, Mn, Mg). The chemical precipitates are primarily metal hydroxides and gypsum.

In the waste repository, natural neutralization of acid tailings liquids may occur through the reaction of tailings liquids with tailings solids. Dissolution of carbonate minerals or amorphous solids and the ion exchange uptake of hydrogen by clay minerals represent possible buffering reactions (Landa 1999). If the acidic tailings liquids react with aluminosilicate minerals, then secondary clay minerals, jarosite, and alunite will be produced. These minerals will plug pore spaces and decrease tailings permeability (Sec. 2.6.4). The minerals will also act as adsorptive sinks for radionuclides and heavy metals (Landa 1999). Such neutralization reactions promote the fixation of radionuclides and metals in tailings solids and reduce the ingress of oxygen and water into the waste. On the other hand, acidic tailings solutions may react with a clay liner placed at the bottom of tailings storage facility. A breach of the clay liner may then be possible.

6.7.4 Tailings Disposal

The specific radioactivity of uranium mill tailings is low (a few Ci kg^{-1}) relative to most other radioactive wastes produced during atomic weapon and nuclear power production. However, large quantities of uranium mill tailings have accumulated and today, there are probably more than 500 Mt of uranium tailings located around the world (Waggitt 1994). Long-term containment of these low-level radioactive wastes is of great environmental concern. Radionuclides can be leached out of these wastes which may result in the contamination of surrounding ground and surface waters, soils, and sediments. Current disposal practices aim: (*a*) to isolate tailings for a long period of time from the surrounding environment; (*b*) to prevent leakage from the repository; and (*c*) to protect ground and surface waters from contamination. The disposal of long-term radioactive, fine-grained, sulfidic or even acidic mine wastes requires special attention.

In the past, uranium mill tailings were often abandoned and left unrehabilitated, or they were discharged into local creek and lake systems. Today, finite disposal options for uranium tailings include: (*a*) placing them under water in a lake, ocean or wetland environment (e.g. MEND 1993b); (*b*) backfilling them into a mined-out open pit (cf. Sec. 4.5); and (*c*) dumping them into a tailings dam. The disposal of uranium mill tailings into open pits and tailings dams are the main waste management strategies:

- *Backfilling.* In-pit disposal of radioactive wastes has numerous advantages compared to conventional tailings dam disposal including: (*a*) the tailings are as secure physically as the original ore; (*b*) there is no hydraulic head due to the position of the tailings below ground, and there is less prospect of contaminants being leached; and (*c*) a greater depth of cover is achieved ensuring radiation safety. Uranium tailings like all other tailings are not simply very fine-grained ore particles (Sec. 4.2.3). Tailings undergo diagenesis, and their mineralogy and chemistry are unlike those of the originally mined ore. In particular, tailings contain soluble salts. These chemical precipitates may dissolve once the tailings become part of the local aquifer. Moreover, the tailings contain interstitial pore waters with reactive process chemicals and dissolved metals, metalloids, and radionuclides. If the open pit is not lined with clay or other impermeable liners, tailings solids and liquids will become part of the local aquifer, and water-rock reactions may lead to the mobilization of contaminants into ground waters. Alternatively, a highly permeable layer is installed around backfilled tailings to allow free ground water circulation (Fig. 6.5). Since the permeability of the tailings is lower than that of the permeable layer, it is likely that there will be no exchange of contaminants between tailings and ground water.
- *Tailings dams.* The disposal of uranium mill tailings into tailings dams has been and will continue to be the main waste management strategy. Deposition of tailings on engineered impermeable clay layers or geotextile liners can limit the seepage into underlying aquifers. However, at many inactive tailings dam sites liners have not been used, and migration of contaminants into the ground water has occurred. Once mining ceases, uranium tailings dams require permanent dry or wet covers, and conventional dry cover designs for uranium tailings are multi-layer barriers. Dry covers are best achieved using a combination of materials including geotextile liners. Tailings are covered with compacted and/or uncompacted waste rock or clay to minimize water ingress and to reduce gamma radiation and radon emanation levels (Fig. 6.6). An impermeable cap of clay and geotextiles can inhibit rainwater inflow and radon escape. Crushed waste rock above the clay will facilitate drainage from

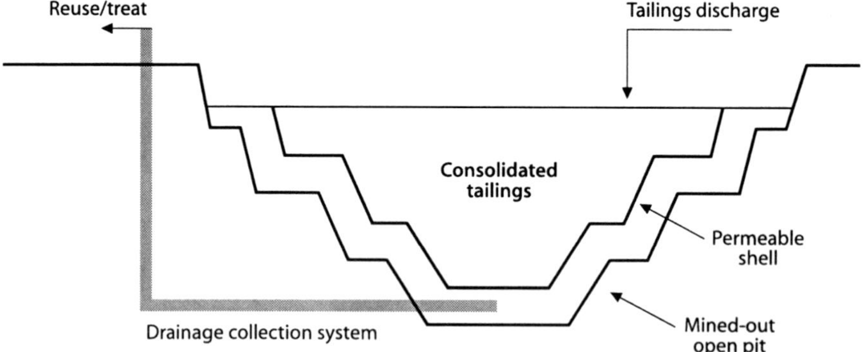

Fig. 6.5. Schematic diagram showing the in-pit disposal concept of uranium mill tailings (after Ripley et al. 1999). Consolidated tailings are discharged into a mined-out open pit that is lined with permeable, benign waste materials. Seepage waters are collected and treated or returned to the processing plant

the waste repository and reduces capillary suction forces. Topping with a suitable substrate, creating a gentle slope, and revegetation using local native species complete the rehabilitation of dry capped tailings dams (Fig. 6.7). The growth of vegetation with long roots should be avoided as this may lead to the disturbance of an intact cover and associated radon escape and radionuclide migration. The placement of a rock cover (so-called "riprap") has been used on tailings repositories in arid regions where rainfall is too low to support vegetation (Rager et al. 1996).

6.7.5 Long-term Stability of Tailings Dams

The design of an uranium tailings dam is important as the repository has to isolate the radionuclides and other contaminants and prevent them from entering the environment. Tailings dams or waste rock dumps are commonly rehabilitated using dry covers. The covers are designed: (*a*) to contain the waste; (*b*) to shed water; (*c*) to convert waste materials into landforms which comply with the final land use; and (*d*) to remain exceptionally stable over a significant period of time.

Fig. 6.6. Erosion and incision into a covered uranium mill tailings repository, Radium Hill, Australia. The tailings (30.6 mSv per year) were originally placed into a depression of crushed waste rock and then capped with soil containing calcrete and rock fragments. The soil cover reduces the radiation levels to 0.91 mSv per year as measured at the cover's surface

Fig. 6.7. Rehabilitated tailings dam at the Mary Kathleen uranium mine, Australia. The tailings dam has been rehabilitated using a multi-barrier system (clay and waste rock layers, vegetation cover). Piezometers allow sampling and monitoring of tailings liquids

The design features of uranium tailings dams are no different to those of other tailings dams. In Australia and the United States, above ground structures for the containment of tailings are required to have an engineered design life of at least 200 years and an effective structural life of up to 1 000 years (e.g. Code of Practice on the Management of Radioactive Wastes from the Mining and Milling of Radioactive Ores). However, uranium tailings contain radioactive isotopes with half-lifes (Th-230: 80 000 years; Ra-226: 1 622 years) that exceed the engineered design and structural life of tailings dams.

The long-term performance of dry covers is dependent on their resistance to wind and water erosion. In regions with high erosion rates (e.g. tropical regions with high intensity rainfalls), the long-term stability of capped uranium tailings dams is particularly important. The slope, thickness and permeability of cover materials and their resistance to erosion become crucial design features of tailings dams and waste rock dumps. Mine waste repositories are commonly designed as raised repositories and represent topographic highs in the surrounding landscape. These artificial landforms are subject to natural erosion once rehabilitation of the repositories has ceased. Erosion will eventually result in the leveling of any topographic expressions. If significantly elevated dumps are constructed in regions with high erosion rates, it has to be established whether the covers will protect the waste repositories from erosion in the long term.

Modeling, based on the universal soil loss equation or other techniques, can indicate whether a cover design or a post-mining landscape will remain viable at a mine site for an adequate number of years (East et al. 1994; Hancock et al. 2006). In addition, erosion studies for surrounding landforms can be used to estimate for how long wastes will be contained in waste repositories. Such modeling attempts are particularly important for the construction and evaluation of uranium tailings dams which are designed to isolate radioactive isotopes from the environment. The relative magnitude of the radioactivity and the long half-lifes of particular radionuclides in uranium tailings present a problem for the very long term (i.e. >1 000 years) (Landa and Gray 1995). Uranium tailings dams have to be secure and stable for over a period of several tens of thousands of years, so the long-term containment of uranium tailings in engineered structures is an important aspect of uranium mine waste management.

Despite their general acceptance and widespread use in the mining industry, the long-term stability and performance of capped repositories are unknown. While modeling and comparative erosion studies of surrounding landforms can indicate the erosive behaviour of waste repositories, only future investigations of capped waste repositories will provide the evidence for their long-term stability. It is a fact that fundamental geological processes such as erosion eventually result in the leveling of any topographic highs. If significantly elevated dumps and dams are constructed in high rainfall, erosive regimes, it is certain that the dry covers will not protect the man-made highlands from erosion in the long term.

6.8 Mine Water

Water management at uranium mine sites follows the general principles of water management at all mine sites (Sec. 3.11). These principles require the containment and disposal of contaminated water in a manner that does not impact on the environment. The release of water from uranium mine sites has to conform with statutory directives. These directives commonly state that the quality of any discharged water has to meet a specified standard which includes maximum acceptable uranium and other radionuclide levels. Water management is to ensure that water courses, aquifers, soils, sediments, irrigation crops, farm animals, and aquatic life are not contaminated by radionuclides, heavy metals, metalloids, sulfate, process chemicals, and other contaminants.

Water management is achieved through collection and disposal. Water is collected from open pit or underground workings, mills, hydrometallurgical plants, waste rock dumps, heap leach piles and ore stockpiles, and channelled into retention ponds (Fig. 6.8). Uncontaminated water can be disposed of by evaporation, land irrigation or discharge whereas contaminated water requires treatment prior to discharge.

6.8.1 Constituents

Waters of uranium mine sites have highly variable compositions, yet they are generally characterized by elevated uranium and thorium levels and their decay products (Ra-226, Rn-222, Pb-210, etc). Some radionuclides found in water such as Ra-226 also originate from natural sources, in particular from the leaching of minerals other than uranium ore minerals. Metal and metalloid concentrations (e.g. As, Be, Cr, Fe, Ni, Se) may be elevated, and oxidation of abundant iron sulfides in mine wastes can lead to AMD development and long-term low pH waters. The geochemical behaviour of radionuclides, metals and metalloids in mine waters is strongly pH and Eh dependent. Enhanced dissolution of radionuclides and heavy metals occurs in acid, oxidized waters.

Fig. 6.8. Water retention pond at the Ranger uranium mine, Australia. Waste rock dumps are present beyond the pond

Various process chemicals are used in the hydrometallurgical extraction of uranium (cf. Sec. 6.5). These include sulfuric acid, sulfates, carbonates, chlorides, calcium oxide, magnesium oxide, sodium hydroxide, potassium permanganate, copper sulfate, cyanides, manganese oxides, nitrogen compounds (e.g. ammonium, ammonia, nitric acid), and organic solvents (e.g. kerosene, alcohols). Hence, spent process waters of processing plants tend to have elevated radionuclide, metal, metalloid as well as sulfate, carbonate, chloride, calcium, magnesium, sodium, manganese and nitrogen levels (e.g. NO_3^-, NH_3, NH_4^+).

6.8.2 Treatment

At uranium mine sites, treatment of surface or ground water may be needed because the water is contaminated with radionuclides, metals, metalloids, acid and/or process chemicals. Various methods are available to remove these contaminants from the mine waters. Land irrigation, neutralization, wetlands, radium removal, pump-and-treat, and in situ bioremediation are some of the techniques available to reduce dissolved contaminant values to acceptable levels:

- *Land irrigation.* Uranium mines in tropical regions with high rainfalls must collect, contain and dispose large volumes of contaminated water in a manner which does not impact on the environment. Excess water at such uranium mining operations has been sprayed onto land application areas (Willett and Bond 1995, 1998; Brown et al. 1998). The soils of the land application areas have to possess adsorptive capacities which retain the contaminants. In particular, soils enriched in iron oxides and organic matter are capable of adsorbing significant concentrations of uranium. Other radionuclides (e.g. Ra-226, Pb-210) are also strongly adsorbed by surface soils in irrigation areas and do not reach the regional aquifers and surface waters (e.g. Willett and Bond 1995, 1998). The radionuclides will remain in the surface soils and do not leach into ground waters. However, such a disposal method has several disadvantages. The topsoil is subject to wind and water erosion, represents the substrate for plants, and may be ingested by animals (Willett and Bond 1998). Furthermore, conservative, non-reactive ions such as sulfate, nitrate, manganese, magnesium, and calcium may migrate into the local aquifer and surface waters (Brown et al. 1998). As a consequence, the application of excess water to an irrigation area will result in soil and ground water contamination and subsequent stress to and dieback of local vegetation.

- *Neutralization.* The acidity of uranium mine waters is either due to AMD development or due to the use of sulfuric acid during uranium extraction. Acid, metal-rich (e.g. Cd, Cu, Mn, Mo, Ni, Zn) mine waters can be treated using the same treatment techniques as applied to AMD waters (Sec. 3.12). The chemical treatment of acid, uranium-rich waters with carbonate and lime will reduce dissolved uranium and heavy metal concentrations and neutralize the acidity (Applegate and Kraatz 1991; Fernandes et al. 1998). Dissolved metals and radionuclides precipitate as sludges, and the sludges need to be disposed of in a waste repository which commonly is the tailings storage facility. On the other hand, chemical treatment of acid, uranium-rich waters with carbonate may promote the undesired desorption, mobilization and leaching of uranium from mine wastes. Dissolution of solid carbonate generates bicarbonate ions, and the uranyl ions are able to complex with the bicarbonate ligands. Consequently, excessive addition of carbonate to uranium mine wastes and contaminated soils can enhance uranium mobility and exacerbate existing contamination problems (Elles and Lee 1998; Wanty et al. 1999) (cf. Sec. 3.12.1).
- *Wetlands.* Wetlands are successfully used to treat waters with elevated uranium and radium concentrations. Wetlands have abundant clays, organic matter, algae, bacteria, fungi, and lichens which are effective in removing dissolved contaminant concentrations (Noller et al. 1997; Haas et al. 1998). Much of the uranium is adsorpted onto the organic-rich wetland sediments, or the reducing conditions convert the mobile U^{6+} to the immobile U^{4+} species and cause the precipitation of uranium in the wetland substrate.
- *Radium removal.* The removal of radium from water is accomplished by precipitation with barium chloride. Barium chloride is introduced into the water circuit, and the radium will coprecipitate with barium sulfate as illustrated by the following reaction (Benes et al. 1983):

$$BaCl_{2(s)} + Ra^{2+}_{(aq)} + SO^{2-}_{4(aq)} \rightarrow 2\,Cl^{-}_{(aq)} + (Ba,Ra)SO_{4(s)} \qquad (6.10)$$

Any calcium present will precipitate as calcium sulfate crystals. Settling and filtration of the precipitated salts follow in constructed ponds. Ponds should be lined to prevent contamination of ground waters with radionuclides. Evaporation of waste waters in ponds lead to the precipitation of salts and sediments. The radioactive salts and evaporation pond sediments cannot be left in the treatment ponds otherwise they redissolve. The salts and sediments are generally placed with the tailings into a tailings storage facility.

Contaminated ground water may be treated using ex situ (e.g. pump-and-treat) or in situ treatment techniques (e.g. bioremediation, permeable reactive barriers):

- *Pump-and-treat.* The pump-and-treat method aims to capture the impacted ground water plume. A number of extraction wells pump the contaminated ground water to the surface. This technique relies on the extraction of uranium and other contaminants with the water and on fast desorption if the contaminants are adsorpted onto the host rock in the polluted aquifer (Abdelouas et al. 1998a,b). Following extraction of the contaminated ground water, the water is treated using separation processes such as ion exchange, reverse osmosis, biosorption, bioreduction, bioaccumulation, and reductive precipitation of uranium (Abdelouas et al. 1999).
- *In situ bioremediation.* In situ bioremediation is based on the injection of nutrients and/or specific bacteria into the aquifer (e.g. *Desulfovibrio, Geobacter metallireducens, Shewanella putrefaciens, Colstridium* sp.). The bacteria precipitate the dissolved uranyl complexes as insoluble uranium oxide through the following processes: (*a*) incorporation of uranium into cell structures; (*b*) adsorption of dissolved uranium from solution; and (*c*) direct or indirect reduction of the mobile U^{6+} to the immobile U^{4+} species which precipitates as insoluble uraninite (Suzuki and Banfield 1999; Abdelouas et al. 1998b, 1999, 2000). This latter property of microorganisms – to be able to reduce U^{6+} to the immobile U^{4+} – is used in the remediation of uranium contaminated ground waters. Direct reduction of the mobile U^{6+} to the immobile U^{4+} species can be performed by particular microorganisms. Indirect reduction of U^{4+} may be caused through the proliferation of sulfate-reducing bacteria, which in turn lead to reducing conditions in the uraniferous ground waters. Both direct and indirect reduction result in the precipitation of dissolved U^{6+} as insoluble U^{4+} oxide.
- *In situ permeable reactive barriers and reactive zones.* In situ permeable reactive barriers (PRB) and reactive zones (RZ) are based on the placement of solid reactive materials into the contaminated ground water. The application of permeable reactive barriers requires excavation of a trench, placement of the reactive materials into the trench, and backfilling; hence, the technique is limited to shallow aquifers. By contrast, the reactive zone technology uses injection points perpendicular to the ground water flow. Reducing materials like metallic iron can be employed to retain uranium and radium on solid particles (Burghardt and Kassahun 2005).

6.9 Monitoring

Waste rock dumps, tailings dams, and heap leach piles are potential sources of radionuclides, acid, salts, heavy metals, and metalloids. These sources need mineralogical, geochemical and radiochemical characterization. Uranium concentrations in solid mine wastes can be determined using a range of analytical methods whereas uranium mineral identification is primarily achieved through X-ray diffraction (Wolf 1999; Hill 1999). Radionuclide concentrations are established using alpha and gamma-ray spectrometry whereas radon gas concentrations are investigated using radon emanometry (Nielson et al. 1991; Sharma 1997).

Kinetic tests are commonly applied to simulate the weathering and oxidation of sulfidic waste samples. The tests expose sulfidic waste over time, from several months to several years, to moisture and air (Sec. 2.7.4.2). The field or laboratory based experiments are also used to investigate the leaching behaviour of uranium mine wastes. The uranium mine wastes are placed into field bins or columns and subjected to periodic or continual leaching in order to simulate the addition of ground or surface waters. The wastes require mineralogical, geochemical and radiochemical characterization prior to and after leaching. The leachates are monitored for water quality parameters (e.g. pH and EC) and analyzed for their chemical and radiochemical composition. Data from such experiments can be used to evaluate acid production and the migratory behaviour of radionuclides, heavy metals, and metalloids over time.

Environmental monitoring of uranium mine sites include periodic chemical and radiochemical measurements of ground and surface waters, stream sediments and soils. In addition, aquatic biota (e.g. fish) are investigated for their radiochemical composition to detect uptake of radionuclides by aquatic organisms. Mosses and lichens are often used to monitor air contaminants as these rootless organisms reflect atmospheric fall-out rather than substrate chemistry (Ripley et al. 1996). Atmospheric measurements of radon gas and dust particles are an integrated part of the monitoring program. In particular, geochemical analyses of waters and stream sediments are very useful for the identification, monitoring, control, and evaluation of environmental impacts in the aquatic system (Noller and Hart 1993). The migratory behaviour of dissolved uranium and other from contamination sources into aquifers can be modelled using computational software, including mass transport models (e.g. MODFLOW) and geochemical speciation programs (e.g. HARPHRQ).

The environmental risks of uranium mine wastes will last until the radioactivity of the mine wastes has decreased to acceptable levels. Therefore, operating and rehabilitated uranium mine sites require monitoring and periodic environmental reviews well beyond mine closure and for at least 1 000 years (i.e. the engineered life-span of tailings repositories).

6.10 Radiation Hazards

A low level of radioactivity is part of the natural environment (Eisenbud and Gesell 1997; Ewing 1999; Gaines 2000). Naturally occurring background levels of radiation originate from cosmic radiation, the radioactive decay products of radon in the air, and the natural radioactivity of the ground. Much of the radioactivity of rocks and minerals is due to the decay of uranium (i.e. U-238, U-234), thorium (i.e. Th-232), their radioactive decay products as well as the decay of the potassium isotope K-40. The abundance of uranium, thorium and potassium varies in different rock types and minerals (Sharma 1997). Hence, the natural radiation dose varies depending on the geological ground above which and altitude at which people live. Radioactivity is also present in water, air, plants, food, and even internally within the human body (Kathren 1998; Ewing 1999). Radiation is unavoidably received by humans and for the great majority of the population, exposure to natural radiation exceeds by far that from artificial, medical and occupational sources.

6.10.1 Radiation Dose and Human Health

Radioactive nuclides of the U-238, U-235 and Th-232 series emit ionizing radiation in the form of alpha and beta particles as well as gamma-rays. Animals and humans get exposed to this radiation through: (*a*) external radiation emitters; and (*b*) ingestion or inhalation of radionuclides. Very high concentrations of radionuclides may be acquired through the foodchain whereby terrestrial or aquatic plants obtain radionuclides from contaminated air, water and soil. The radionuclides get into organisms using the transport mechanisms of nutrient ions, so the radionuclides become concentrated in the foodchain. For example, Ra-226 follows the pathways of calcium. The radionuclide accumulates in calcium sinks such as bones and continues to emit ionizing radiation.

It is established that work with radioactive materials carries with it an increased risk of tissue damage including radiation induced cancers, leukemia, and genetic mutations. Radiation protection standards assume that any dose of radiation – no matter how small – involves the possibility of risk to human health. Ionizing radiation can be particularly harmful if a large enough dose is received (Table 6.5). While small radiation doses such as medical X-rays are not capable of causing radiation sickness or death, they slightly increase the risk of cancer several years after the exposure has occurred. As more is known about the effects of ionizing radiation, the dose levels considered to be safe have been revised. Radiation exposures for the general public and in the workplace have been lowered over the years, and the occupational maximum permissible radiation exposure level has dropped to about one tenth of that considered safe in the early 1950s.

The health effects of radon exposure are still debatable, and the effects of low level radiation dose on cells are still to be determined. While exposure to excessive radiation is a health hazard and may even lead to death, a small radiation dose may be beneficial for the cure of various ailments. There is a practice of radon consumption and inhalation by the general public in certain parts of the world. In Germany, Austria and Poland, mineral waters and underground mines with known high levels of radon are successfully used for the treatment of various ailments (Kathren 1998). The consumption of radon bearing mineral waters and the inhalation of radon-rich air appear to have a beneficial effect. Will decommissioned uranium mine sites and other non-uranium mines with high radon levels have a future as sanatoriums and health spas?

The radiation dose to members of the public from operating uranium mines is in most cases negligible as access to the mine sites is controlled and occupation of the mining areas by the general public is zero. In populated areas, however, environmental dust and radon monitoring programs have to be implemented around uranium mine sites. Direct radiation and ingestion as well as inhalation pathways need to be evaluated as well. The potential for significant radiation exposure especially exists for mine staff.

6.10.2 Occupational Radiation Exposure

Much of the occupational radiation exposure at uranium mine sites does not originate from the decay of uranium isotopes but from the decay products of U-238 and

Table 6.5. Comparative radiation doses and likely effects of radiation doses to the human body (Ripley et al. 1996; Uranium Information Centre 2001b,c; Major 2001)

Radiation dose		Effects of radiation dose
10 000	mSv in a short term dose	Immediate illness and subsequent death
1 000	mSv in a short term dose	Temporary illness; likelyhood of cancer
750	mSv per year	Mean annual radiation dose to East German miners from 1946 to 1954, resulting in thousands of cases of lung cancer (cf. Case Study 6.1)
350	mSv in lifetime	Criterion for relocating people after the 1986 Chernobyl accident
100	mSv per year	No evidence for increased cancer risk or immediate effects below this level
20	mSv per year av. over 5 years	Occupational maximum permissible exposure level for nuclear industry employees and uranium miners (International Commission on Radiological Protection)
5.2	mSv per year	Average dose to employees at the Ranger uranium mine, Australia
4–5	mSv per year	Average dose to Australian uranium miners
5	mSv	Abdominal X-ray
3	mSv per year	Typical background radiation to North American public
2.9	mSv per year	Average occupational dose to United States nuclear industry employees
2.7	mSv per year	Average dose to underground employees of the Olympic Dam uranium mine, South Australia
2	mSv per year	Typical background radiation to Australian public
1.1	mSv per year	Average dose to Canadian uranium refinery workers
1	mSv per year	Maximum permissible exposure level for general public above that received from background and medical exposure (International Commission on Radiological Protection)

U-235 (i.e. Ra-226, Rn-222). While uranium isotopes are radioactive, the isotopes U-238 and U-235 have very long half-lifes and uranium is not strongly radioactive. On the other hand, both uranium isotopes emit alpha particles. The release of alpha particles is a potential hazard if the uranium occurs in fine-grained material. The finer grain size results in larger surface areas and greater release of alpha particles from the material. Alpha particles cannot penetrate the skin and are only a potential hazard if they are inside the human body. Consequently, the handling of solid uranium concentrate at mine sites is not so much a radioactive hazard but a potential inhalation or ingestion hazard. Uranium itself is a suspected human carcinogen and has a chemical toxicity (Wanty et al. 1999). The inhalation or ingestion of uranium can be prevented using appropriate ventilation, protective clothing, and strict hygiene standards.

The prime health hazard – during uranium mining and after mine closure – relates to ionizing radiation and the carcinogenic properties of radionuclides (e.g. Ra-226) and radon (Rn-222) gas. Employees of uranium mines can be exposed to radiation and radon gas via four principal exposure pathways (Table 6.6):

- *Direct external radiation from uranium ores, tailings and waste rocks.* Employees will be exposed to external radiation in the form of gamma-rays as well as alpha and beta particles emitted by isotopes of the U-238, U-235 and Th-232 series.
- *Ingestion of contaminated water.* Leaching of wastes can carry radioactive nuclides, metals, and metalloids to surface and ground waters. Thus, ground and surface water within or near the mine site can contain elevated levels of dissolved radioactive isotopes and should not be used as drinking water.
- *Inhalation of radon.* Radon-222 gas emanates from ores, wastes, and tailings and may be inhaled by employees. Airborne radionuclides, particularly Rn-222, may impact on the health of humans, because Rn-222 and its radioactive decay products deliver an internal radiation dose to the lungs.
- *Inhalation and ingestion of dust containing radioactive isotopes.* At uranium mine sites, atmospheric emissions are not restricted to radon gas. All mining, tipping, handling, transport, crushing and milling operations produce dust. In addition, dry uncovered tailings can be a source of radioactive dust particles. The ore or tailings dust contains radioactive isotopes of the U-238, U-235 and Th-232 series, and the dust may be inhaled and ingested by employees.

Uranium mine wastes emit very low to elevated levels of radiation and variable levels of radon gas. Today, low level radiation doses for employees can be achieved through (Table 6.6):

- *Installation of covers.* The release of Rn-222 to the atmosphere from tailings is major pathway of radon exposure at uranium mine sites. A reduction in radon emanation from tailings repositories is generally achieved by installing cover materials. At operating uranium mines, tailings are commonly covered with water which reduces radiation levels. Where tailings are collected and kept wet under permanent water cover, the radon flux is reduced to only 1% of the radon flux from dry tailings (Davy and Levins 1984). Wet or dry covers on tailings and other wastes need to be of at least 30 cm thickness in order to reduce radiation levels to background levels.
- *Dust suppression.* The inhalation and ingestion of radionuclides contained in dust need to be minimized as radionuclides emit gamma-rays and alpha particles. Dust suppression can be achieved by wetting and covering the potential dust sources.
- *Appropriate ventilation.* Uranium mine sites are always monitored for levels of radon gas that is released during the natural radioactive decay of uranium. Ventilation generally reduces radon gas levels, however, in some mines the radiation and radon levels are so high that mining must be done with remote control equipment and robots. Natural ventilation in an open cut mine removes radon gas whereas an underground uranium mine requires an engineered ventilation system. Forced ventilation reduces radon and radon daughter exposure.
- *Radiation dose monitoring.* Personal radiation doses are monitored to limit the radiation exposure of individual employees.
- *Strict hygiene standards, ventilation, and protective clothing.* Employees handling uranium concentrate have to adhere to strict hygiene standards and have to work in protective clothing in appropriately vented environments. This will minimize the ingestion, adsorption and inhalation of uranium.

Table 6.6. Potential occupational hazards at uranium mining and processing sites and their management (after Davy and Levins 1994; OECD 1999)

Hazards	Sources	Possible contaminants	Management
External radiation	Mine workings, ore stockpiles, tailings repository, waste rock dumps, processing plant	Gamma-rays, alpha and beta particles	Monitoring of personal radiation doses; engineered ventilation system
Ingestion of surface or ground water	Water within or near the mine site	Dissolved radionuclides	Inappropriate drinking water supply
Inhalation of radon	Mine workings, ore stockpiles, waste rock dumps, processing plant	Radon-222	Monitoring of personal radiation doses; engineered ventilation system
Inhalation of radon	Tailings repository	Radon-222	Wet cover during operation; dry or wet cover once disposal ceases
Inhalation of dust containing radioactive isotopes	Working areas with mining, blasting, crushing, and screening; wind erosion of ore and waste rock piles and tailings	Radionuclides, silica dust	Forced ventilation in underground mines; wetting of working areas; wind resistant covers for ore and waste piles; collection of dust using filters; dust suppression using wetting and covering of potential dust sources
Inhalation, skin adsorption or ingestion of uranium	Packaging of uranium concentrate	Uranium	Ventilation; air hoods; filters; protective clothing; hygiene

The above protection measures ensure that radiation doses are below recommended dose limits and are as low as reasonably achievable. In the past, little attention was paid towards proper ventilation of underground mines, dust suppression, and the radiation exposure of mine workers. An example for this is the Radium Hill uranium mine in South Australia. The mine operated from 1952 to 1961 (Fig. 6.6). The frequency of lung cancer deaths in the former Radium Hill workforce is associated with duration of work underground and cumulative exposure to radon (Woodward and Mylvaganam 1993). In contrast, modern underground uranium mines are far ore strictly controlled and far better ventilated. Furthermore, in industrialized nations, health standards have been set for the exposure to gamma radiation and radon gas.

6.11 Environmental Impacts

Many uranium ores were exposed to the surface prior to mining. At these sites, there was natural release of radioactivity into the environment as well as natural dispersion of uranium and other elements into soils, plants and creeks prior to mining. Plants may adapt to uranium-rich soils. In fact, plants growing on naturally radioactive places are known to absorb so much Ra-228 that the plants produce an image when placed on photographic film. Also, ground waters in contact with uranium ores are often naturally radioactive and too saline for human and animal consumption. Thus, natural dispersion of radionuclides, metals and metalloids is known to occur from uranium ores into surrounding soils, waters, plants, and stream sediments.

Mining, processing and metallurgical extraction of uranium ore may impact on air, soil, sediment as well as surface and ground water unless suitable waste disposal and rehabilitation strategies are adopted (Table 6.7, Case Study 6.2). The following environmental problems are encountered with uranium mine wastes:

- *Excessive radioactivity levels and radon emissions.* The escape of radon and radionuclides from waste repositories may be possible due to non-existent or inadequate covers. Uncovered tailings are exposed to erosion, and windblown radioactive dust and erosive products are dispersed from the storage facility into the surrounding soils, sediments, and waters. Intrusion of plant roots and burrowing animals through cover materials into the wastes can provide pathways for infiltrating surface water and cause the upward migration of radionuclides and the release of radon.
- *Inappropriate use of tailings and waste rocks.* At some historic uranium mine sites, tailings and crushed wastes have been used for landscaping purposes, for the production of gravel and cement, and for the construction of houses, roads and railway beds. As a consequence, wastes with elevated levels of uranium and radioactivity have been dispersed over large areas and along transport corridors, and homes constructed with uranium mine wastes exhibit high radon and gamma-ray radiation levels.
- *Failure of tailings dams.* Rehabilitated or operating tailings dams may fail to contain the tailings and release them to the environment through: liquefaction; rapid increase in dam wall height; foundation failure; excessive water levels; or excessive seepage (cf. Sec. 4.3.3). There have been several uranium tailings dam leakages, discharges and failures since the 1950s. The largest single release of uranium tailings ever occurred in the region of the upper Puerco River of New Mexico, United States (Landa and Gray 1995). In this area, uranium mines were dewatered, and the effluent was released to the local river for over 22 years. In 1979, a tailings dam pond failed, resulting in the release of uranium tailings. An estimated 360 000 m^3 of tailings liquid and 1 000 t of tailings solids were discharged. The tailings liquid had a pH value of 1.6, and the total gross-alpha activity was estimated to be 130 000 pCi l^{-1} which is approximately 10 000 times higher than drinking water quality guidelines. Twenty years after the spillage, the tailings pond spill and the uncontrolled release of mine waters were still discernible as indicated by uranium contamination of the Puerco River as far as 70 km downstream from the mines (Landa and Gray 1995).
- *Soil and sediment contamination.* Failure of tailings dams, unconstrained erosion of waste repositories, and improper disposal of contaminated water from mining, mineral processing, and metallurgical operations may release contaminants into local environments (Case Study 6.3). If mine waters are released into local stream systems, the environmental impact will depend on the quality of the released effluent. The release may cause soils, floodplain sediments, and stream sediments to be become contaminated with radionuclides, metals, metalloids, and salts (Pinto et al. 2004). The metals and metalloids may be contained in various sediment fractions, including the

Table 6.7. Potential environmental impacts at uranium mining and processing sites and their treatment and management (after Davy and Levins 1994; OECD 1999)

Environmental impacts	Sources	Possible contaminants	Treatment and management
Excessive radioactivity levels	Waste rock dumps, tailings repository	Radon, radionuclides	Construction of engineered waste repositories to reduce radon release and radioactivity levels
Erosion of tailings cover materials and wastes	Waste rock dumps, tailings dams	Radionuclides, metals, metalloids, acid, process chemicals	Wet cover during operation; dry or wet cover once disposal ceases
Failure of tailings dams	Tailings dams	Radionuclides, metals, metalloids, acid, process chemicals	Construction of waste repositories for the long term
Ground and surface water contamination (drainages and seepages)	Ore stockpiles, waste rock dumps, tailings repository, processing plant	Radionuclides, metals, metalloids, acid, process chemicals	Collection of seepage and run-off waters for treatment or reuse in mineral processing; construction of waste repositories to limit water infiltration: (*a*) clay or synthetic liners to seal the floor of the waste storage facility; and (*b*) clay to cap tailings storage facility
Ground and surface water contamination (process water)	Processing plant	Radionuclides, metals, metalloids, acid, process chemicals	Neutralization of acidity and precipitation of heavy metals by adding lime or limestone; precipitation of Ra-226 by adding barium chloride; recycling to reduce volume
Ground and surface water contamination (AMD)	Sulfidic ore stockpiles, waste rock dumps, tailings dams	Acidity, metals, metalloids, radionuclides	Clay or synthetic liners to seal the floor of the waste storage facility; wet or dry cover on waste storage facility; collection of AMD water for treatment
Atmospheric emissions	Sulfuric acid production plant	Sulfur oxides, sulfuric acid mist	Demister, scubbers, tall stacks

adsorptive, iron and manganese hydroxide, carbonate, organic, and residual silicate fractions.

- *Ground and surface water contamination.* In the past, untreated mine waters have been released accidentally or deliberately into local rivers and lakes. In addition, unconstrained seepage and run-off waters have mobilized and transported contaminants from waste rock dumps, tailings dams or ore stockpiles into local creeks and aquifers. Thus, ground and surface water contamination may originate from tailings dams, waste rock piles, evaporation ponds, and contaminated soils (Pinto et al. 2004). In particular, water may leak from tailings impoundments into underlying aquifers if the waste repositories are uncapped, unlined and permeable at their base (Zielinski et al. 1997). At such sites, significant concentrations of uranium and other contaminants have been found in ground water plumes migrating from the uranium tailings impoundments. If not rectified, the plumes of contaminated water will migrate over time downgradient, spreading beyond the waste repositories, surfacing at seepage points, and contaminating surface waters. The migration rate of plumes is highly

Case Study 6.2. Environmental Review of the Rehabilitated Mary Kathleen Uranium Mine, Australia

Mining and Rehabilitation at Mary Kathleen

The Mary Kathleen uranium deposit was discovered in 1954, mined in 1956–1963 and 1976–1982, and subsequently rehabilitated (Flanagan et al. 1983; Ward et al. 1983, 1984; Ward and Cox 1985). The aim of rehabilitation was to leave the site in a safe and satisfactory condition consistent with future land use in the area (i.e. cattle grazing), requiring no foreseeable on-going maintenance and a minimum of precautionary monitoring (Ward et al. 1983, 1984; Ward and Cox 1985). All areas were to be made safe for public access, and radiation and radon daughter levels were to be within acceptable limits. Rehabilitation of the site was completed in 1985 at a cost of some A$19 million. The work won an award for environmental excellence from the Institute of Engineers Australia in 1986. In 1999 and 2000, gamma-ray data, plus stream sediment, soil, rock chip, mineral efflorescence, vegetation and water samples were collected from selected sites to assist in the examination of the current environmental status of the rehabilitated area (Lottermoser and Ashley 2005a; Lottermoser et al. 2005).

Open Pit

Elevated gamma-ray readings in the open pit correspond to exposed ore lenses, the former haul road, and abandoned ore stockpiles (up to 16 mSv per year). Surficial oxidation of ore and adjacent sulfide-bearing rocks has led to contemporary precipitation of mineral efflorescences on the pit walls. Wallrock oxidation of reactive sulfides (mainly pyrrhotite breakdown) produces acidic solutions, yet buffering reactions of these fluids with gangue calc-silicates and carbonate phases prevent low pH conditions from developing. The open pit lake contains saline surface waters at a pH of 6.11.

Waste Rock Piles

Waste rock piles are up to 30 m thick and have been covered by a thin veneer of benign waste. However, there are high radiation levels on several waste rock piles (up to 20 mSv per year) which were only partly covered or were ripped for seeding after the thin cover was put into place. Biogeochemical analyses indicate that *Enneapogon lindleyanus* (grass), *Cymbopogon bombycinus* (grass), *Aerva javanica* (kapok bush), *Aristida longicollis* (poaceae) and *Acacia chisholmii* (wattle) accumulate certain elements (As, Ce, Cu, La, Ni, Pb, Th, U, Y, Zn) at mined and disturbed areas compared to background sites. Stream sediments – accumulating below waste rock piles – are enriched in copper, zinc, arsenic, nickel, lanthanum, cerium, and uranium, indicating active weathering and erosion of waste materials into the local drainage system.

Tailings Dam

The tailings disposal system consisted of a tailings dam (0.28 km^2) and a series of evaporation ponds (0.6 km^2). The tailings dam contains 7.1 Mt of un-neutralized tailings. The dam was formed by constructing a rolled earth fill and metasedimentary waste rock embankment across a narrow section of a small valley. It was constructed without a basal liner. In 1982, kinetic testing of the tailings indicated that the leachate from the tailings dam would contain low radionuclide, heavy metal and acidity levels. Studies concluded that the composition of leachates from the tailings dam would largely be controlled by gypsum dissolution. Sulfate contamination of the ground water system would be the main environmental impact (Flanagan et al. 1983). The former tailings dam was subsequently rehabilitated using a multi-barrier system (i.e. clay and waste rock layers).

variable and dependent on the physical and chemical characteristics of the aquifer or waste material. The contaminated ground water may not only contain elevated uranium but also high sulfate, nitrate, heavy metal and metalloid concentrations as well as elevated salinities and TDS values (Lawrence et al. 1997; Abdelouas et al. 1998a).

Case Study 6.2. ***Continued***

Today, gamma-ray measurements demonstrate an intact tailings dam cover. Nevertheless, seepage of acid (pH 5.71), saline (0.31%) waters occurs from the toe of the tailings dam into the evaporation ponds and local drainage system. The sulfate concentration of the seepage water ($4.2\ g\,l^{-1}$) is well below that of the discharge water from the former mill ($21\ g\,l^{-1}$) and that of the tailings liquid before rehabilitation ($24\ g\,l^{-1}$). Hence, acid-producing reactions are sufficiently buffered within the tailings storage area, yet there is dissolution of solid phases within the repository. Seepage waters are calcium-, sulfate-rich with elevated iron, manganese, nickel, uranium and zinc values. The seepage waters precipitate abundant sulfate efflorescences and iron oxyhydroxide ochres. These precipitates exhibit elevated radiation and major (i.e. >1 wt.%) concentrations of Fe, S and Ca, minor contents (i.e. >1 000 ppm) of Ce, La, K, Mg, Mn, and Na, elevated (i.e. >100 ppm) contents of U, and traces (i.e. <100 ppm) of As, Ba, Cu, Ni, P, Pb, Sr, an Th.

Drainage

Seepage of saline waters occurs from the tailings dam and evaporation ponds into the local creek via surface and subsurface flows as indicated by salt-encrusted creek banks. This stream and its tributaries are usually active only in the wet season. At other times, few permanent water holes exist although there is some sustenance of flow from the tailings dam seepage. When sampled during the dry season, pools in the creek system are shallow, saline (up to 3%), alkaline (up to pH 8.6), and strongly enriched in sulfate (up to $26\ g\,l^{-1}$) and uranium (up to $5\ mg\,l^{-1}$). The pools locally sustain fish and reeds in lower salinity regimes.

Conclusions

The environmental assessment of the rehabilitated Mary Kathleen uranium mine has revealed the following features:

1. Radiation levels are below acceptable limits.
2. The tailings dam structure and its cover are intact.
3. No exposure pathway to radiation has been identified that could pose a threat to human health or other life forms.
4. Uncovered or partly covered, steeply sided, thick waste rock dumps continue to shed dissolved and particulate contaminants into local soils, sediments, and waters.
5. In 1984, at the time of mine closure and mine site rehabilitation, scientists of the Australian Nuclear Science and Technology Organisation predicted that AMD and radionuclide mobility from the tailings dam would not be a problem and that the tailings had a high capacity to neutralize acid. This investigation has confirmed that AMD generation does not occur. However, there is dissolution of solid phases within the tailings dam, and radionuclides are mobilized from the waste repository into surface water seepages.
6. Similarly at the time of closure, it was predicted that uranium and heavy metals would be adsorbed within the tailings. They would not be leached from the tailings dam into the environment. However, metal (Fe, Mn, Pb, Th, U, Y), metalloid (As), alkali (Ca, Mg, Na, K, Sr, Ba), rare earth element (La, Ce), radionuclide and sulfate mobilization occurs from the tailings storage area into surface water seepages.
7. Saline subsurface and surface waters flow from the tailings dam and the evaporation ponds into the local creek system causing its salinization during the dry season. The saline water is accessible to cattle.

Natural attenuation processes in aquifers can reduce the constituent concentrations to background levels in the pathway of the subsurface drainage (Erskine et al. 1997; Schramke et al. 1997) (cf. Sec. 3.12.6). For example, reducing environments in the ground water will limit the migration of redox sensitive elements (e.g. U, Se, Cr). Also, neutralizing minerals including carbonates may be contained in the aquifers, and these minerals buffer acid ground waters (Fig. 3.9). The neutralizing minerals will

Case Study 6.3. Environmental Review of the Rehabilitated Radium Hill Uranium Mine, South Australia

Introduction

The evaluation of rehabilitated mine sites is a challenging task because most rehabilitated mine sites represent very young landforms (i.e. less than 20 years old), which have not achieved equilibrium with the surrounding environment. Leaching processes, AMD development and physical erosion are all kinetic processes; processes which require time. Therefore, rehabilitated mine sites should not be evaluated immediately post mining and post remediation.

Recent research on rehabilitated uranium mine sites located in wet climates has revealed the varied success of the applied rehabilitation efforts (e.g. Menzies and Mulligan 2000; Peacey et al. 2002; Ritchie and Bennett 2003). In wet and seasonally wet climates, AMD development and the leaching of waste repositories are dominant pathways of contaminants into surrounding environments. In comparison, there is little knowledge of the status and environmental impacts of rehabilitated uranium mines in dry climates. To date, the few studies conducted on rehabilitated uranium mine sites in semi-arid and arid climates have largely focussed on uranium mill tailings repositories and related ground and surface water investigations.

The Radium Hill Mine

The Radium Hill uranium deposit, in semi-arid eastern South Australia, was discovered in 1906 and mined for radium between 1906 and 1931 and for uranium between 1954 and 1961 (production of 969300 t of davidite ore averaging 0.12% U_3O_8). In the 1980s, rehabilitation was limited to removal of mine facilities, sealing of underground workings and capping of selected waste repositories. In 2002, gamma-ray data and samples of tailings, uncrushed and crushed waste rock, stream sediment, topsoil and vegetation were collected to assist in examination of the current environmental status of the mine site (Lottermoser and Ashley 2005b).

Physical Dispersion

The data indicate that capping of tailings storage facilities did not ensure the long-term containment of the low-level radioactive wastes due to the erosion of sides of the impoundments. Moreover, wind erosion of waste fines (phyllosilicates, ore minerals) from various, physically unstable waste repositories has caused increasing radiochemical (from a background dose of 35–70 $nSv\,hr^{-1}$ to max. 940 $nSv\,hr^{-1}$) and geochemical (Ce, Cr, La, Lu, Rb, Sc, Th, U, V, Y, Yb) impacts on local soils. Plants (saltbush, pepper tree) growing on waste dumps display evidence of biological uptake of lithophile elements, with values being up to 1–2 orders of magnitude above values for plants of the same species at background sites. However, radiation doses associated with the mine and processing site average 670 $nSv\,hr^{-1}$; hence, visitors to the Radium Hill site will not be exposed to excessive radiation levels. Although rehabilitation procedures have been partly successful in reducing dispersion of U and related elements into the surrounding environment, it is apparent that 20 years after rehabilitation, there is significant physical and limited chemical mobility, including transfer into plants. Additional capping and landform design of the crushed waste and tailings repositories are required in order to minimize erosion and impacts on surrounding soils and sediments.

be eventually consumed, and the acid ground water plume will migrate further downgradient. Such contaminated ground water requires treatment (cf. Sec. 6.8.2).

- *Acid mine drainage.* Uranium ores which contain abundant iron-rich sulfides may oxidize upon exposure to the atmosphere, and AMD may develop in mine workings or mining, processing and metallurgical wastes. The problems of AMD and elevated levels of dissolved radionuclides are linked. Low pH, oxygenated waters dissolve uranium ore minerals and sulfides, and radionuclides and metals are mobilized into ground and surface waters.

6.12 Summary

Radioactivity is the issue which sets uranium mine wastes apart from other mine wastes. Uranium commonly occurs in the form of uraninite, brannerite, coffinite, and pitchblende. Much of the radioactivity of uranium ores and wastes is due to the decay of the uranium isotopes U-238 and U-235 and their radioactive decay products. These radionuclides serve as a long-term source of ionizing radiation. Radium-226 and Rn-222 are intermediate daughter nuclides of the U-238 series. They are of most concern in uranium mining, processing, metallurgical extraction, and waste management. Radium-226 is a radionuclide that has: (*a*) a significant half-life of 1 622 years; (*b*) a geochemical behaviour similar to calcium; (*c*) a high radiotoxicity; (*d*) an affinity for accumulating in bones; and (*e*) a high solubility in ground and surface waters. Radium-226 is the source of Rn-222. Radon-222 is a radionuclide that: (*a*) has a half-life of 3.8 days; (*b*) is soluble in water; (*c*) is highly mobile due to its gaseous state and inert chemical properties; and (*d*) decays to radioactive solid daughter products that can be deposited in human lungs.

Mineral processing of uranium ores relies on acid or alkaline leaching. Conventional uranium extraction is based on the leaching of the ore in hydrometallurgical plants or heap leach piles. In situ leach techniques leave the ore in the ground and inject the process chemicals via drill holes into the ore horizon. The process chemicals produce a solution from which uranium is obtained.

Mining, mineral processing and metallurgical extraction of uranium ores result in the production of tailings, mine water, waste rocks, and heap leach residues. Most wastes of uranium mining and processing operations can be classified in terms of radioactivity as non-radioactive or low-level radioactive waste.

Conventional extraction of uranium in a hydrometallurgical plant produces uranium mill tailings. Once the uranium (U-238, U-235, U-234) is recovered from the ore, most of the radioactive daughter nuclides end up in uranium mill tailings. Thus, uranium tailings have special containment requirements because of their radioactivity. Tailings carry about 85% of the radioactivity originally present in the uranium ore.

Uranium tailings contain radionuclides, heavy metals and metalloids as: (*a*) dissolved species; and (*b*) insoluble and potentially soluble solid forms. The potential release of radionuclides, heavy metals, and metalloids from tailings solids and the presence of radionuclides in solution are undesirable. Pore waters laden with contaminants may migrate into aquifers or surface at seepage points. The mobilization of radionuclides, heavy metals and metalloids from tailings solids into the tailings liquids may occur after tailings deposition. Such mobilization can be induced through several factors: (*a*) AMD development; (*b*) presence of process chemicals; (*c*) acid leaching or reduction of iron and manganese oxyhydroxides; (*d*) bacterial reduction; and (*e*) presence of clay minerals. Acid leached tailings tend to contain high concentrations of soluble and dissolved radionuclides, heavy metals and sulfate unless the tailings have been neutralized prior to discharge to the tailings storage facility.

Uranium tailings need to be covered during operation in order to reduce Rn-222 emanation. Finite disposal options for uranium tailings include: (*a*) placing them under water in a lake, ocean or wetland environment; (*b*) backfilling them into a mined-out open pit; and more commonly (*c*) dumping them into a tailings dam. Once mining ceases, uranium tailings repositories require covers, and conventional cover designs are multi-layer barriers. Uranium tailings contain radioactive isotopes with half-lifes exceeding the engineered design and structural life of tailings dams. Therefore, the long-term stability of uranium tailings dams is important in regions with high erosion rates.

Mine water at uranium mine sites is invariably radioactive arising from dissolved U-238, U-235, Th-230, Ra-226 and Pb-210 ions. The water may also contain various heavy metals and metalloids, and some of the heavy metal load can be the result of sulfide oxidation and resulting AMD. Dissolved uranium in oxidized mine water is mainly in the form of the uranyl ion (UO_2^{2+}) which forms complexes with carbonate, sulfate, and other ligands. Uranium can be effectively leached from mine wastes under acid, oxidizing weathering conditions. The uranium will remain in solution until: (*a*) the pH is raised to higher levels; (*b*) coprecipitation or adsorption reactions occur; or (*c*) reducing environments are met. Surface and ground water can be contaminated with radionuclides, metals, metalloids, acid, and process chemicals. Treatment techniques of such contaminated waters include land irrigation, neutralization, wetlands, and radium removal. Contaminated ground water may be treated using ex situ (pump-and-treat) or in situ (bioremediation, permeable reactive barriers) techniques.

Environmental concerns with uranium mining include radiation hazards, in particular to the workforce. Employees of uranium mines can be exposed to ionizing radiation and radon gas via: direct external radiation from uranium ores, tailings and waste rocks; ingestion of contaminated water; inhalation of radon; and inhalation and ingestion of dust containing radioactive isotopes. These radiation hazards can be minimized and low level radiation doses can be achieved through: installation of covers; appropriate ventilation; dust suppression; radiation dose monitoring; strict hygiene standards, ventilation, and protective clothing; and installation of covers on tailings. Environmental impacts of uranium mine wastes include: excessive radioactivity levels and radon emissions; inappropriate use of tailings and waste rocks; failure of tailings dams; soil, sediment, and ground and surface water contamination; and acid mine drainage.

Further information on uranium mine wastes can be obtained from web sites shown in Table 6.8.

Table 6.8. Web sites covering aspects of uranium mine wastes

Organization	Web address and description
US Environmental Protection Agency (EPA)	*http://www.epa.gov/iaq/radon/* *http://www.epa.gov/radiation/* *http://www.epa.gov/radiation/docs/radwaste/* *http://www.epa.gov/radiation/tenorm/sources.htm* Health effects of radiation and radon exposure; factsheet on uranium mill tailings
US Department of Energy (DOE), Uranium Mill Tailings Remedial Actiont	*http://www.em.doe.gov/bemr96/umtra.htm* *http://www.em.doe.gov/emprimer/erumtra.html* Details on uranium mill tailings remediation programme
US Nuclear Regulatory Commission	*http://www.nrc.gov/reading-rm/doc-collections/fact-sheets/mill-tailings.html* Factsheet on uranium mill tailings
Department of the Environment and Heritage, Supervising Scientist, Australia	*http://www.deh.gov.au/ssd/uranium-mining/index.html* Publications and information on uranium mining in the Alligator Rivers region
Wyoming Mining Association	*http://www.wma-minelife.com/uranium/uranium.html* Uranium mining and milling in Wyoming
Uranium Information Centre	*http://www.uic.com.au* Information on uranium mining and nuclear energy; information sheet on mineral sands
World Nuclear Association	*http://www.world-nuclear.org* Information on nuclear energy and uranium mining
World Information Service on Energy (WISE) – Uranium Project	*http://www.wise-uranium.org* Information on environmental impacts of uranium mining
Radwaste	*http://radwaste.org* Reference source for radioactive waste management

Chapter 7

Wastes of Phosphate and Potash Ores

7.1 Introduction

Plants and agricultural crops require phosphorus, potassium, and nitrogen as macronutrients. In order to maintain agricultural crop yields, these elements must be added to replace those lost from the soil. In most cases, the nutrients are added to agricultural land as mineral fertilizers (UNEP/IFA 2001). Nitrogen fertilizers are generally produced from atmospheric nitrogen, water, and energy. In contrast, the production of phosphate and potassium fertilizers relies on the provision of phosphate rock and potash ores. The majority of mined phosphate rock and potash ore is processed to fertilizer; a minor proportion of the mined material is used for other purposes. World population growth and the necessity to provide adequate food supplies have resulted in the significant growth of phosphate and potash mining, and fertilizer consumption over the last 100 years. This growth has also led to the ever increasing volume of phosphate and potash mine wastes.

In this chapter, potash and phosphate deposits will be introduced prior to the documentation of the different wastes accumulating at potash and phosphate mines. This is because an understanding of phosphate and potash mine wastes requires a knowledge of their ore deposits since ore properties influence the characteristics of mine wastes. As many of the waste disposal problems and environmental impacts of phosphate mining and fertilizer manufacturing are centered on phosphogypsum, particular emphasis is placed on this topic.

7.2 Potash Mine Wastes

Potash is a generic term which refers to a number of potassium salts including carbonate, sulfate and chloride compounds. The production of potash and other salts principally relies on the mining of evaporitic salt deposits. As with all mining, mineral processing and metallurgical activities, the exploitation of salt ores not only generates mineral resources but also mine wastes. Salt ores have the specific issue of salinity and hence, potash mine wastes are invariably saline. This property differentiates potash mine wastes from other mine wastes. Appropriate disposal and rehabilitation strategies of potash mine wastes should ensure that these wastes do not release salinity into the environment and cause significant environmental harm.

7.2.1 Potash Ores

Potash is mined from potash deposits and as a by-product from rock salt (i.e. halite) ores. The nature of these deposits largely stipulates the mining method. Potash and rock salt mines are underground operations and less commonly, in situ solution mining operations. Gypsum, anhydrite, and halite are the main minerals in potash and rock salt ores. In rock salt deposits, halite is extracted as a primary product whereas potassium (sylvite: KCl; carnallite: $KMgCl_3 \cdot 6\,H_2O$; polyhalite: $K_2Ca_2Mg(SO_4)_4 \cdot 2\,H_2O$) and magnesium salts (kieserite: $MgSO_4$; tachyhydrite: $CaMgCl_4 \cdot 12\,H_2O$) are of lesser abundance. Nonetheless, the concentrations of these less common potassium minerals in rock salt deposits provide important potash resources. In potash deposits, the ores contain sylvite and/or carnallite as dominant potassium minerals, and halite is a by-product. Gangue minerals in these deposits include clay minerals, sulfides, carbonates, iron oxides, and numerous evaporative salts. The potassium ore minerals contain variable K_2O concentrations (sylvite 63.2 wt.% K_2O; carnallite 16.9 wt.% K_2O). Potassium ore grades are expressed in terms of potassium oxide (K_2O), and currently mined potassium ores have 8 to 30 wt.% K_2O (Rauche et al. 2001).

7.2.2 Mining and Processing Wastes

Mineral processing of potash and rock salt ores involves flotation of the crushed salt ore. The flotation technique aims to concentrate the salt minerals and to reject the gangue phases. Alternatively, dissolution of the entire crude salt occurs by hot aqueous solutions, and the salts are precipitated (Rauche et al. 2001). Consequently, potash and rock salt mining produces very little mining waste whereas mineral processing results in the rejection of the majority of the mined ore as liquid and solid wastes (Ripley et al. 1996; UNEP/IFA 2001). In particular, potash ores generate more waste than any other salt ores. The major waste products of potash and rock salt processing include:

- *Brines.* The liquid waste from potash and rock salt operations is a saline solution. This brine can be enriched in one or more of the ore elements/compounds including calcium, potassium, sodium, magnesium, chloride, and sulfate. Disposal techniques for brines vary (Fig. 7.1). The solution may be disposed of by: (*a*) reinjection into deep aquifers below the orebodies; (*b*) discharge into the ocean; (*c*) collection in large ponds, treatment, and release into local rivers; or (*d*) pumping – with or without the solid residues – back into the underground workings and emplacement as hydraulic backfill (Ripley et al. 1996; Rauche et al. 2001; UNEP/IFA 2001). At shallow underground salt mines, the backfill disposal practice prevents surface subsidence which could otherwise occur in these highly ductile and leachable ore environments.
- *Tailings.* Solid processing wastes of potash and rock salt deposits contain rejected gangue minerals and mineral processing salts. These tailings may be backfilled into underground workings or are stacked near the mine site into large piles (Fig. 7.2) (UNEP/IFA 2001). The relative ease with which salt minerals dissolve requires that

Fig. 7.1. Potash mining operations at Unterbreizbach, Germany. Dry salt tailings have been stacked into piles, and retention ponds allow the controlled release of brines into the local river

Fig. 7.2. Potash tailings stack at Röhrigshof, Germany

piled tailings are covered with appropriate impermeable barriers (e.g. topsoils, clay seals). Alternatively, the tailings may be topped with a leached layer that acts as a suitable substrate for vegetation (Ripley et al. 1996).

7.3 Phosphate Mine Wastes

7.3.1 Phosphate Rock

Phosphate rock is defined herein as a rock containing phosphate minerals in high enough concentrations to be mined commercially. Phosphate resources exploited commonly exceed 20 wt.% P_2O_5. Phosphate rock occurs in various deposit types, and the principal industrial mineral of all deposits is apatite in the form of fluorapatite ($Ca_5(PO_4)_3(F)$) and/or carbonate fluorapatite ($Ca_5(PO_4,CO_3)_3(F)$). Exploitable phosphate deposits can be divided into three types: phosphorites; carbonatites and alkaline igneous rocks; guano deposits. Of these deposits, phosphorites constitute the predominantly mined ore.

1. *Phosphorites.* Phosphorites or marine sedimentary phosphate deposits contain carbonate fluorapatite with 1 wt.% fluorine and appreciable amounts of carbonate as the phosphate mineral (e.g. USA, Morocco, Togo, Egypt, eastern Australia). This apatite form is also referred to as francolite. The francolite is present as nodules or pellets up to several millimeters in diameter which may contain various impurities. The impurities occur as physical inclusions (e.g. quartz, clays, iron oxyhydroxides, organic matter, carbonates) or as crystallographic substitutions within the phosphate mineral. Ions which will substitute for calcium, phosphate and fluorine within the francolite lattice are diverse and mainly include sodium, magnesium, strontium, carbonate, and sulfate as well as various trace elements (e.g. U, Th, REE, Y, Cd, Zn) (Jarvis et al. 1994; Rutherford et al. 1994). Trace heavy metal and metalloid enrichments (e.g. Ag, As, Cd, Cu, Mo, Ni, Sb, Se, V, Zn) of phosphorites are associated with abundant organic matter. The enrichment is likely due to the adsorption or incorporation of these elements in organic substances or sulfides.
2. *Carbonatites and alkaline igneous rocks.* Carbonatites and other alkaline igneous rocks commonly contain elevated phosphate concentrations in the form of fluorapatite (e.g. Russia, Canada). Weathering of these igneous rocks removes soluble carbonate minerals and leads to the natural concentration of fluorapatite and other weathering resistant minerals (e.g. magnetite, pyrochlore). Deep weathering profiles – overlying and derived from carbonatites – are known to contain distinctly elevated phosphate concentrations in the form of fluorapatite (e.g. western Australia, Brazil). Secondary aluminophosphate minerals such as crandallite ($CaAl_3(PO_4)_2(OH)_5H_2O$) may form during weathering of the apatite. These crandallite-rich horizons are not suitable for phosphate extraction.
3. *Guano deposits.* Guano deposits on tropical coral islands represent originally bird excrements which have been leached, oxidized, and redeposited as amorphous and crystalline phosphates. Commercial guano deposits have largely been mined out, and there has been no production of mining or processing wastes on site due to the nature of the deposits. Some of the mined deposits require extensive rehabilitation (e.g. Christmas Island, Nauru, Ocean Island). For example, phosphate mining on the island of Nauru has disturbed land to such an extent that only the coastal fringe of the island remains habitable.

The presence of elevated radionuclide, arsenic, cadmium, selenium and thallium concentrations in some guano and phosphorite deposits can limit the use of these deposits. Fertilizers manufactured from such deposits will contain elevated concentrations of these elements (Lottermoser and Schomberg 1993; Jarvis et al. 1994). Cadmium, in particular, can accumulate in soils and plants through repeated fertilizer use, and fertilizer application to agricultural land may result in the transfer of cadmium into the food chain. Cadmium-rich phosphorites are considered to be unsuitable for fertilizer production, and the application of cadmium enriched fertilizers has been restricted in some countries. Also, fertilizers contain trace quantities of uranium, and the application of fertilizer to agricultural land and soil erosion add uranium to local waterways and oceans. For instance, the Everglade wetlands and the Mediterranean Sea contain fertilizer-derived uranium due to phosphate fertilizer inputs (Ragnarsdottir and Charlet 2000).

7.3.1.1 *Mineralogy and Geochemistry*

The principal mineral of all ores is apatite while the gangue mineralogy of phosphate deposits is deposit specific. Gangue minerals include framework, ortho, ring, chain and sheet silicates as well as sulfides, oxides, hydroxides, sulfates, and carbonates. The variation in gangue mineralogy causes the major and trace element geochemistry of phosphate deposits to be variable. In particular, phosphorite deposits tend to possess elevated uranium, thorium, rare earth element, yttrium, heavy metal and metalloid values (Jarvis et al. 1994). Carbonatite and alkaline igneous rock related deposits display high uranium, thorium, rare earth element, yttrium, niobium, tantalum, titanium and zirconium concentrations.

Some phosphate deposits have elevated radioactivity and radon levels (Rutherford et al. 1994). The radioactivity and radon originate from abundant uranium and thorium, and the daughter nuclides of the U-238, U-235 and Th-232 decay series (Table 6.2). Uranium and thorium concentrations of phosphate deposits are highly variable, and igneous deposits tend to contain lower uranium and higher thorium levels than sedimentary phosphorites. Beneficiation of phosphate rocks not only leads to concentration of the phosphate minerals but also to the concentration of the specific radioactivity ($Ci\ kg^{-1}$) contained in the phosphate rock. Most of the radioactive nuclides are likely hosted by the apatite crystal lattice or are adsorbed onto surfaces of clays and organics (Burnett et al. 1995). On the other hand, some radionuclides, such as Ra-226, have ionic radii and charges which are not commensurate with the size and charge of the host calcium cation of apatite. As a consequence, Ra-226 may be present in separate barium-strontium sulfate phases (Rutherford et al. 1994).

7.3.2 Mining, Processing and Hydrometallurgical Wastes

Due to the nature of many phosphate deposits, *waste rocks* or *overburden* must be removed to extract the ore. The mined rock is then processed with water to remove unwanted gangue minerals and to concentrate the phosphate minerals to a raw material for phosphoric acid production. The processing generates a phosphate mineral concentrate and unwanted fine-grained rock and mineral particles. These *tailings* are either discarded in tailings ponds or are discharged into rivers and oceans. The production of phosphoric acid is commonly achieved by dissolving the washed and concentrated phosphate minerals in sulfuric acid. The unwanted by-product of such fertilizer manufacturing is referred to as "*phosphogypsum*". In many cases, fertilizer plants have been built adjacent to, or in the vicinity of, phosphate mines. As a result, vast quantities of phosphogypsum accumulate at phosphate mine sites. Wastes of phosphate mining, processing and metallurgical extraction can be grouped into the following four main categories (UNEP/IFA 2001) (Fig. 7.3):

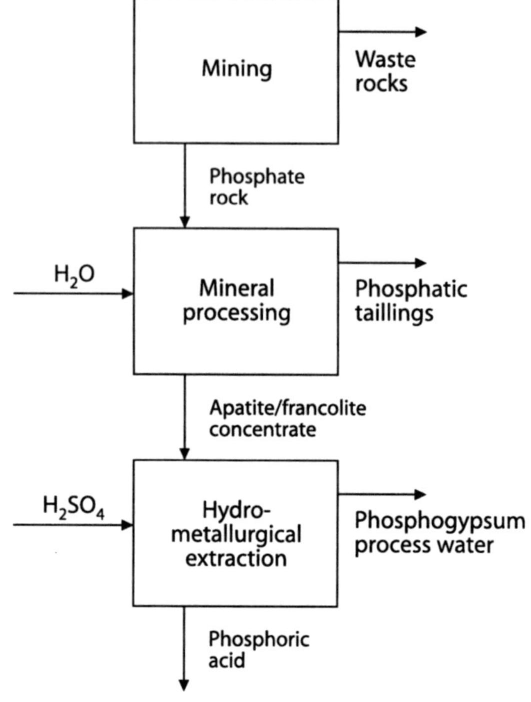

Fig. 7.3. Simplified flow-chart of a phosphate mine and phosphoric acid plant. Phosphate rock is mined, processed and metallurgically treated to yield phosphoric acid and mine wastes (i.e. waste rocks, phosphatic tailings, phosphogypsum, process water)

1. *Waste rocks.* Mining of phosphate deposits produces phosphate rock which is either used as rock phosphate fertilizer or is further processed to phosphoric acid for fertilizer manufacturing. The mining activities invariably generate *waste rocks.* These are country rocks enclosing phosphate deposits, and they need to be mined to access the deposit. The wastes possess no or subeconomic amounts of apatite and display low phosphate concentrations and thus, low radioactivity and radon levels. Sulfide contents are highly variable and deposit specific. Highly sulfidic waste rocks may generate AMD, and such waste materials require appropriate characterization, prediction, monitoring, and control (Chap. 2). Mined waste rocks are commonly disposed of in piles near the mine. The long-term stability of waste rock dumps is of prime concern, especially for those piles constructed in areas with high erosion rates. Such dumps require monitoring for erosional stability; the non-sulfidic wastes do not need other monitoring strategies, treatment of seepages or installation of covers. If wastes contain significant quantities of metals and metalloids (e.g. Ag, As, Cd, Cu, Mo, Ni, Sb, Se, V, Zn), leachates with significant metal and metalloid concentrations may be formed (Vance 2000). In most cases, however, the waste rocks are benign wastes and do not pose an environmental threat. They contain naturally elevated nutrient concentrations, so the revegetation of non-sulfidic waste rock dumps is easily achieved. Phosphatic waste rocks may even be put to good use and be consumed in the rehabilitation of mine sites and waste repositories (e.g. landscaping, capping, revegetation).
2. *Phosphatic tailings.* The mined phosphate rock is commonly treated prior to phosphoric acid manufacturing. Mineral processing of phosphorites usually involves a

combination of various treatment techniques. The beneficiation aims: (*a*) to remove fine-grained rock and mineral particles as slimes; (*b*) to remove coarse-grained quartz sand particles as sands; and (*c*) to concentrate the apatite. Organic-rich phosphate rock may have to be heated prior to processing in order to reduce the organic matter content to acceptable levels. The rejected gangue minerals include calcite, dolomite, quartz, clays, iron oxides, and aluminium and iron phosphates. The wastes are in the sand to clay size range and are termed "*phosphatic tailings*". Thus, phosphatic clays and sands accumulate during the beneficiation of phosphorites. The sand particles may be pumped to storage impoundments, or they may be reclaimed and used as a backfill of mine workings. The clay particles – liberated during benefication – are either disposed of to rivers, mined-out areas or engineered storage impoundments. Beneficiation of particularly clay-rich phosphorites results in the production of exceptionally fine-grained particles and colloids. Such phosphatic clays are also referred to as "*fines*" or "*slimes*". The production of slimes is, of course, minimal for those deposits where the clay content is volumetrically insignificant. Phosphatic clays have poor settling characteristics and require long periods to settle without any treatment. Settling of these slimes may require construction of large ponds covering vast areas, and even larger settling ponds are needed for those deposits containing large quantities of clays (e.g. Florida). The settling time of the slime particles and colloids can be accelerated to some degree using flocculants, which are added to the slurry upon discharge to the ponds. Sludge ponds may, therefore, contain chemicals used as reagents in the flotation process. These chemicals include soda ash, diesel, fatty acid, ammonium hydroxide, sodium silicate, sulfuric acid, and amine.

3. *Phosphogypsum.* The greatest volume of waste products is generated in the phosphoric acid plant where the extraction of phosphorus from beneficiated phosphate rock is conducted (Fig. 7.4). The production of phosphoric acid is achieved by thermal reduction of the phosphate rock in an electric furnace to produce elemental phosphorus (i.e. pyrometallurgy), or more commonly by the chemical reaction of the phosphate rock with sulfuric acid (i.e. hydrometallurgy). The latter, so-called "wet process", not only results in the production of phosphoric acid but also in the generation of *process waters* and *phosphogypsum* waste. The most common manufacturing process for phosphoric acid is based on the use of sulfuric acid, which results in the production of phosphogypsum. In the process of phosphoric acid manufacture, about 3 to 6 t of phosphogypsum are generated for every tonne of phosphoric acid produced. The phosphogypsum is a substantial waste product of the fertilizer producing industry and is commonly stored in very large piles near the processing plants.
4. *Process water.* Spent process waters are disposed of with the phosphogypsum and are commonly pumped to the phosphogypsum repository. Hence, phosphogypsum dumps contain very low pH (as low pH 1), high TDS process waters with potentially elevated fluoride, sulfate, phosphate, ammonia, radionuclide, heavy metal and metalloid concentrations. Such waters require isolation within the waste repository or collection in lined ponds. Excess process waters may need to be neutralized with lime prior to their discharge to receiving streams. Sludges generated from the neutralization process generally require isolation, for example, in the phosphogypsum stack (Ericson et al. 1997).

Fig. 7.4. View of parts of the fertilizer production facility at Phosphate Hill, Australia. Ammonia plant (*far right*), phosphoric acid facility (*centre*), granulation plant (*left*), and rail loadout facility (*far left*). A conveyer belt delivers phosphogypsum to the stack

7.3.3 Phosphogypsum

In the wet process, the ground phosphate rock concentrate ($Ca_5(PO_4)_3(F)$) is reacted with sulfuric acid (H_2SO_4) to produce phosphoric acid (H_3PO_4) and calcium sulfate crystals ($CaSO_4 \cdot 2\,H_2O$). These calcium sulfate solids are referred to as phosphogypsum. Other reaction products include hydrofluoric acid (HF). The reaction of phosphate rock with sulfuric acid can be written as follows:

$$Ca_5(PO_4)_3(F)_{(s)} + 5\,H_2SO_{4(aq)} + 10\,H_2O_{(l)} \rightarrow 3\,H_3PO_{4(aq)} + 5\,CaSO_4 \cdot 2\,H_2O_{(s)} + HF_{(aq)} + \text{heat} \tag{7.1}$$

7.3.3.1 *Mineralogy and Geochemistry*

The above chemical reaction (Reaction 7.1) illustrates the manufacturing process of phosphoric acid and the production of calcium sulfate (i.e. phosphogypsum), yet it does not adequately account for other reaction products. The calcium sulfate can be produced in different mineralogical forms. Depending on the manufacturing process, the phosphoric acid production can result in the formation of phosphogypsum in the dihydrate (gypsum: $CaSO_4 \cdot 2\,H_2O$) or hemihydrate (bassanite: $CaSO_4 \cdot 0.5\,H_2O$) form (Rutherford et al. 1994). The dihydrate form is the most common waste product; hemihydrate is metastable in water and may convert in time to dihydrate.

Phosphogypsum dominantly consists of calcium sulfate crystals. It also contains other small solid phases. These include reaction products of the wet process (e.g. alkali fluosilicates, fluorides) and unreacted phosphate rock and gangue mineral particles (e.g. quartz, phosphates, organic matter, feldspars) (Luther et al. 1993; Rutherford et al. 1995; Arocena et al. 1995a). In addition, the calcium sulfate crystals contain liquid inclusions and process waters trapped in the interstices of mineral particles. The pore liquids are variably enriched in phosphoric acid, sulfuric acid, fluorine, nitrate, heavy metals, metalloids, and radionuclides (Luther et al. 1993). The term "*phosphogypsum*" is, therefore, a collective term for a waste mixture comprising major solid and minor liquid waste components.

Phosphogypsum has physical and chemical properties broadly similar to natural gypsum and bassanite. Phosphogypsum readily dissolves in rainwater in a similar way to gypsum, and its dissolution is independent of pH. In contrast, the properties of an entire phosphogypsum pile are dissimilar to those of a natural gypsum mass of the same size. A phosphogypsum pile contains numerous very fine-grained phosphogypsum particles. The total specific surface area (i.e. $m^2 g^{-1}$) of all phosphogypsum particles is exceptionally large and as a result, the phosphogypsum pile dissolves more rapidly than a gypsum mass of the same size. Thus, mineralogical properties influence the performance of phosphogypsum dumps. If a closed, uncapped phosphogypsum pile is continuously exposed to rainwater infiltration, the dump will develop karst features such as solution channels and cavities.

The chemical composition of phosphogypsum can vary greatly depending on: (*a*) the nature of the phosphate rock used in phosphoric acid manufacturing; (*b*) the type of wet phosphoric acid process used; (*c*) the efficiency of the plant operation; and (*d*) any contaminants which may be introduced into the phosphogypsum during manufacturing (Rutherford et al. 1994). Major compounds of phosphogypsum are calcium and sulfate (Table 7.1). Significant trace constituents include fluorine, rare earth elements, heavy metals, metalloids, and radionuclides (Table 7.2). The material is acidic due to the presence of residual phosphoric, sulfuric and fluoride acids. Prolonged leaching of phosphogypsum in stacks leads to the flushing of trapped acids, heavy metals, and metalloids. Hence, aged and leached materials display near neutral pH and lower trace element values.

The siting of trace elements, metals and metalloids in phosphogypsum is variable. For instance, it has been suggested that: (*a*) cadmium and strontium may substitute for calcium in the calcium sulfate crystal lattice; (*b*) uranium may be adsorbed onto the surface of calcium sulfate crystals; (*c*) selenium may be adsorbed onto iron oxyhydroxides; and (*d*) silver may form discrete halide minerals (Rutherford et al. 1994; Arocena et al. 1995a). The finer particle fraction (<20 µm) of phosphogypsum tends to have higher fluorine, heavy metal, metalloid and radionuclide concentrations than coarser size fractions (Rutherford et al. 1994; Arocena et al. 1995a).

7.3.3.2
Radiochemistry

During the wet phosphoric acid process, the radionuclides within the phosphate rock are liberated from their host phases and released into solution. The individual radionuclides are partitioned into the phosphoric acid or the phosphogypsum, according

Table 7.1. Major element composition (wt.%) of phosphogypsum produced at various phosphoric acid plants

Component	Florida[a] hemihydrate	Florida[a] dihydrate	Australia[b] dihydrate	Iraq[c] dihydrate	Morocco[d] dihydrate	Senegal[d] dihydrate
	Type of analysis					
	Bulk geochemical analysis				Mineral chemical analysis	
CaO	36.9	32.5	32.8	32.94	42.24	33.58
SO_3	50.3	44.0	45.2	44.94	59.85	59.86
SiO_2	0.7	0.5	0.21	0.45	0.84	3.49
Al_2O_3	0.3	0.1	0.17	1.05	0.13	0.28
$Fe_2O_3^{total}$	0.1	0.1	0.02	0.4	<0.07	0.09
MgO	n.a.[e]	0.1	0.05	0.46	n.a.	n.a.
P_2O_5	1.5	0.65	0.5	0.18	2.57	0.99
F	0.8	1.2	1.24	0.6	n.a.	n.a.
$H_2O_{(crystalline)}$	9.0	19.0	20.07	19.18	n.a.	n.a.

[a] Kouloheris (1980).
[b] Beretka (1980)
[c] Khalil et al. (1990).
[d] Martin et al. (1999).
[e] Not analysed.

Table 7.2. Trace element content (ppm) of phosphogypsum produced at various phosphoric acid plants

Trace element	Phosphate rock source						
	Florida[a]	Florida[a]	Idaho[a]	Tunisia[a]	South Africa[a]	Morocco[b]	Senegal[b]
Cu	8	n.a.[c]	10–42	6	103	<6	9
Pb	3–7	n.a.	3–7	n.a.	n.a.	<18	<11
Zn	9	n.a.	18–112	315	6	18	6
Cd	7	9–28	n.a.	40	n.a.	n.a.	n.a.
Mo	16	n.a.	<1–2	5	n.a.	n.a.	n.a.
Ni	2	n.a.	3–15	15	13	13	13
Cr	n.a.	1.5	<10–70	n.a.	n.a.	n.a.	n.a.
Hg	n.a.	n.a.	n.a.	14	<0.05	n.a.	n.a.
As	40	0.25	<1–2	n.a.	n.a.	<5	<6
Sb	100	n.a.	0.3–0.8	n.a.	n.a.	n.a.	n.a.
Se	n.a.	n.a.	4–67	n.a.	n.a.	n.a.	n.a.
U	n.a.	18	6–13	n.a.	n.a.	n.a.	n.a.

[a] Rutherford et al. (1994).
[b] Martin et al. (1999).
[c] Not analysed.

to their solubility (Rutherford et al. 1994). Uranium and thorium radionuclides and Pb-210 concentrate in the phosphoric acid whereas most of the Ra-226 and Po-210 are concentrated in the phosphogypsum. The latter radionuclides are not contained in calcium sulfate crystals but are hosted in separate phases (Jarvis et al. 1994; Rutherford et al. 1994). Radium-226 of the U-238 decay series has a charge (4^+) and ionic radius (0.152 nm) which makes it an unlikely substitution for calcium (2^+, 0.099 nm) in the calcium sulfate lattice. A proportion of the Ra-226 is associated with extremely fine-grained water insoluble particles. The water insoluble phases are possibly barium or strontium sulfates, fluorides, phosphates and/or aluminium phosphates resembling the mineral crandallite (Jarvis et al. 1994; Rutherford et al. 1994; Burnett et al. 1995). The remaining percentage of Ra-226 is likely adsorpted onto organics, associated with colloids, and present in water soluble solids (Burnett et al. 1995). Thus, radionuclides in phosphogypsum are contained in solid crystal lattices and on adsorption sites. In addition, phosphogypsum stacks contain stack fluids in the interstices of the phosphogypsum solids which contain acids, including phosphoric acid. These pore fluids have been observed to be very high in uranium and Pb-210 with moderate concentrations of Ra-226 (Burnett and Elzerman 2001).

Radium-226 activity (about 500 to 2 000 Bq kg^{-1}) is usually the largest source of radioactivity in phosphogypsum although high activities (>1 000 Bq kg^{-1}) of Pb-210 and Po-210 have also been reported (Rutherford et al. 1994). Radium-226 decays to the important radon isotope Rn-222 which is of significant environmental concern (cf. Sec. 6.4.2.2).

7.3.4 Disposal of Phosphogypsum

The amount of phosphogypsum produced by the fertilizer industry on a worldwide basis is in the order of 100 Mt per year (Wissa and Fuleihan 2000). Such large quantities of phosphogypsum create a major disposal problem. Disposal options for this waste are either discharging it into the sea, backfilling it into mined-out open pits, or stacking it in large heaps. In Florida alone, phosphogypsum is generated at a rate of 40 Mt per year, and phosphate mining operations have produced 1 000 Mt of stockpiled phosphogypsum covering over 2 000 ha of land (Burnett and Elzerman 2001).

7.3.4.1 *Marine Disposal*

The dumping of phosphogypsum at sea and into rivers has been pursued for many years at various operations around the world. Such a disposal practice has a number of advantages:

1. Insoluble mineral particles like quartz settle on the stream or sea bed and become incorporated into marine sediments without any environmental impacts.
2. Soluble and sparingly soluble constituents (i.e. calcium sulfate crystals, free phosphoric and sulfuric acids, fluorine compounds, heavy metals, metalloids and radionuclides) are dispersed in a very large volume of water to background concentrations (i.e. dilution is the solution to pollution).
3. There is no contamination threat to ground and surface waters on land.

Despite these advantages, the disposal practice may cause elevated phosphate, cadmium and radionuclide levels in coastal seawater, and discharged spoils could end up being dredged and dumped on coastal land. Increased phosphate concentrations stimulate the growth of algae, and cadmium and radionuclides may bioaccumulate in coastal marine life (van der Heijde et al. 1990). Hence, while sea dumping is an effective disposal option, there are question marks about the risks of environmental impact and contaminant transfer into the local foodchain. Nowadays, only a few phosphoric acid plants discharge their phosphogypsum directly into the sea. Furthermore, fertilizer manufacturing plants are rarely situated close to the open sea into which phosphogypsum can be discharged. This leaves land disposal as the only option.

7.3.4.2
Backfilling

Backfilling mined-out open cuts or underground workings with phosphogypsum is a possible disposal option (Wissa and Fuleihan 2000). However, phosphogypsum contains acids in liquid inclusions and pores. This acidity requires that backfilling is based on the blending of phosphogypsum with sufficient acid buffering materials. The phosphatic clays produced during beneficiation may be suitable for blending as they can have a high calcareous content. Mixing these calcareous materials with phosphogypsum can neutralize the acidity remaining in the phosphogypsum (cf. Sec. 2.10.4).

Once backfilling has occurred, the ground water will eventually return, approximating the pre-mining ground water table. If the open pit is not lined with clay or other impermeable liners, the phosphogypsum will become part of the local aquifer (cf. Sec. 4.5). Phosphogypsum is similar to gypsum in its solubility, and phosphogypsum can be host to elevated radionuclide, heavy metal and metalloid concentrations. Consequently, the disposal of phosphogypsum into mining voids may lead to: (*a*) the dissolution of phosphogypsum; (*b*) the mobilization of contaminants; and (*c*) the contamination of ground water with acid, radionuclides, heavy metals, and metalloids. Because of such concerns, phosphogypsum is generally not recommended as backfill (Wissa and Fuleihan 2000).

7.3.4.3
Phosphogypsum Stacks

The most widespread practice in the phosphate industry is to stack the phosphogypsum near the production plant. The use of tailings dams is less frequent. Prior to stacking, the phosphogypsum is filtered and/or washed at the processing plant to remove any soluble phosphate. Two different stacking procedures are employed, wet-stacking and dry-stacking. Dry-stacking is applied in arid areas with limited water supplies. Wet-stacking is the most common method employed and is based on the pumping of the waste slurry, containing about 20 to 25% solids, from the plant to a repository. The phosphogypsum is slurried with process, sea or fresh water and pumped to the top of an impoundment known as stack where a pond-and-pile system is operating (Fig. 7.5). At the top of the stack, the sand- to silt-sized phosphogypsum solids settle. The water is removed from the settling pond and sent to a nearby collection pond. Alternatively, the water is indirectly removed – after it seeps through the stack – and collected in

Fig. 7.5. Phosphogypsum stack at the Phosphate Hill mine, Australia. The dry phosphogypsum is reslurried prior to disposal and discharged onto the top of the stack where the solids settle. Waters (pH 1.5) seep through the stack and are collected in circumscribing plastic lined ditches

ditches or ponds that circumscribe the stack (Fig. 7.5). If acidic, phosphatic process water has been used to slurry the phosphogypsum, the process water is usually recycled. If sea water has been used, the process water is discharged into the ocean. The stack grows as the dikes that form the impoundment at the top are built up with phosphogypsum. The phosphogypsum stacks may reach several hundred hectares in size and 100 m in height.

Phosphogypsum stacks may exhibit distinct chemical gradients similar to sulfidic waste rock piles and tailings repositories (cf. Secs. 3.9.1, 4.3.2). It is likely that acidic, aerobic conditions are prevalent in the upper part of the stack. Neutral, anaerobic conditions can be expected to develop near its base. Aged waste material, which has been leached for a considerable period of time, would possess near neutral pH values. These aerobic and anaerobic conditions will influence the mobility of radionuclides.

A proportion of Ra-226 in phosphogypsum is water soluble while the remaining percentage may be sorpted onto organics, associated with colloids, and incorporated into extremely fine-grained water insoluble particles (Burnett et al. 1995). The release of Ra-226 from phosphogypsum solids into stack pore waters is possible. Dissolution of the water insoluble phases may occur under anaerobic conditions at the base of the stack. At the base, sulfate-reducing bacteria may not only interact with the calcium sulfate crystals but also with water insoluble minerals. Under anaerobic conditions and upon bacterial reduction, the Ra-226 is released from the sulfates and becomes available for mobilization. Similarly, the highly toxic Po-210 is highly soluble in reducing environments, may be leached from stack bases, and could migrate into aquifer systems (Bennett et al. 1995). Thus, it appears that phosphogypsum wastes share common characteristics with uranium mill tailings (cf. Sec. 6.7). Uranium mill tailings and phosphogypsum may exhibit elevated radioactivity levels due to abundant Ra-226, and some of this Ra-226 is hosted by solid phases (possibly sulfates, fluorides, phosphates and/or aluminophosphates). Ra-226 and Po-210 that are contained in sulfates, may be released by sulfate-reducing bacteria into porewaters, and may subsequently be mobilized into ground and surface waters.

In order to prevent ground water contamination, any dry or wet stack of phosphogypsum needs to be constructed with a composite liner (i.e. compacted soil or compacted phosphogypsum with a high density polyethylene membrane) and a leachate collection system (US EPA 1997). Otherwise, process water or trapped liquids within phosphogypsum may be leached during rainfall events, and the ground water can become contaminated with acid, sulfate, phosphate, fluorine, radionuclides, and heavy metals. If a phosphogypsum stack has been constructed on calcareous or clay-rich subsoil or with a clay liner, the fluorine-rich process water may react with the carbonate or clay minerals, causing their alteration and dissolution (Arocena et al. 1995b). The dissolution of these minerals would greatly increase permeability and would lead to the transfer of contaminants into aquifers. Therefore, engineering designs for phosphogypsum waste impoundments advocate plastic liners (US EPA 1997).

Upon closure, the stack should be capped with an impermeable layer because the dissolution of phosphogypsum and the leaching of contaminants from the waste need to be avoided. If closed stacks are left uncapped, the phosphogypsum will dissolve and karst-type features will form in the stack (Wissa and Fuleihan 2000). Phosphatic clays with their long-term acid buffering capacity and alkaline pH may be useful materials for the capping of closed phosphogypsum stacks. In addition, compacted phosphogypsum may be suitable capping material. Vegetation can be established directly on the phosphogypsum with or without suitable amendments, including limestone, lime or phosphatic clay (Richardson 1995). These amendments are to raise the phosphogypsum pH and to ameliorate the tendency for crusting and cracking of phosphogypsum. Capping the phosphogypsum with materials, such as phosphate waste rocks or phosphatic clays, not only reduces leaching of contaminants and atmospheric emissions from the stack, it also creates suitable substrates for vegetation.

7.3.4.4
Recycling

Much effort has been put into finding alternative uses for phosphogypsum. These possible uses require either reprocessing (e.g. for the production of sulfur or pure calcium sulfate) or minimal/nil reprocessing for large-scale applications (e.g. in agriculture, mine and landfill reclamation, earthworks and construction) (Korcak 1998; Johnson and Traub 1996).

Reprocessing aims to extract pure elements or pure calcium sulfate solids. For example, sulfur can be obtained for sulfuric acid production, and calcium sulfate can be extracted for building materials (e.g. gypsum plaster, boards, tiles, cement, hydraulic binder, mineralizer, artifical marble, fibre boards, glass, glass-ceramics). Some phosphate rocks contain elevated uranium, yttrium, and/or rare earth element concentrations, and their extraction is possible (Scientific Issue 7.1). However, the recovery of

Scientific Issue 7.1. Recovery of Trace Constituents from Phosphate Rock

Resources within Phosphate Rock

Phosphate rock contains on average 0.015% uranium, 0.035% yttrium, and 0.055% rare earth element oxides (Weterings and Janssen 1985; Habashi and Awadalla 1986). Phosphate rock from carbonatite deposits may contain even higher rare earth element concentrations. About 140 Mt of phosphate rock are treated annually for fertilizer production. In the wet phosphoric acid process, much of the uranium, yttrium and rare earth elements present in the phosphate rock ends up in solution or in the solid phosphogypsum particles. The elements are not extracted and are disposed of with the phosphogypsum waste. In total, approximately 21 000 t of uranium, 49 000 t of yttrium and 77 000 t of rare earth element oxides are lost annually to phosphogypsum process waters and solid wastes. The waste of such resources is extreme. The amount of yttrium and rare earth elements discarded by phosphoric acid plants around the world is much greater than the annual global production of these elements from other sources.

Recovery of Elements

Extensive research shows that recovery of uranium, thorium, scandium, vanadium, yttrium, and rare earth elements is feasible during the manufacture of phosphoric acid. In fact, commercial recovery of rare earth elements during fertilizer production was formerly undertaken in Finland. Whether the extraction of these trace constituents is economically warranted depends on the element content of the phosphate rock, the phosphoric acid production volume of the acid plant, the processing technique, and the market price of the element (Weterings and Janssen 1985). To date, many of the proposed and validated extraction procedures are too expensive to pursue, and despite considerable research efforts, none of the elements have been recovered. There are over 1 000 Mt of phosphogypsum waste stockpiled around the world. These enormous waste stacks could be useful resources of uranium, yttrium, thorium, scandium, vanadium, and rare earth elements in the future.

pure elements or pure calcium sulfate from phosphogypsum is not economically viable at this stage. Phosphogypsum has to compete with mined natural gypsum which is mined at considerably higher purity. Moreover, the use of phosphogypsum as raw material for building purposes can be limited due to residual phosphate, high water content, fine particle size, and inconsistent composition. While these characteristics hinder the use of phosphogypsum in building products, it has been found that compacted phosphogypsum can form a high strength material for road beds, making it a suitable road base material.

Phosphogypsum has the potential to be used as agricultural amendment. Phosphogypsum is particularly useful as an amendment for highly weathered, nutrient-poor soils as well as saline, sodic, acid and calcareous soils (Rutherford et al. 1994). Crop yields and pasture quality have been found to improve on phosphogypsum amended soils. Phosphogypsum application can counter crusting of soils, reducing soil runoff and erosion rates. However, elevated radionuclide (Ra-226, Pa-231, Pb-210, Po-210), fluorine and heavy metal levels as well as contained acid water with dissolved phosphate and heavy metals all represent problems for possible agricultural applications (El-Mrabet et al. 2003). Eutrophication of waterways is possible, and radionuclides, heavy metals, and metalloids may be taken up by plants.

In the United States, radioactivity limits imposed by the US EPA restrict the use of phosphogypsum as building material and soil amendment. If phosphogypsum is used for building and agricultural applications, purification of phosphogypsum may be necessary. Separation of radionuclides, heavy metals and metalloids from phosphogypsum would have to be achieved through chemical or physical means. The distinct chemical and mineralogical properties of different size fractions form the basis of proposed purification procedures. The purification is designed to lower the contaminant load and radioactivity of phosphogypsum. A reduction in contaminant load is possible to some extent through the removal of the finer size fraction which is enriched in radionuclides and heavy metals (Arocena et al. 1995a). Despite the research efforts into alternative uses of phosphogypsum, there are no large-scale applications of phosphogypsum. As a result, the amount of phosphogypsum piled up around the world is steadily increasing.

7.3.5 Potential Hazards and Environmental Impacts

7.3.5.1 *Phosphogypsum*

The most widespread disposal practice in the phosphate industry is to stack the phosphogypsum near the production plant. Improper phosphogypsum stacking causes most of the environmental impacts of phosphate mining and processing. The principal potential hazards and environmental impacts of phosphogypsum stacks are:

- *Failure of stacks.* In the past, several failures of phosphatic tailings ponds and phosphogypsum stacks and dams have occurred. These failures released wastes into waterways and resulted in damage to the local ecosystem (Table 7.3). Phosphorite deposits are commonly hosted or enclosed by carbonate rocks which are susceptible to chemical weathering, leaching, and karst development. Hence, there is a karst sinkhole potential, and in Florida, for example, an entire phosphogypsum stack collapsed into a karst sinkhole.
- *Atmospheric emissions.* Phosphogypsum stacks may emit gaseous radon and fluorine compounds (SiF_4, HF) in significant amounts into the atmosphere. While radon emissions from phosphogypsum stacks are generally significantly lower than those of uranium mill tailings (cf. Sec. 6.10.2), the US EPA has set radon exhalation limits for phosphogypsum stacks. Phosphate processing plants may also emit sulfur dioxide, radon, and fluorine compounds (SiF_4, HF) in significant amounts into the atmosphere. The technology exists to limit atmospheric emissions of fluorine from phosphoric acid manufacturing plants to environmentally acceptable levels. Few operations recover fluorine by absorption in water using scrubbers, which would minimize environmental impacts and prevent potential health impacts on workers.
- *Ground water and surface water contamination.* Discharge of phosphopgypsum into local rivers, seepage of process water from stacks, and leaching of stacks – via rainwater percolating through the heaps – have the potential to contaminate local streams and aquifers. Ground water contamination has occurred at several phosphogypsum stacks where the repositories have not been lined and leachates were not collected. At these sites, ground water has elevated total dissolved solids and conductivity values as well as high acid, phosphate, fluorine, sulfate, heavy metal, metalloid, and

Table 7.3. Examples of phosphate waste repository failures (WISE Uranium Project 2006, reprinted from *www.wise-uranium.org*, with permission from P. Diehl)

Date	Location	Incident	Release	Impact
14.04.2005	Bangs Lake, Florida	Phosphogypsum stack failure	~65 000 m^3	Spillage into marshland; plant death
05.09.2004	Riverview, Florida	Phosphogypsum stack failure	227 000 m^3 of acid liquid	Spillage into local creek
31.12.1998	Huelva, Spain	Dam failure during storm	50 000 m^3 of phosphogypsum with pH 1.5	Spillage into local river
07.12.1997	Mulberry, Florida	Phosphogypsum stack failure	200 000 m^3 of phosphogypsum and process water	Spillage into local river; destruction of ecosystem
19.11.1994	Hopewell, Florida	Dam failure	1.9 million m^3 of water from a phosphatic clay pond	Spillage into local wetland and river
02.10.1994	Payne Creek, Florida	Dam failure	6.8 million m^3 of water from a phosphatic clay settling pond	Mostly contained in mine area
October 1994	Fort Meade, Florida	?	76 000 m^3 of water	Spillage into local river
June 1994	IMC-Agrico, Florida	Sinkhole opens in phosphogypsum stack	?	?
October 1993	Gibsonton, Florida	?	?	Spillage into local river, fish kill
1988	Riverview, Florida	?	Acidic spill	Spillage into local river, fish kill
03.12.1971	Fort Meade, Florida	Clay pond dam failure	9 million m^3 of phosphatic clay water	Spillage into local river, fish kill
March 1967	Fort Meade, Florida	Dam failure	250 000 m^3 of phosphatic clay slimes, 1.8 million m^3 of water	Spillage into local river, fish kill

radionuclide concentrations (especially Ra-226, Rn-222, U-238 and Po-210). Polonium-210 is considered to be extremely radiotoxic and has been found well above recommended guidelines in aquifers below Florida phosphate mining areas (Bennett et al. 1995). Elevated TDS and conductivity values with high fluorine, heavy metal, metalloid and radionuclide levels in ground waters can be due to natural water-rock interactions and the natural leaching of undeveloped phosphate resources. Consequently, aquifers in phosphate mining districts can have naturally elevated contaminant loads. For instance, high fluorine concentrations are known to cause health problems in humans obtaining their water supplies from ground water bores in such areas (e.g. Senegal).

- *Enhanced radioactivity.* High uranium and thorium concentrations in some phosphate rocks lead to significant radioactive decay of radionuclides in process waters, phosphatic clays, and phosphogypsum. In fact, the amount of uranium contained within some phosphate rocks, particularly phosphorite deposits, is so high – several tens to hundreds of parts per million – that uranium can be recovered as a by-product. Liquid and solid wastes may then emit radioactivity levels high enough to require environmental monitoring and safety measures. Thus, improper disposal practices may lead to enhanced radioactivity levels in soils, sediments and waters surrounding a phosphate plant (Bolivar et al. 1995).

From a radiation perspective, inhalation of dust containing radioactive isotopes, direct external radiation, ingestion of surface water, and inhalation of radon represent possible risks to human health (cf. Sec. 6.10.2). Radionuclides are emitted from phosphate rock mining and processing operations in particulate (e.g. U-238), aqueous (e.g. Ra-226) and gaseous forms (e.g. Rn-222), which primarily cause increased radiation exposures in the work force. Phosphogypsum stacks do not pose a dust hazard, because the phosphogypsum has the tendency to form crusts and active phosphogypsum waste repositories are wet enough to prevent dust generation. Radon emissions are the main risk and therefore, Ra-226 – the parent nuclide of Rn-222 – is of prime concern. During mineral processing and phosphoric acid manufacturing, much of the Ra-226 may be incorporated into the phosphogypsum, which may then contain the bulk of Ra-226 and emit high levels of radiation (several tens of pCi g^{-1}). The emission of Rn-222 and its progeny, from phosphogypsum stacks and phosphoric acid plants, is variable and has attracted US EPA regulatory emission standards since there are perceived radiation risks. Moreover, regulations have been put into place on the maximum permissable radiation levels of phosphogypsum (10 pCi g^{-1}) when used in construction and agriculture. This in turn reduces the recycling potential of phosphogypsum. Some studies have indicated, however, that there is no significant risk to human health as the additional radiation dose from phosphogypsum is minimal in comparison to natural levels and background variations (Rutherford et al. 1994).

7.3.5.2
Waste Rocks and Tailings

Environmental concerns are not exclusively related to phosphogypsum stacks. Waste rocks removed to access the phosphate rock are generally piled up next to the mine and hence, become exposed to atmospheric leaching. The leaching may generate seepages with elevated metal, metalloid or radionuclide concentrations. In some cases, leaching of waste repositories led to the contamination of soils, sediments, waters, plants and aquatic organisms with selenium which in turn caused selenium poisoning in grazing animals (Vance 2000; Hamilton and Buhl 2004).

Beneficiation of phosphate rocks results in the production of tailings, which in some instances are discharged into local rivers and coastal marine settings. At such sites, elevated phosphate, fluorine, metal, cadmium and radionuclide levels are found in sediments, waters and marine organisms including seafood (Gnandi and Tobschall 1999; Gnandi et al. 2006). Increased phosphate concentrations stimulate the growth of algae, and cadmium and radionuclides may bioaccumulate in aquatic life which then enter the foodchain (van der Heijde et al. 1990).

7.4 Summary

The production of potash and other salts principally relies on the mining of evaporitic salt deposits. Mineral processing of potash ores involves flotation of the crushed salt which results in the concentration of the salt minerals and rejection of the gangue phases. Alternatively, dissolution of the entire crude salt occurs by hot aqueous solutions, and various salts are precipitated. Currently mined potassium ores have 8 to 30 wt.% K_2O and consequently, potash mineral processing leads to the rejection of the majority of the mined ore as liquid and solid wastes. The major waste products of potash processing include brines and tailings. Brines may be disposed of by: (*a*) reinjection into deep aquifers below the orebodies; (*b*) discharge into the ocean; (*c*) collection in large ponds and release into local rivers; and (*d*) pumping – with or without the solid residues – back into the underground workings and emplacement as hydraulic backfill. Tailings largely consist of mineral processing salts and rejected primary gangue minerals, including clay minerals, sulfides, carbonates, iron oxides, and numerous evaporative salts. The tailings may be backfilled into underground workings or are stacked near the mine site into large piles.

Phosphate rock is mined from sedimentary and igneous deposits, with phosphorites being the predominantly mined deposit type. The principle mineral of all deposit types is apatite in the form of fluorapatite and/or carbonate fluorapatite. Some phosphate rocks display elevated uranium, thorium, heavy metal, metalloid and rare earth element levels. Abundant nuclides of the U-238, U-235 and Th-232 decay series can cause elevated radioactivity and radon levels of the phosphate rock and waste products.

Mining, beneficiation and hydrometallurgical processing of phosphate rock generate waste rocks, phosphatic clays, phosphogypsum, and spent process waters. In the wet process, the beneficiated phosphate rock is reacted with sulfuric acid to produce phosphoric acid. The product of this chemical reaction is a slurry that consists of phosphoric acid and suspended solid crystals. The solids are removed from the phosphoric acid and are termed "phosphogypsum". The term phosphogypsum is a collective term for a waste mixture, comprising major solid and minor liquid waste components. The solid crystals are mainly calcium sulfate (gypsum or bassanite), minor reaction products, and unreacted phosphate rock particles. The calcium sulfate crystals contain liquid inclusions and process water trapped in the interstices of mineral particles. The chemical composition of phosphogypsum is characterized by variably elevated heavy metal, metalloid and radionuclide concentrations. The material is acid due to the presence of residual phosphoric, sulfuric and fluoride acids. Prolonged leaching of stacked phosphogypsum results in the flushing of trapped acids, metals and metalloids; therefore, aged and leached materials display near neutral pH values.

The amount of phosphogypsum produced on a worldwide basis is in the order of 100 Mt per year, and for every tonne of phosphoric acid produced, there are 3 to 6 t of phosphogypsum generated. These large quantities of phosphogypsum create a major disposal problem. The dumping of phosphogypsum at sea has been pursued for many years around the globe. Today, land disposal has become the dominant disposal option. Backfilling phosphogypsum into mine voids is not recommended since it may

Table 7.4. Web sites covering aspects of phosphate and potash mine wastes

Organization	Web address and description
Florida Institute of Phosphate Research (FIPR)	*http://www.fipr.state.fl.us* Publications and information on phosphate mining, mine wastes, and mine site rehabilitation
US Environmental Protection Agency (EPA)	*http://www.epa.gov/radiation/neshaps/subpartr/index.html* *http://www.epa.gov/radiation/tenorm/sources.htm* Radiation and Radon emission from phosphogypsum stacks
World Information Service on Energy (WISE) – Phosphate tailings	*http://www.wise-uranium.org* Case studies and information on phosphate mine wastes
International Fertilizer Industry Association (IFA)	*http://www.fertilizer.org/ifa/* Publications on fertilizer mining, manufacture, trade, and use

bring about mineral dissolution and associated mobilization of contaminants into local ground waters. The preferred disposal option is the stacking of phosphogypsum in large piles near the production plant. A pond-and-pile system allows the sequential built-up of the solid waste and the recycling of process water. Lining and capping of phosphogypsum stacks limit atmospheric emissions of radon and fluorine, and prevent contamination of ground water by acid, sulfate, phosphate, fluorine, radionuclides, heavy metals, and metalloids.

The ever increasing volume of phosphogypsum has stimulated much research into recycling potentials. Numerous alternative uses have been proposed including the use of phosphogypsum as agricultural amendment, as a source of uranium, yttrium and rare earth elements, and as earthworks, construction and building material. However, elevated radionuclide levels, and inefficient and costly extraction procedures, have so far prevented such alternative uses.

Further information on phosphate and potash mine wastes can be obtained from web sites shown in Table 7.4.

References

Aachid M, Mbonimpa M, Aubertin M (2004) Measurement and prediction of the oxygen diffusion coefficient in unsaturated media, with applications to soil covers. Water Air Soil Poll 156:163–193

Abdelouas A, Lutze W, Nuttall E (1998a) Chemical reactions of uranium in ground water at a mill tailings site. J Contam Hydrol 34:343–361

Abdelouas A, Nuttall HE, Lutze W, Lu Y (1998b) In situ removal of uranium from ground water. In: Tailings and mine waste '98. Balkema, Rotterdam, pp 669–677

Abdelouas A, Lutze W, Nuttall HE (1999) Uranium contamination in the subsurface: Characterization and remediation. In: Burns PC, Finch R (eds) Uranium: Mineralogy, geochemistry and the environment. Mineralogical Society of America, Washington (Reviews in mineralogy and geochemistry, vol 38, pp 433–473)

Abdelouas A, Lutze W, Gong W, Nuttall EH, Strietelmeier BA, Travis BJ (2000) Biological reduction of uranium in groundwater and subsurface soil. Sci Total Environ 250:21–35

Abraitis PK, Pattrick RAD, Kelsall GH, Vaughan DJ (2004) Acid leaching and dissolution of major sulphide ore minerals: processes and galvanic effects in complex systems. Min Mag 68:343–351

Accornero M, Marini L, Ottonello G, Zuccolini MV (2005) The fate of major constituents and chromium and other trace elements when acid waters from the derelict Libiola mine (Italy) are mixed with stream waters. Appl Geochem 20:1368–1390

Acero P, Ayora C, Torrento C, Nieto JM (2006) The behavior of trace elements during schwertmannite precipitation and subsequent transformation into goethite and jarosite. Geochim Cosmochim Acta 70:4130–4139

Ackman TE (2003) An introduction to the use of airborne technologies for watershed characterization in mined areas. Mine Water Environ 22:62–68

Agricola G (1556) De re metallica. Translated by Hoover HC, Hoover LH (1950). Dover Publications, New York

Ahn JS, Chon CM, Moon HS, Kim KW (2003) Arsenic removal using steel manufacturing byproducts as permeable reactive materials in mine tailing containment systems. Water Res 37:2478–2488

Al TA, Martin CJ, Blowes DW (2000) Carbonate-mineral/water interactions in sulphide-rich mine tailings. Geochim Cosmochim Acta 64:3933–3948

Alakangas L, Öhlander B (2006) Formation and composition of cemented layers in low-sulphide mine tailings, Laver, northern Sweden. Environ Geol 50:809–819

Allan RJ (1995) Impact of mining activities on the terrestrial and aquatic environment with emphasis on mitigation and remedial measures. In: Salomons W, Förstner U, Mader P (eds) Heavy metals: problems and solutions. Springer, Berlin Heidelberg New York, pp 119–140

Alpers CN, Blowes DW (eds) (1994) Environmental geochemistry of sulfide oxidation. American Chemical Society, Washington DC (Symposium Series 550)

Alpers CN, Nordstrom DK (1999) Geochemical modeling of water-rock interactions in mining environments. In: Plumlee GS, Logsdon MS (eds) The environmental geochemistry of mineral deposits. Part A: Processes, techniques and health issues. Society of Economic Geologists, Littleton (Reviews in economic geology, vol 6A, pp 289–323)

Alpers CN, Blowes DW, Nordstrom DK, Jambor JL (1994) Secondary minerals and acid mine-water chemistry. In: Jambor JL, Blowes DW (eds) Environmental geochemistry of sulfide mine-wastes. Mineralogical Association of Canada, Nepean (Short course handbook, vol 22, pp 247–270)

Ambers RKR, Hygelund BN (2001) Contamination of two Oregon reservoirs by cinnabar mining and mercury amalgamation. Environ Geol 40:699–707

Amos PW, Younger PL (2003) Substrate characterization for a subsurface reactive barrier to treat colliery spoil leachate. Water Res 37:108–120

Anderson CWN, Brooks RR, Stewart RB, Simcock R (1998) Harvesting a crop of gold in plants. Nature 395:553–554

Anderson CWN, Brooks RR, Chiarucci A, LaCoste CJ, Leblanc M, Robinson BH, Simcock R, Stewart RB (1999) Phytomining for nickel, thallium and gold. J Geochem Explor 67:407–415

Appelo CAJ, Postma D (1999) Geochemistry, groundwater and pollution. Balkema, Rotterdam

Applegate RJ, Kraatz M (1991) Rehabilitation of Rum Jungle uranium mine. In: Proceedings of the 2nd international conference on the abatement of acidic drainage. MEND, pp 155–169

Apte SC, McNamara J, Bowles KC (1996) Fate of trace metals in the riverine environment downstream of the Porgera gold mine, PNG. In: Proceedings of 3rd international and 21st annual Minerals Council of Australia environmental workshop. Minerals Council of Australia, Dickson, pp 38–49

Ariza LM (1998) River of vitriol. Sci Amer 279(3):15–18

Arocena JM, Rutherford PM, Dudas MJ (1995a) Heterogeneous distribution of trace elements and fluorine in phosphogypsum by-product. Sci Total Environ 162:149–160

Arocena JM, Dudas MJ, Poulsen L, Rutherford PM (1995b) Weathering of clay minerals induced by fluoride-containing solutions from phosphogypsum by-product. Can J Soil Sci 75:219–226

Asher CJ, Bell LC (eds) (1999) Proceedings of the workshop on future directions in tailings environmental management. Australian Centre for Mining Environmental Research, Brisbane

Ashley PM, Lottermoser BG (1999a) Geochemical, mineralogical and biogeochemical characterisation of abandoned metalliferous mine sites, southern New England Orogen. In: Proceedings of the NEO'99 Conference. Armidale, Division of Earth Sciences, University of New England, pp 409–418

Ashley PM, Lottermoser BG (1999b) Arsenic contamination at the Mole River mine, northeastern New South Wales, Australia. Austral J Earth Sci 46:861–874

Ashley PM, Craw D, Graham BP, Chappell DA (2003) Environmental mobility of antimony around mesothermal stibnite deposits, New South Wales, Australia and southern New Zealand. J Geochem Explor 77:1–14

Asmund G, Broman PG, Lindgren G (1994) Managing the environment at the Black Angel mine, Greenland. Int J Surface Min Reclam Environ 8:37–40

Attendorn HG, Bowen RNC (1997) Radioactive and stable isotope geology. Chapman & Hall, London

Bailey EA, Gray JE, Theodorakos PM (2002) Mercury in vegetation and soils at abandoned mercury mines in southwestern Alaska, USA. Geochem Explor Environ Anal 2:275–285

Banks D, Younger PL, Arnesen R-T, Iversen ER, Banks SB (1997) Mine-water chemistry: the good, the bad and the ugly. Environ Geol 32:157–174

Banks D, Skarphagen H, Wiltshire R, Jessop C (2004) Heat pumps as a tool for energy recovery from mining wastes. In: Gieré R, Stille P (eds) Energy, waste, and the Environment: a geochemical perspective. Geological Society London (Special Publications 236, pp 499–513)

Barbour AK, Shaw IC (2000) Ecotoxicological impacts of the extractive industries. In: Warhurst A, Noronha L (eds) Environmental policy in mining. Lewis Publishers, Boca Raton, pp 57–80

Barnett MO, Turner RR, Singer PC (2001) Oxidative dissolution of metacinnabar (β-HgS) by dissolved oxygen. Appl Geochem 16:1499–1512

Barriga FJAS, Carbalho D (eds) (1997) Geology and VMS deposits of the Iberian pyrite belt. Society of Economic Geologists, Littleton (Guidebook series, vol 27)

Barton CD, Karathanasis AD (1999) Renovation of a failed constructed wetland treating acid mine drainage. Environ Geol 39:39–50

Bayless ER, Olyphant GA (1993) Acid generating salts and their relationship to the chemistry of groundwater and storm runoff at an abandoned mine site in southwestern Indiana, USA. J Contam Hydrol 12:313–328

Bednar AJ, Garbarino JR, Ranville JF, Wildeman TR (2005) Effects of iron on arsenic speciation and redox chemistry in acid mine water. J Geochem Explor 85:55–62

Belzile N, Chen YW, Cai MF, Li Y (2004) A review on pyrrhotite oxidation. J Geochem Explor 84:65–76

Benes P, Sebesta F, Sedlacek J, Obdrzalek M, Sandrik R (1983) Particulate forms of radium and barium in uranium mine waste waters and receiving river waters. Water Res 17:619–624

Benito G, Benito-Calvo A, Gallart F, Martin-Vide JP, Regües D, Bladé E (2001) Hydrological and geomorphological criteria to evaluate the dispersion risk of waste sludge generated by the Aznalcollar mine spill (SW Spain). Environ Geol 40:417–428

Benner SG, Blowes DW, Gould WD, Herbert RB Jr, Ptacek CJ (1999) Geochemistry of a permeable reactive barrier for metals and acid mine drainage. Environ Sci Technol 33:2793–2799

Benner SG, Blowes DW, Ptacek CJ, Mayer KU (2002) Rates of sulfate reduction and metal sulfide precipitation in a permeable reactive barrier. Appl Geochem 17:301–320

Bennett JW, Ritchie AIM (1993) Bio-oxidation of pyrite in mine wastes. Mechanisms which govern pollutant generation. In: Biomine '93, International conference and workshop applications of biotechnology to the minerals industry. Australian Mineral Foundation, Glenside, pp 12.1–12.9

Bennett JW, Gibson DK, Ritchie AIM, Tan Y, Broman PG, Jönsson H (1994) Oxidation rates and pollution loads in drainage: correlation of measurements in a pyritic waste rock dump. In: Proceedings of the international land reclamation and mine drainage conference and 3rd international conference on the abatement of acidic drainage, vol 1. United States Department of the Interior, Bureau of Mines (Special Publication SP06A-94, pp 401–409)

Bennett MW, Kempton JH, Maley PJ (1997) Applications of geological block models to environmental management. In: Proceedings from the 4th international conference on acid rock drainage, vol 1. Vancouver, pp 293–303

Bennett JW, Timms GP, Ritchie AIM (1999) The effectiveness of the covers on waste rock dumps at Rum Jungle and the impact in the long term. In: Proceedings of the 24th annual Minerals Council of Australia environmental workshop. Minerals Council of Australia, Dickson, pp 379–388

Benson AK (1995) An integration of geophysical methods and geochemical analysis to map acid mine drainage – a case study. Explo Min Geol 4:411–419

Benzaazoua M, Bussiere B, Dagenais AM, Archambault M (2004) Kinetic test comparison and interpretation for prediction of the Joutel tailings acid generation potential. Environ Geol 46:1086–1101

Beretka J (1980) Properties and utilization of by-product gypsum in Australia. In: Borris DP, Boody PW (eds) Proceedings of the international symposium on phosphogypsum. Florida Institute of Phosphate Research (Publication no 01-001-017)

Berger AC, Bethke CM, Krumhansl JL (2000) A process model of natural attentuation in drainage from a historic mining district. Appl Geochem 15:655–666

Berkun M (2005) Submarine tailings placement by a copper mine in the deep anoxic zone of the Black Sea. Water Res 39:5005–5016

Bernier LR (2005) The potential use of serpentinite in the passive treatment of acid mine drainage: batch experiments. Environ Geol 47:670–684

Bertrand VJ, Monroy MG, Lawrence RW (2000) Weathering characteristics of cemented paste backfill: mineralogy and solid phase chemistry. In: Proceedings from the 5th international conference on acid rock drainage, vol 2. Society for Mining, Metallurgy, and Exploration, Littleton, pp 863–876

Bethune KJ, Lockington DA, Williams DJ (1997) Acid mine drainage: comparison of laboratory testing to mine site conditions. In: Proceedings from the 4th international conference on acid rock drainage, vol 1. Vancouver, pp 306–318

Bews BE, Barbour SL, Wickland B (1999) Lysimeter design in theory and practice. In: Tailings and mine waste '99. Balkema, Rotterdam, pp 13–21

Bigham JM (1994) Mineralogy of ochre deposits formed by sulfide oxidation. In: Jambor JL, Blowes DW (eds) Environmental geochemistry of sulfide mine-wastes. Mineralogical Association of Canada, Nepean (Short course handbook, vol 22, pp 103–132)

Bigham JM, Nordstrom DK (2000) Iron and aluminium hydroxysulfates from acid sulfate waters. In: Alpers CN, Jambor JL, Nordstrom DK (eds) Sulfate minerals: crystallography, geochemistry and environmental significance. Mineralogical Society of America, Washington (Reviews in mineralogy and geochemistry, vol 40, pp 351–403)

Bigham JM, Schwertmann U, Traina SJ, Winland RL, Wolf M (1996) Schwertmannite and the chemical modeling of iron in acid sulfate waters. Geochim Cosmochim Acta 60:2111–2121

Bilek F (2006) Column tests to enhance sulphide precipitation with liquid organic electron donators to remediate AMD-influenced groundwater. Environ Geol 49:674–683

Black A, Craw D, Younson JH, Karubaba J (2004) Natural recovery rates of a river system impacted by mine tailing discharge: Shag River, East Otago, New Zealand. J Geochem Explor 84:21–34

Blainey G (1991) Gold – a lesson or two? In: Blainey G (ed) Eye on Australia. Schwartz & Wilkinson, Melbourne, pp 181–188

Blanchette MC, Hynes TP, Kwong YTJ, Anderson MR, Veinott G, Payne JF, Stirling C, Sylvester PJ (2001) A chemical and ecotoxicological assessment of the impact of marine tailings disposal. In: Tailings and mine waste '01. Balkema, Rotterdam, pp 323–331

Blowes DW, Ptacek CJ (1994) Acid-neutralization mechanisms in inactive mine tailings. In: Jambor JL, Blowes DW (eds) Environmental geochemistry of sulfide mine-wastes. Mineralogical Association of Canada, Nepean (Short course handbook, vol 22, pp 271–292)

Blowes DW, Reardon EJ, Jambor JL, Cherry JA (1991) The formation and potential importance of cemented layers in inactive sulfide mine tailings. Geochim Cosmochim Acta 55:965–978

Blowes DW, Ptacek CJ, Jambor JL (1994) Remediation and prevention of low-quality drainage from tailings impoundments. In: Jambor JL, Blowes DW (eds) Environmental geochemistry of sulfide mine-wastes. Mineralogical Association of Canada, Nepean (Short course handbook, vol 22, pp 365–380)

Blowes DW, Jambor JL, Hanton-Fong CJ, Lortie L, Gould WD (1998) Geochemical, mineralogical and microbiological characterization of a sulphide-bearing carbonate-rich gold-mine tailings impoundment, Joutel, Québec. Appl Geochem 13:687–705

Blowes DW, Bain JG, Smyth DJ, Ptacek CJ (2003) Treatment of mine drainage using permeable reactive materials. In: Jambor JL, Blowes DW, Ritchie AIM (eds) Environmental aspects of mine wastes. Mineralogical Association of Canada, Nepean (Short course handbook, vol 31, pp 361–376)

Bobos I, Duraes N, Noronha F (2006) Mineralogy and geochemistry of mill tailings impoundments from Algares (Aljustrel), Portugal: implications for acid sulfate mine waters formation. J Geochem Explor 88:1–5

Bodénan F, Baranger P, Piantone P, Lassin A, Azaroual M, Gaucher E, Braibant G (2004) Arsenic behaviour in gold-ore mill tailings, Massif Central, France: hydrogeochemical study and investigation of in situ redox signatures. Appl Geochem 19:1785–1800

Boger DV (1998) Environmental rheology and the mining industry. In: Proceedings of AusIMM '98 The Mining Cycle. Australasian Institute of Mining and Metallurgy, Melbourne, pp 459–461

Bolivar JP, Garcia-Tenorio R, Garcia-Leon M (1995) Enhancement of natural radioactivity in soils and salt-marshes surrounding a non-nuclear industrial complex. Sci Total Environ 173/174:125–136

Bond PL, Druschel GK, Banfield JE (2000) Comparison of acid mine drainage microbial communities in physically and geochemically distinct ecosystems. Appl Environ Microbiol 66:4962–4971

Boon M, Snijder M, Hansfrod GS, Heijnen JJ (1998) The oxidation kinetics of zinc sulphide with *Thiobacillus ferrooxidans*. Hydrometall 48:171–186

Boorman RS, Watson DM (1976) Chemical processes in abandoned sulphide tailings dumps and environmental implication for northeastern New Brunswick. CIM Bulletin 69:86–96

Borden R (2001) Geochemical evolution of sulphide-bearing waste rock soils at the Bingham Canyon mine, Utah. Geochem Explor Environ Anal 1:15–22

Bordon RK, Black R (2005) Volunteer revegetation of waste rock surfaces at the Bingham Canyon mine, Utah. J Environ Qual 34:2234–2242

Botz M (2001) Cyanide treatment methods. Mining Environmental Management, May 2001:28–30

Boulet MP, Larocque ACL (1998) A comparative mineralogical and geochemical study of sulfide mine tailings at two sites in New Mexico, USA. Environ Geol 33:130–142

Bowell RJ, Bruce I (1995) Geochemistry of iron ochres and mine waters from Levant mine, Cornwall. Appl Geochem 10:237–250

Braithwaite RSW, Kampf AR, Pritchard RG, Lamb RPH (1993) The occurrence of thiosulfates and other unstable sulfur species as natural weathering products of smelting slags. Mineral Petrol 47:255–261

Brake SS, Dannelly HK, Connors KA (2001a) Controls on the nature and distribution of an alga in coal mine-waste environments and its potential impact on water quality. Environ Geol 40:458–469

Brake SS, Dannelly HK, Connors KA, Hasiotis ST (2001b) Influence of water chemistry on the distribution of an acidophilic protozoan in an acid mine drainage system at the abandoned Green Valley coal mine, Indiana, USA. Appl Geochem 16:1641–1652

Brant DL, Ziemkiewicz PF, Robbins EI (1999) Passive removal of manganese from acid mine drainage at the Shade mining site, Somerset County, PA (USA). In: Goldsack D, Belzile N, Yearwood P, Hall G (eds) Proceedings of Sudbury '99 Mining and the Environment, vol 3. Sudbury Environmental, Sudbury, Canada, pp 1241–1249

Braungardt CB, Achterberg EP, Elbaz-Poulichet F, Morley NH (2003) Metal geochemistry in a mine-polluted estuarine system in Spain. Appl Geochem 18:1757–1771

BRGM (2000) Management of mining, quarrying and ore-processing waste in the European Union. European Commission, Directorate-General Environment, ENV.E.3 – Waste Management (BRGM/RP-50319-FR, *http://europa.eu.int/comm/environment/waste/mining_report.pdf*)

Briggs TJ, Kelso IJ (2003) Ammonium nitrate-sulfide reactivity at the Century Zn-Pb-Ag mine, northwest Queensland, Australia. Explor Mining Geol 10:177–190

Brookins DG (1988) Eh-pH diagrams for geochemistry. Springer, Berlin Heidelberg New York

Brooks RR (1998) Plants that hyperaccumulate heavy metals: their role in phytoremediation, microbiology, archaeology, mineral exploration and phytomining. CAB International, Wallingord

Brown PL, Guerin M, Hankin SI, Lowson RT (1998) Uranium and other contaminant migration in groundwater at a tropical Australian uranium mine. J Contam Hydrol 35:295–303

Brown M, Barley B, Wood H (2002) Minewater treatment: technology, application and policy. International Water Association Publishing

Brownlow AH (1996) Geochemistry, 2nd edn. Prentice Hall, Upper Saddle River

Brzezinski LS (2001) Surface disposal of thickened tailings. Mining Environmental Management 9:7–12

Buckby T, Black S, Coleman ML, Hodson ME (2003) Fe-sulphate-rich evaporative mineral precipitates from the Rio Tinto, southwest Spain. Min Mag 67:263–278

Bullock SET, Bell FG (1997) Some problems associated with past mining at a mine in the Witbank coalfield, South Africa. Environ Geol 33:61–71

Burghardt D, Kassahun A (2005) Development of a reactive zone technology for simultaneous in situ immobilisation of radium and uranium. Environ Geol 49:314–320

Burnett WC, Elzerman AW (2001) Nuclide migration and the environmental radiochemistry of Florida phosphogypsum. J Environ Rad 54:27–51

Burnett WC, Cowart JB, LaRock P, Hull CD (1995) Microbiology and radiochemistry of phosphogypsum. Florida Institute of Phosphate Research, publication no 05-035-115

Burns PC (1999) The crystal chemistry of uranium. In: Burns PC, Finch R (eds) Uranium: mineralogy, geochemistry and the environment. Mineralogical Society of America, Washington (Reviews in mineralogy and geochemistry, vol 38, pp 23–90)

Bussière B, Benzaazoua M, Aubertin M, Mbonimpa M (2004) A laboratory study of covers made of low-sulphide tailings to prevent acid mine drainage. Environ Geol 45:609–622

Butler TW, Seitz JC (2006) Apparent seasonal variations in iron photoreduction in acidic discharge from a former pyrite mine, Oakland, California. Appl Geochem 21:1109–1122

Bycroft BM, Coller BAW, Deacon GB, Coleman DJ, Lake PS (1982) Mercury contamination of the Lerderderg River, Victoria, Australia, from an abandoned gold field. Environ Poll Series A 28:135–147

Cabral A, Lefebvre G, Proulx ME, Audet C, Labbé M, Michaud C (1997) Use of deinking residues as cover material in the prevention of AMD generation at an abandoned mine site. In: Tailings and mine waste '97. Balkema, Rotterdam, pp 257–266

Caldwell E, Johnson J (1997) Use of risk-based standards for restoration of ground water at in situ uranium mines. In: Tailings and mine waste '97. Balkema, Rotterdam, pp 21–25

Campbell DL, Fittermann DV (2000) Geoelectrical methods for investigating mine dumps. In: Proceedings from the 5th international conference on acid rock drainage, vol 2. Society for Mining, Metallurgy, and Exploration, Littleton, pp 1513–1523

Carlson L, Kumpulainen S (2000) Mineralogy of precipitates formed from mine effluents in Finland. In: Tailings and mine waste '00. Balkema, Rotterdam, pp 269–275

Carroll SA, O'Day PA, Piechowski M (1998) Rock-water interactions controlling zinc, cadmium, and lead concentrations in surface waters and sediments, US tri-state mining district. 2. Geochemical interpretation. Environ Sci Technol 32:956–965

Casagrande DJ, Finkelman RB, Caruccio FT (1989) The nonparticipation of organic sulfur in acid mine drainage. Environ Geochem Health 11:187–192

Casiot C, Morin G, Juillot F, Bruneel O, Personné JC, Leblanc M, Duquesne K, Bonnefoy V, Elbaz-Poulichet F (2003) Bacterial immobilization and oxidation of arsenic in acid mine drainage (Carnoulès creek, France). Water Res 37:2929–2936

Casiot C, Bruneel O, Personné JC, Leblanc M, Elbaz-Poulichet F (2004) Arsenic oxidation and bioaccumulation by the acidophilic protozoan, *Euglena mutabilis*, in acid mine drainage (Carnoulès, France). Sci Total Environ 320:259–267

Castelo-Branco MA, Santos J, Moreira O, Oliveira A, Pereira Pires F, Magalhaes I, Dias S, Fernandes LM, Gama J, Vieira e Silva JM, Ramalho Ribeiro J (1999) Potential use of pyrite as an amendment for calcareous soil. J Geochem Explor 66:363–367

Castendyk DN, Mauk JL, Webster JG (2005) A mineral quantification method for wall rocks at open pit mines, and application to the Martha Au-Ag mine, Waihi, New Zealand. Appl Geochem 20:135–156

Castro JM, Moore JN (2000) Pit lakes: their characteristics and the potential for their remediation. Environ Geol 39:1254–1260

Cathles LM (1994) Attempts to model the industrial-scale leaching of copper-bearing mine waste. In: Alpers CN, Blowes DW (eds) Environmental geochemistry of sulfide oxidation. American Chemical Society, Washington DC (Symposium Series 550, pp 123–131)

Chapman BM, Jones DR, Jung RF (1983) Processes controlling metal ion attenuation in acid mine drainage streams. Geochim Cosmochim Acta 47 1957–1973

Chermak JA, Runnells DD (1996) Self-sealing hardpan barriers to minimize infiltration of water into sulfide-bearing overburden, ore, and tailings piles. In: Tailings and mine waste '96. Balkema, Rotterdam, pp 265–273

Chermak JA, Runnells DD (1997) Development of chemical caps in acid drainage environments. Mining Eng 49:93–97

Cidu R, Fanfani L (2002) Overview of the environmental geochemistry of mining districts in southwestern Sardinia, Italy. Geochem Explor Environ Anal 2:243–251

Cleary D, Thornton I (1994) The environmental impact of gold mining in the Brazilian Amazon. In: Hester RE, Harrison RM (eds) Mining and its environmental impact. Issues in Environ Sci Technol 1. Royal Society of Chemistry, Cambridge, pp 17–29

Coggans CJ, Blowes DW, Robertson WD, Jambor JL (1999) The hydrogeochemistry of a nickel-mine tailings impoundment – Copper Cliff, Ontario. In: Filipek LH, Plumlee GS (eds) The environmental geochemistry of mineral deposits. Part B: Case studies and research topics. Society of Economic Geologists, Littleton (Reviews in economic geology, vol 6B, pp 447–465)

Cornell RM, Schwertmann U (1996) The iron minerals: structure, properties, reactions, occurrence and uses. VCH, Weinheim

Costa MC, Duarte JC (2005) Bioremediation of acid mine drainage using acidic soil and organic wastes for promoting sulphate-reducing bacteria activity on a column reactor. Water Air Soil Poll 165:325–345

Courtin-Nomade A, Grosbois C, Bril H, Roussel C (2005) Spatial variability of arsenic in some iron-rich deposits generated by acid mine drainage. Appl Geochem 20:383–396

Covelli S, Faganeli J, Horvat M, Brambati A (2001) Mercury contamination of coastal sediments as the result of long-term cinnabar mining activity (Gulf of Trieste, northern Adriatic sea). Appl Geochem 16:541–558

Cravotta CA (1994) Secondary iron-sulfate minerals as sources of sulfate and acidity: geochemical evolution of acidic groundwater at a reclaimed surface coal mine in Pennsylvania. In: Alpers CN, Blowes DW (eds) Environmental geochemistry of sulfide oxidation. American Chemical Society, Washington DC (Symposium Series 550, pp 345–364)

Cravotta CA (2003) Size and performance of anoxic limestone drains to neutralize acidic mine drainage. J Environ Qual 32:1277–1289

Cravotta CA, Bilger MD (2001) Water-quality trends for a stream draining the Southern Anthracite Field, Pennsylvania. Geochem Explor Environ Anal 1:33–50

Cravotta CA, Trahan MK (1999) Limestone drains to increase pH and remove dissolved metals from acidic mine drainage. Appl Geochem 14:581–606

Cravotta CA, Dugas DL, Brady KBC, Kovalchuk TE (1994) Effects of selective handling pyritic, acid forming materials on the chemistry of pore gas and ground water at a reclaimed surface coal mine, Clarion County, PA, USA In: Proceedings of the international land reclamation and mine drainage conference and 3rd international conference on the abatement of acidic drainage. US Department of the Interior, Bureau of Mines (Special Publication SPO6A-94, pp 365–374)

Craw D (2001) Tectonic controls on gold deposits and their environmental impact, New Zealand. J Geochem Explor 73:43–56

Craw D (2003) Geochemical changes in mine tailings during a transition to pressure-oxidation process discharge, Macraes mine, New Zealand. J Geochem Explor 80:81–94

Craw D, Pacheco (2002) Mobilisation and bioavailability of arsenic around mesothermal gold deposits in a semiarid environment, Otago, New Zealand. The Sci World J 2:308–319

Craw D, Chappell D, Nelson M, Walrond M (1999) Consolidation and incipient oxidation of alkaline arsenopyrite-bearing mine tailings, Macreas Mine, New Zealand. Appl Geochem 14:485–498

Craw D, Wilson N, Ashley PM (2004) Geochemical controls on the environmental mobility of Sb and As at mesothermal antimony and gold deposits. Trans Inst Min Metall B 113:B3–B10

Crock JG, Arbogast BF, Lamothe PJ (1999) Laboratory methods for the analysis of environmental samples. In: Plumlee GS, Logsdon MS (eds) The environmental geochemistry of mineral deposits. Part A: Processes, techniques and health Issues. Society of Economic Geologists, Littleton (Reviews in economic geology, vol 6A, pp 265–287)

Cruz R, Bertrand V, Monroy M, Ganzalez I (2001a) Effect of sulfide impurities on the reactivity of pyrite and pyritic concentrates: a multi-tool approach. Appl Geochem 16:803–819

Cruz R, Mendez BA, Monroy M, Gonzalez I (2001b) Cyclic voltammetry applied to evaluate reactivity in sulfide mining residues. Appl Geochem 16:1631–1640

Currey NA, Ritchie PJ, Durham AJP, Wilson GW (1999) Field performance and optimisation of two low flux soil cover systems for the prevention of acid mine drainage in a semi arid environment. In: Proceedings of the 24th annual Minerals Council of Australia environmental workshop. Minerals Council of Australia, Dickson, pp 458–465

Dalton JB, King TVV, Bove DJ, Kokaly RF, Clark RN, Vance JS, Swayze GA (2000) Distribution of acid-generating and acid-buffering minerals in the Animas River watershed as determined by AVIRIS spectroscopy. In: Proceedings from the 5th international conference on acid rock drainage, vol 2. Society for Mining, Metallurgy, and Exploration, Littleton, pp 1541–1550

Davies MP, Martin TE (2000) Upstream constructed tailings dams – a review of the basics. In: Tailings and mine waste '00. Balkema, Rotterdam, pp 3–15

Davis A, Ashenberg D (1989) The aqueous geochemistry of the Berkeley Pit, Butte, Montana, USA. Appl Geochem 4:23–36

Davis RA Jr, Welty AT, Borrego J, Morales JA, Pendon JG, Ryan JG (2000) Rio Tinto estuary (Spain): 5 000 years of pollution. Environ Geol 39:1107–1116

Davy DR, Levins DM (1984) Air and water borne pollutants from uranium mines and mills. Mining Magazine March 1984:283–291

Demchak J, Morrow T, Skousen J (2001) Treatment of acid mine drainage by four vertical flow wetlands in Pennsylvania. Geochem Explor Environ Anal 1:71–80

Demchak J, Skousen J, McDonald LM (2004) Longevity of acid discharges from underground mines located above regional water table. J Environ Qual 33:656–668

Dempsey BA, Jeon B-H (2001) Characteristics of sludge produced from passive treatment of mine drainage. Geochem Explor Environ Anal 1:89–94

Denimal S, Bertrand C, Mudry J, Paquette Y, Hochart M, Steinmann M (2005) Evolution of the aqueous geochemistry of mine pit lakes – Blanzy-Montceau-les-Mines coal basins (Massif Central, France): origin of sulfate contents; effects of stratification on water quality. Appl Geochem 20:825–839

Descostes M, Vitorge P, Beaucaire C (2004) Pyrite oxidation in acidic media. Geochim Cosmochim Acta 68:4559–4569

Deutsch WJ (1997) Groundwater geochemistry: fundamentals and applications to contamination. Lewis Publishers, Boca Raton

Dinelli E, Tateo F (2001) Sheet silicates as effective carriers of heavy metals in the ophiolitic mine area of Vigonzano (northern Italy). Min Mag 65:121–132

Dinelli E, Lucchini F, Fabbri M, Cortecci G (2001) Metal distribution and environmental problems related to sulfide oxidation in the Libiola copper mine area (Ligurian Apennines, Italy). J Geochem Explor 74:141–152

Dobos SK (2000) Potential problems with geologically uncontrolled sampling and the interpretation of chemical tests for waste characterisation and AMD prediction. 4th Australian workshop on acid mine drainage. Australian Centre for Mining Environmental Research, Brisbane

Dold B (2003) Speciation of the most soluble phases in a sequential extraction procedure adapted for geochemical studies of copper sulfide mine waste. J Geochem Explor 80:55–68

Dold B, Fontbote L (2001) Element cycling and secondary mineralogy in porphyry copper tailings as a function of climate, primary mineralogy, and mineral processing. J Geochem Explor 74:3–55

Domagalski JL (1998) Occurrence and transport of total mercury and methyl mercury in the Sacramento River Basin, California. J Geochem Explor 64:277–291

Domagalski JL (2001) Mercury and methylmercury in water and sediment of the Sacramento River Basin, California. Appl Geochem 16:1677–1691
Domvile SJ, Li MG, Sollner DD, Nesbitt W (1994) Weathering behaviour of mine tailings and waste rock: a surface investigation. In: Proceedings of the international land reclamation and mine drainage conference and 3rd international conference on the abatement of acidic drainage, vol 1. United States Department of the Interior, Bureau of Mines (Special Publication SP06A-94, pp 167–176)
Donahue R, Hendry MJ (2003) Geochemistry of arsenic in uranium mine mill tailings, Saskatchewan, Canada. Appl Geochem 18:1733–1750
Donahue R, Hendry MJ, Landine P (2000) Distribution of arsenic and nickel in uranium mill tailings, Rabbit Lake, Saskatchewan, Canada. Appl Geochem 15:1097–1119
Doyle GA, Runnells DD (1997) Physical limnology of existing mine pit lakes. Mining Eng 49:76–80
Drever JI (1997) The geochemistry of natural waters: surface and groundwater environments. Prentice Hall, Upper Saddle River
Eary LE (1998) Predicting the effects of evapoconcentration on water quality in mine pit lakes. J Geochem Explor 64:223–236
Eary LE (1999) Geochemical and equilibrium tends in mine pit lakes. Appl Geochem 14:963–987
Eary LE, Runnells DD, Esposito KJ (2003) Geochemical controls on ground water composition at the Cripple Creek mining district, Cripple Creek, Colorado. Appl Geochem 18:1–24
East TJ, Uren CJ, Noller BN, Cull RF, Curley PM, Unger CJ (1994) Erosional stability of rehabilitated uranium mine structures incorporating natural landform characteristics, northern tropical Australia. Zeitschrift für Geomorphologie 38:283–298
Edwards KJ, Bond PL, Gihring TM, Banfield JF (2000) An Archaeal iron-oxidizing extreme acidophile important in acid mine drainage. Science 287:1796–1799
Eisenbud M, Gesell T (1997) Environmental radioactivity from natural, industrial and military sources. Academic Press, San Diego
Elbaz-Poulichet F, Leblanc M (1996) High metal fluxes from a major mining province to ocean through acidic rivers (Rio Tinto, Spain). CR Acad Sci II A 322:1047–1052
Elberling B, Schippers A, Sand W (2000) Bacterial and chemical oxidation of pyritic mine tailings at low temperatures. J Contam Hydrol 41:225–238
Elberling B, Asmund G, Kunzendorf H, Krogstad EJ (2002) Geochemical trends in metal-contaminated fiord sediments near a former lead-zinc mine in West Greenland. Appl Geochem 17:493–502
Elless MP, Lee SY (1998) Uranium solubility of carbonate-rich uranium-contaminated soils. Water Air Soil Poll 107:147–162
Elliott LCM, Liu L, Stogran SW (1997) Evaluation of single layer organic and inorganic cover materials for oxidized tailings. In: Tailings and mine waste '97. Balkema, Rotterdam, pp 247–256
El-Mrabet R, Abril J-M, Perianez R, Manjon G, Garcia-Tenorio R, Delgado A, Andreu L (2003) Phosphogypsum amendment effect on radionuclide content in drainage water and marsh soils from southwestern Spain. J Environ Qual 32:1262–1268
Environment Australia (1995) Tailings containment. Best practice environmental management in mining. Environment Australia, Canberra
Environment Australia (1997) Managing sulphidic mine wastes and acid drainage. Best practice environmental management in mining. Environment Australia, Canberra
Environment Australia (1998) Cyanide management. Best practice environmental management in mining. Environment Australia, Canberra
Environment Australia (1999) Water management. Best practice environmental management in mining. Environment Australia, Canberra
Ericson WA, Patel SK, Marek AC, Brown CM (1997) Proposed design for fertilizer gypsum stack and acid pond closure. In: Tailings and mine waste '97. Balkema, Rotterdam, pp 69–77
Eriksson N (1998) Effect of water flow on mobilisation and transport of weathering products in mine waste rock heaps. In: McLean RW, Bell LC (1998) Proceedings of the 3rd Australian acid mine drainage workshop. Australian Centre for Mine site Rehabilitation Research, Brisbane, pp 59–67
Ernst WHO (1998) The origin and ecology of contaminated, stabilized and non-pristine soils. In: Vangronsveld J, Cunningham S (eds) Metal contaminated soils: in situ inactivation and phytorestoration. Springer, Berlin Heidelberg New York, pp 17–29
Erskine DW, Yancey CL, Lawrence EP (1997) Natural attenuation of hazardous constituents in groundwater at uranium mill tailings sites. In: Tailings and mine waste '97. Balkema, Rotterdam, pp 489–498

España JS, Pamo EL, Santofimia E, Aduvire O, Reyes J, Barettino D (2005) Acid mine drainage in the Iberian Pyrite Belt (Odiel river watershed, Huelva, SW Spain): geochemistry, mineralogy and environmental implications. Appl Geochem 20:1320–1356

España JS, Pamo EL, Pastor ES, Andrés JR, Rubi JAM (2006) The impact of acid mine drainage on the water quality of the Odiel River (Huelva, Spain): evolution of precipitate mineralogy and aqueous geochemistry along the Concepcion-Tintillo segment. Water Air Soil Poll 173:121–149

Ettler V, Legendre O, Bodénan F, Touray JC (2001) Primary phases and natural weathering of old lead-zinc pyrometallurgical slag from Pribram, Czech Republic. Can Mineral 39:873–888

Ettler V, Piantone P, Touray JC (2003) Mineralogical control on inorganic contaminant mobility in leachate from lead-zinc metallurgical slag: experimental approach and long-term assessment. Min Mag 67:1269–1283

Ettner DC, Braastad G (1999) Induced hardpan formation in a historic tailings impoundment, Røros, Norway. In: Tailings and mine waste '99. Balkema, Rotterdam, pp 457–464

Evangelou VP (1995) Pyrite oxidation and its control. CRC Press, Boca Raton

Evangelou VP (1998) Pyrite chemistry: the key for abatement of acid mine drainage. In: Geller W, Klapper H, Salomons W (eds) Acidic mining lakes: acid mine drainage, limnology and reclamation. Springer, Berlin Heidelberg New York, pp 197–222

Evangelou VP (2001) Pyrite microencapsulation technologies: principles and potential field applications. Ecol Eng 17:165–178

Evangelou VP, Zhang YL (1995) A review: pyrite oxidation mechanisms and acid mine drainage prevention. Crit Rev Environ Sci Technol 25:141–199

Ewing RC (1999) Radioactivity and the 20th century. In: Burns PC, Finch R (eds) Uranium: mineralogy, geochemistry and the environment. Mineralogical Society of America, Washington (Reviews in mineralogy and geochemistry, vol 38, pp 1–21)

Fala O, Molson J, Aubertin M, Bussiere B (2005) Numerical modeling of flow and capillary barrier effects in unsaturated waste rock piles. Mine Water Environ 24:172–185

Falk H, Lavergren U, Bergbäck B (2006) Metal mobility in alum shale from Öland, Sweden. J Geochem Explor 90:157–165

Fennemore GG, Neller WC, Davis A (1998) Modeling pyrite oxidation in arid environments. Environ Sci Technol 32:2680–2687

Ferguson KD, Erickson PM (1988) Pre-mine prediction of acid mine drainage. In: Salomons W, Förstner U (eds) Environmental management of solid waste, dredged material and mine tailings. Springer, Berlin Heidelberg New York, pp 24–43

Fernandes HM, Franklin MR, Veiga LH (1998) Acid rock drainage and radiological environmental impacts. A study case of the uranium mining and milling facilities at Pocos de Caldas. Waste Management 18:169–181

Ferrara R (1999) Mercury mines in Europe: assessment of emissions and environmental contamination. In: Ebinghaus R, Turner RR, de Lacerda LD, Vasiliev O, Salomons W (eds) Mercury contaminated sites. Springer, Berlin Heidelberg New York, pp 51–72

Ferris FG, Tazaki K, Fyfe WS (1989) Iron oxides in acid mine drainage environments and their association with bacteria. Chem Geol 74:321–330

Ficklin WH, Mosier EL (1999) Field methods for sampling and analysis of environmental samples for unstable and selected stable constituents. In: Plumlee GS, Logsdon MS (eds) The environmental geochemistry of mineral deposits. Part A: Processes, techniques and health issues. Society of Economic Geologists, Littleton (Reviews in Economic Geology, vol 6A, pp 249–264)

Ficklin WH, Plumlee GS, Smith KS, McHugh JB (1992) Geochemical classification of mine drainages and natural drainages in mineralized areas. In: Kharaka YK, Maest AS (eds) Proceedings of water-rock interaction no 7. Balkema, Rotterdam, pp 381–384

Finch R, Murakami T (1999) Systematics and paragenesis of uranium minerals. In: Burns PC, Finch R (eds) Uranium: mineralogy, geochemistry and the environment. Mineralogical Society of America, Washington (Reviews in mineralogy and geochemistry, vol 38, pp 91–179)

Fischer RC, Viskocil KH, Thompson LC, Caldwell CS, Seep AE, Williams L, Hubbard J (1996) Pilot-scale biotreatment of cyanide and metal-containing wastewater at the Summitville Mine Superfund Site. In: Tailings and mine waste '96. Balkema, Rotterdam, pp 535–545

Flanagan JC, Morton WH, Ward TA (1983) Groundwater management around uranium mine waste areas, Mary Kathleen, Australia. In: International conference on groundwater and man. Australian Water Resources Council Conference Series no 8, pp 81–88

Förstner U (1999) Introduction. In: Azcue JM (ed) Environmental impacts of mining activities: emphasis on mitigation and remedial measures. Springer, Berlin Heidelberg New York, pp 1–3
Foos AM (1997) Geochemical modeling of coal mine drainage, Summit County, Ohio. Environ Geol 31:205–210
Ford RC (2000) Closure of copper heap leach facilities in semi-arid and arid climates: challenges for strategic management of acid drainage. In: Proceedings from the 5th international conference on acid rock drainage, vol 2. Society for Mining, Metallurgy, and Exploration, Littleton, pp 1283–1289
Ford RC, Clapper JL, Romig BR (1998) Stability of water treatment sludge, Climax Mine, Colorado. Part I: Chemical and mineralogical properties. In: Tailings and mine waste '98. Balkema, Rotterdam, pp 645–656
Forsberg LS, Ledin S (2003) Effects of iron precipitation and organic amendments on porosity and penetrability in sulphide mine tailings. Water Air Soil Poll 142:395–408
Forsberg LS, Ledin S (2006) Effects of sewage sludge on pH and plant availability of metals in oxidizing sulphide mine tailings. Sci Total Environ 358:21–35
Foster AL, Brown Jr GE, Tingle TN, Parks GA (1998) Quantitative arsenic speciation in mine tailings using X-ray absorption spectroscopy. Amer Mineral 83:553–568
Fowler TA, Holmes PR, Crundwell FK (1999) Mechanism of pyrite dissolution in the presence of *Thiobacillus ferrooxidans*. Appl Environ Microbiol 65:2987–2993
Frau F (2000) The formation-dissolution-precipitation cycle of melanterite at the abandoned pyrite mine of Genna Luas in Sardinia, Italy: environmental implications. Min Mag 64:995–1006
Frostad S, Klein B, Lawrence RW (2002) Evaluation of laboratory kinetic test methods for measuring rates of weathering. Mine Water Environ 21:183–192
Fuge R, Pearce FM, Pearce NJG, Perkins WT (1994) Acid mine drainage in Wales and influence of ochre precipitation on water chemistry. In: Alpers CN, Blowes DW (eds) Environmental geochemistry of sulfide oxidation. American Chemical Society, Washington DC (Symposium Series 550, pp 261–274)
Fukuschi K, Sasaki M, Sato T, Yanase N, Amano H, Ikeda H (2003) A natural attenuation of arsenic in drainage from an abandoned arsenic mine dump. Appl Geochem 18:1267–1278
Furniss G, Hinman NW, Doyle GA, Runnells DD (1999) Radiocarbon-dated ferricrete provides a record of natural acid rock drainage and paleoclimatic changes. Environ Geol 37:102–106
Fyfe WS (1981) The environmental crisis: quantifying geosphere interactions. Science 213:105–110
Fytas K, Evangelou B (1998) Phosphate coating on pyrite to prevent acid mine drainage. Int J Surface Min Reclam Environ 12:101–104
Gaines M (2000) Radiation and risk. New Scientist 2230 (Inside Science, 4 pp)
Gammons CH, Nimick DA, Parker SR, Cleasby TE, McCleskey RB (2005) Diel behaviour of iron and other heavy metals in a mountain stream with acidic to neutral pH: Fisher Creek, Montana, USA. Geochim Cosmochim Acta 69:2505–2516
Ganguli PM, Mason RP, Abu-Saba KE, Anderson RS, Flegal AR (2000) Mercury speciation in drainage from the New Idria mercury mine, California. Environ Sci Technol 34:4773–4779
Ganne P, Cappuyns V, Vervoort A, Buvé, Swennen R (2006) Leachability of heavy metals and arsenic from slags of metal extraction industry at Angleur (eastern Belgium). Sci Total Environ 356:69–85
Gault AG, Cooke DR, Townsend AT, Charnock JM, Polya DA (2005) Mechanisms of arsenic attentuation in acid mine drainage from Mount Bischoff, western Tasmania. Sci Total Environ 345:219–228
Gazeba B, Adam K, Kontopoulos A (1996) A review of passive systems for the treatment of acid mine drainage. Mineral Eng 9:23–42
Geldenhuis S, Bell FG (1998) Acid mine drainage at a coal mine in the eastern Tranvaal, South Africa. Environ Geol 34:234–242
Genevois R, Tecca PR (1993) The tailings dams of Stava (northern Italy): an analysis of the disaster. In: Chowdhury RN, Sivakumar SM (eds) Environmental management geo-water and engineering aspects. Balkema, Rotterdam, pp 23–36
Georgopoulou ZJ, Fytas K, Soto H, Evangelou B (1996) Feasibility and cost of creating an iron-phosphate coating on pyrrhotite to prevent oxidation. Environ Geol 28:61–69
Gerhardt A, Janssens de Bisthoven L, Soares AMVM (2004) Macroinvertebrate response to acid mine drainage: community metrics and on-line behavioural toxicity bioassay. Environ Poll 130:263–274
Gerke HH, Molson JW, Frind EO (2001) Modelling the impact of physical and chemical heterogeneity on solute leaching in pyritic overburden mine spoils. Ecol Eng 17:91–101

Gessner TP, Kadlec RH, Reaves RP (2005) Wetland remediation of cyanide and hydrocarbons. Ecol Eng 25:457–469
Getaneh W, Alemayehu T (2006) Metal contamination of the environment by placer and primary gold mining in the Adola region of southern Ethiopia. Environ Geol 50:339–352
Ghomshei MM, Allen DM (2000) Hydrochemical and stable isotope assessment of tailings pond leakage, Nickel Plate Mine, British Columbia. Environ Geol 39:937–944
Gibert O, de Pablo J, Cortina JL, Ayora C (2005) Municipal compost-based mixture for acid mine drainage bioremediation: metal retention mechanisms. Appl Geochem 20:1648–1657
Gibson DK, Ritchie AIM (1991) Options to control acid generation in existing pyritic mine waste dumps. In: Randol Gold Forum Cairns '91, pp 109–111
Gieré R, Sidenko NV, Lavareva EV (2003) The role of secondary minerals in controlling the migration of arsenic and metals from high-sulfide wastes (Berikul gold mine, Siberia). Appl Geochem 18:1347–1359
Gleisner M, Herbert Jr RB, Frogner Kockum PC (2006) Pyrite oxidation by *Acidithiobacillus ferrooxidans* at various concentrations of dissolved oxygen. Chem Geol 225:16–29
Gnandi K, Tobschall HJ (1999) The pollution of marine sediments by trace elements in the coastal region of Togo caused by dumping of cadmium-rich phosphorite tailing into the sea. Environ Geol 38:13–24
Gnandi K, Tchangbedji G, Killi K, Baba G, Abbe K (2006) The impact of phosphate mine tailings on the bioaccumulation of heavy metals in marine fish and crustaceans from the coastal zone of Togo. Mine Water Environ 25:56–62
Goh SW, Buckley AN, Lamb RN, Rosenberg RA, Moran D (2006) The oxidation states of copper and iron in mineral sulfides, and the oxides formed on initial exposure of chalcopyrite and bornite to air. Geochim Cosmochim Acta 70:2210–2228
Gould WD, Kapoor A (2003) The microbiology of acid mine drainage. In: Jambor JL, Blowes DW, Ritchie AIM (eds) Environmental aspects of mine wastes. Mineralogical Association of Canada, Nepean (Short course handbook, vol 31, pp 203–226)
Gould WD, Bechard G, Lortie L (1994) The nature and role of microorganisms in the tailings environment. In: Jambor JL, Blowes DW (eds) Environmental geochemistry of sulfide mine-wastes. Mineralogical Association of Canada, Nepean (Short course handbook, vol 22, pp 185–199)
Goulden WD, Hendry MJ, Clifton AW, Barbour SL (1998) Characterization of radium-226 in uranium mill tailings. In: Tailings and mine waste '98. Balkema, Rotterdam, pp 561–570
Gray NF (1997) Environmental impact and remediation of acid mine drainage: a management problem. Environ Geol 30:62–71
Gray NF (1998) Acid mine drainage composition and the implications for its impact on lotic systems. Water Res 32:2122–2134
Gray JE, Crock JG, Fey DL (2002a) Environmental geochemistry of abandoned mercury mines in West-Central Nevada, USA. Appl Geochem 17:1069–1079
Gray JE, Crock JG, Lasorsa BK (2002b) Mercury methylation at mercury mines in the Humboldt River Basin, Nevada, USA. Geochem Explor Environ Anal 2:143–149
Greben HA, Maree JP (2005) Removal of sulphate, metals, and acidity from a nickel and copper mine effluent in a laboratory scale bioreactor. Mine Water Environ 24:194–198
Haas JR, Bailey EH, Purvis OW (1998) Bioaccumulation of metals by lichens: uptake of aqueous uranium by *Peltigera membranacea* as a function of time and pH. Amer Mineral 83:1494–1502
Habashi F, Awadalla FT (1986) The recovery of uranium and lanthanides during the production of nitrophosphate fertilisers using tertiary amyl alcohol. J Chem Technol Biotechnol 36:1–6
Hallberg KB, Johnson DB (2005) Mine water microbiology. Mine Water Environ 24:28–37
Hamilton SJ, Buhl KJ (2004) Selenium in water, sediment, plants, invertebrates, and fish in the Blackfoot River drainage. Water Air Soil Poll 159:3–34
Hammack RW, Sams JI, Veloski GA, Mabie JS (2003a) Geophysical investigation of the Sulphur Bank Mercury Mine superfund site, Lake county, California. Mine Water Environ 22:69–79
Hammack RW, Love EI, Veloski GA, Ackman TE, Harbert W (2003b) Using helicopter electromagnetic surveys to identify environmental problems at coal mines. Mine Water Environ 22:80–84
Hammarstrom JM, Sibrell PL, Belkin HE (2003) Characterization of limestone reacted with acid-mine drainage in a pulsed limestone bed treatment system at the Friendship Hill National Historical Site, Pennsylvania, USA. Appl Geochem 18705–1721

Hancock GR, Grabham MK, Martin P, Evans KG, Bollhöfer A (2006) A methodology for the assessment of rehabilitation success of post mining landscapes – sediment and radionuclide transport at the former Nabarlek uranium mine, Northern Territory, Australia. Sci Total Environ 354:103–119

Harries JR (1997) Acid mine drainage in Australia: its extent and potential future liability. Supervising Scientist report 125. Supervising Scientist, Canberra

Harries JR, Ritchie AIM (1987) The effect of rehabilitation on the rate of oxidation of pyrite in a mine waste rock dump. Environ Geochem Health 9:27–36

Harries JR, Ritchie AIM (1988) Rehabilitation measures at the Rum Jungle mine site. In: Salomons W, Förstner U (eds) Environmental management of solid waste, dredged material and mine tailings. Springer, Berlin Heidelberg New York, pp 131–151

Harris DL, Lottermoser BG (2006a) Phosphate stabilization of polyminerallic mine wastes. Min Mag 70:1–13

Harris DL, Lottermoser BG (2006b) Evaluation of phosphate fertilizers for ameliorating acid mine waste. Appl Geochem 21:1216–1225

Harrison D (2002) Minesite remediation with seawater neutralised red mud. Groundwork 5:32–33

Hassinger BW (1997) Erosion. In: Marcus JJ (ed) Mining environmental handbook: effects of mining on the environment and American environmental controls on mining. Imperial College Press, London, pp 136–140

Hawkins JW (1999) Hydrogeologic characteristics of surface-mine spoil. In: Coal mine drainage prediction and pollution prevention in Pennsylvania. The Pennsylvania Department of Environmental Protection, pp 3-1 to 3-11 (*http://www.dep.state.pa.us/dep/deputate/minres/districts/cmdp/main.htm*)

Hedin RS, Nairn RW, Kleinmann RLP (1994a) Passive treatment of coal mine drainage. US Bureau of Mines Information Circular 9389, US Department of the Interior, Bureau of Mines, Pittsburgh

Hedin RS, Watzlaff GR, Nairn RW (1994b) Passive treatment of acid mine drainage with limestone. J Environ Qual 23:1338–1345

Heijde HB van der, Klijn PJ, Duursma K, Eisma D, de Groot AJ, Hagel P, Köster HW, Nooyen JL (1990) Environmental aspects of phosphate fertilizer production in the Netherlands with particular reference to the disposal of phosphogypsum. Sci Total Environ 90:203–225

Herbert RB (2003) Zinc immobilization by zerovalent Fe: surface chemistry and mineralogy of reaction products. Min Mag 67:1285–1298

Herbert RB (2006) Seasonal variations in the composition of mine drainage-contaminated groundwater in Dalarna, Sweden. J Geochem Explor 90:197–214

Herr C, Gray NF (1997) Sampling riverine sediments impacted by acid mine drainage: problems and solutions. Environ Geol 29:37–45

Hesse CA, Ellis DV (1995) Quartz Hill, Alaska: a case history of engineering and environmental requirements for STD in the USA. Marine Geores Geotechnol 13:135–182

Hettler J, Lehmann B (1995) Environmental impact of large-scale mining in Papua New Guinea: mining residue disposal by the Ok Tedi copper-gold mine. A176-II, Berliner Geowissenschaftliche Abhandlungen, Freie Universität Berlin

Hettler J, Irion G, Lehmann B (1997) Environmental impact of mining waste disposal on a tropical lowland river system: a case study on the Ok Tedi mine, Papua New Guinea. Mineralium Deposita 32:280–291

Higueras P, Oyarzun R, Oyarzun J, Maturana H, Lillo J, Morata D (2004) Environmental assessmemt of copper-gold-mercury mining in the Andacollo and Punitaqui districts, northern Chile. Appl Geochem 19:1855–1864

Hill FC (1999) Identification of selected uranium-bearing minerals and inorganic phases by X-ray powder diffraction. In: Burns PC, Finch R (eds) Uranium: mineralogy, geochemistry and the environment. Mineralogical Society of America, Washington (Reviews in mineralogy and geochemistry, vol 38, pp 653–679)

Hinton JJ, Veiga MM (2002) Earthworms as bioindicators of mercury pollution from mining and other industrial activities. Geochem Explor Environ Anal 2:269–274

Hodson ME (2004) Heavy metals: geochemical bogey men? Environ Poll 129:341–343

Hodson ME, Valsami-Jones E, Cotter-Howells JD (2000) Metal phosphates and remediation of contaminated land. In: Cotter-Howells JD, Campbell LS, Valsami-Jones E, Batchelder M (eds) Environmental mineralogy: microbial interactions, anthropogenic influences, contaminated land and waste management. Mineralogical Society, London (Mineralogical Society Series no 9, pp 291–311)

Hollings P, Hendry MJ, Nicholson RV, Kirkland RA (2001) Quantification of oxygen consumption and sulphate release rates for waste rock piles using kinetic cells: Cluff lake uranium mine, northern Saskatchewan, Canada. Appl Geochem 16:1215–1230
Holmström H, Öhlander B (1999) Oxygen penetration and subsequent reactions in flooded sulphidic mine tailings: a study at Stekenjokk, northern Sweden. Appl Geochem 14:747–759
Holmström H, Öhlander B (2001) Layers rich in Fe- and Mn-oxyhydroxides formed at the tailings-pond water interface, a possible trap for trace metals in flooded mine tailings. J Geochem Explor 74:189–203
Holmström H, Ljungberg J, Ekström M, Öhlander B (1999) Secondary copper enrichment in tailings at the Laver mine, northern Sweden. Environ Geol 38:327–342
Hong S, Candelone JP, Patterson CC, Boutron CF (1996) History of ancient copper smelting pollution during Roman and Medieval times recorded in Greenland ice. Science 272:246–249
Hossner LR, Hons FM (1992) Reclamation of mine tailings. Adv Soil Sci 17:311–350
Hozhina EI, Khramov AA, Gerasimov PA, Kumarkov AA (2001) Uptake of heavy metals, arsenic, and antimony by aquatic plants in the vicinity of ore mining and processing industries. J Geochem Explor 74:153–162
Huang X, Evangelou VP (1994) Suppression of pyrite oxidation rate by phosphate addition. In: Alpers CN, Blowes DW (eds) Environmental geochemistry of sulfide oxidation. American Chemical Society, Washington DC (Symposium Series 550, pp 562–573)
Hudson-Edwards KA, Edwards SJ (2005) Mineralogical controls on storage of As, Cu, Pb and Zn at the abandoned Mathiatis massive sulphide mine, Cyprus. Min Mag 69:695–706
Hudson-Edwards KA, Schell C, Macklin MG (1999) Mineralogy and geochemistry of alluvium contaminated by metal mining in the Rio Tinto area, southwest Spain. Appl Geochem 14:1015–1030
Hudson-Edwards KA, Macklin MG, Jamieson HE, Brewer PA, Coulthard TJ, Howard AJ, Turner JN (2003) The impact of tailings dam spills and clean-up operations on sediment and water quality in river systems: the Rios Agrio-Guadiamar, Aznalcollar, Spain. Appl Geochem 18:221–239
Hulshof AHM, Blowes DW, Gould WD (2006) Evaluation of in situ layers for treatment of acid mine drainage: a field comparison. Water Res 40:1816–1826
Hutchison I, Ellison R (1992) Mine waste management. Lewis Publishers, Boca Raton
Hynes TP, Harrison J, Bonitenko E, Doronina TM, Baikowitz H, James M, Zinck JM (1998) The International scientific commission's assessment of the impact of the cyanide spill at Barskaun, Kyrgyz Republic May 20, 1998. Mining and Mineral Sciences Laboratories Report MMSL 98-039(CR), CANMET (*http://envirolab.nrcan.gc.ca/spill-e.htm*)
Iribar V, Izco F, Tames P, Antigüedad I, da Silva A (2000) Water contamination and remedial measures at the Troya abandoned Pb-Zn mine (The Basque country, northern Spain). Environ Geol 39:800–806
Jacobson G, Sparksman GF (1988) Acid mine drainage at Captains Flat, New South Wales. BMR Journal of Australian Geology and Geophysics 10:391–393
Jambor JL (1994) Mineralogy of sulfide-rich tailings and their oxidation products. In: Jambor JL, Blowes DW (eds) Environmental geochemistry of sulfide mine-wastes. Mineralogical Association of Canada, Nepean (Short course handbook, vol 22, pp 59–102)
Jambor JL (2000) The relationship of mineralogy to acid- and neutralization-potential values in ARD. In: Cotter-Howells JD, Campbell LS, Valsami-Jones E, Batchelder M (eds) Environmental mineralogy: microbial interactions, anthropogenic influences, contaminated land and waste management. Mineralogical Society, London (Mineralogical Society Series no 9, pp 141–159)
Jambor JL (2003) Mine-waste mineralogy and mineralogical perspectives of acid-base accounting. In: Jambor JL, Blowes DW, Ritchie AIM (eds) Environmental aspects of mine wastes. Mineralogical Association of Canada, Nepean (Short course handbook, vol 31, pp 117–145)
Jambor JL, Nordstrom DK, Alpers CN (2000a) Metal sulfate salts from sulfide mineral oxidation. In: Alpers CN, Jambor JL, Nordstrom (eds) Sulfate minerals: crystallography, geochemistry and environmental significance. Mineralogical Society of America, Washington (Reviews in mineralogy and geochemistry, vol 40, pp 303–350)
Jambor JL, Blowes DW, Ptacek CJ (2000b) Mineralogy of mine wastes and strategies for remediation. In: Vaughan DJ, Wogelius RA (eds) Environmental mineralogy. EMU Notes in Mineralogy, vol 2, pp 255–290
Jambor JL, Dutrizac JE, Chen TT (2000c) Contribution of specific minerals to the neutralization potential in static tests. In: Proceedings from the 5th international conference on acid rock drainage, vol 1. Society for Mining, Metallurgy, and Exploration, Littleton, pp 551–565

Jambor JL, Dutrizac JE, Raudsepp M, Groat LA (2003) Effect of peroxide on neutralization-potential values of siderite and other carbonate minerals. J Environ Qual 32:2373–2378
James LP (1994) The mercury "tromol" mill: an innovative gold recovery technique, and a possible environmental concern. J Geochem Explor 50:493–500
Janzen MP, Nicholson RV, Scharper JM (2000) Pyrrhotite reaction kinetics: reaction rates for oxidation by oxygen, ferric iron, and for nonoxidative dissolution. Geochim Cosmochim Acta 64:1511–1522
Jarvis I, Burnett WC, Nathan Y, Almbaydin FSM, Attia AKM, Castro LN, Flicoteaux R, Hilmy ME, Husain V, Qutawnah AA, Serjani A, Zanin YN (1994) Phosphorite geochemistry: state-of-the-art and environmental concerns. Ecl Geol Helv 87:643–700
Jarvis AP, Moustafa M, Orme PHA, Younger PL (2006) Effective remediation of grossly polluted acidic, and metal-rich, spoil heap drainage using a novel, low-cost, permeable reactive barrier in Northumberland, UK. Environ Poll 143:261–268
Jaxel R, Gelaude P (1986) New mineral occurrences from the Laurion slags. Mineral Rec 17:183–190
Jeffrey J, Marshman N, Salomons W (1988) Behaviour of trace metals in a tropical river system affected by mining. In: Salomons W, Förstner U (eds) Chemistry and biology of solid waste: dredged materials and mine tailings. Springer, Berlin Heidelberg New York, pp 259–274
Jennings SR, Dollhopf DJ, Inskeep WP (2000) Acid production from sulfide minerals using hydrogen peroxide weathering. Appl Geochem 15:235–243
Jerz JK, Rimstidt JD (2003) Efflorescent iron sulfate minerals: paragenesis, relative stability, and environmental impact. Amer Mineral 88:1919–1932
Jönsson J, Lövgren L (2000) Sorption properties of secondary iron precipitates in oxidized mining waste. In: Proceedings from the 5th international conference on acid rock drainage, vol 1. Society for Mining, Metallurgy, and Exploration, Littleton, pp 115–123
Jönsson J, Persson P, Sjöberg S, Lövgren L (2005) Schwertmannite precipitated from acid mine drainage: phase transformation, sulphate release and surface properties. Appl Geochem 20:179–191
Jönsson J, Jönsson J, Lövgren L (2006) Precipitation of secondary Fe(III) minerals from acid mine drainage. Appl Geochem 21:437–445
Johansson L, Xydas C, Messios N, Stoltz E, Greger M (2005) Growth and Cu accumulation by plants grown on Cu containing mine tailings in Cyprus. Appl Geochem 20:101–107
Johnson DB (1998a) Biological abatement of acid mine drainage: the role of acidophilic protozoa and other indigenous microflora. In: Geller W, Klapper H, Salomons W (eds) Acidic mining lakes. Springer, Berlin Heidelberg New York, pp 285–301
Johnson DB (1998b) Biodiversity and ecology of acidophilic microorganisms. FEMS Microbiol Ecol 27:307–317
Johnson DB, Hallberg KB (2005a) Biogeochemistry of the compost bioreactor components of a composite acid mine drainage passive remediation system. Sci Total Environ 338:81–93
Johnson DB, Hallberg KB (2005b) Acid mine drainage remediation options: a review. Sci Total Environ 338:3–14
Johnson JR, Traub RJ (1996) Risk estimates for uses of phosphogypsum. Florida Institute of Phosphate Research (Publication no 05-041-124)
Johnson KL, Younger PL (2005) Rapid manganese removal from mine waters using an aerated packed-bed bioreactor. J Environ Qual 34:987–993
Johnson SW, Rice SD, Moles DA (1998) Effects of submarine mine tailings disposal on juvenile yellowfin sole (*Pleuronectes asper*): a laboratory study. Marine Poll Bull 36:278–287
Johnson RH, Blowes DW, Robertson WD, Jambor JL (2000) The hydrogeochemistry of the Nickel Rim mine tailings impoundment, Sudbury, Ontario. J Contam Hydrol 41:49–80
Johnson CA, Leinz RW, Grimes DJ, Rye RO (2002) Photochemical changes in cyanide speciation in drainage from a precious metal ore heap. Environ Sci Technol 36:840–845
Jones DR, Chapman BM (1995) Wetlands to treat AMD – facts and fallacies. In: Grundon NJ, Bell LC (eds) Proceedings of the 2nd Australian acid mine drainage workshop. Australian Centre for Minesite Rehabilitation Research, Brisbane, pp 127–145
Jones SG, Ellis DV (1995) Deep water STD at the Misima gold and silver mine, Papua New Guinea. Marine Geores Geotechnol 13:183–200

Jones DR, Laszcyk H, Ellerbroek (1998) Wetlands for control of acid mine drainage in tailings. In: McLean RW, Bell LC (eds) Proceedings of the 3rd Australian acid mine drainage workshop. Australian Centre for Minesite Rehabilitation Research, Brisbane, pp 161–170

Jong de JM, Bennett R, McPhail DC, Dunkerley DL (2000) Abandoned alluvial tin mines in NE Tasmania: characterisation for successful remediation. Geological Society of Australia, Sydney (15th Australian geological convention, abstracts no 59, p 124)

Jurjovec J, Ptacek CJ, Blowes DW (2002) Acid neutralization mechanisms and metal release in mine tailings: a laboratory column experiment. Geochim Cosmochim Acta 66:1511–1523

Kafwembe BS, Veasey TJ (2001) The problems of artisan mining and mineral processing. Mining Environmental Management, November 2001:17–21

Kairies CL, Capo RC, Watzlaf GR (2005) Chemical and physical properties of iron hydroxide precipitates associated with passively treated coal mine drainage in the Bituminous Region of Pennsylvania and Maryland. Appl Geochem 20:1445–1460

Kalin M (2004) Passive mine water treatment: the correct approach? Ecol Eng 22:299–304

Kalin M, Fyson A, Wheeler WN (2006) The chemistry of conventional and alternative treatment systems for the neutralization of acid mine drainage. Sci Total Environ 366:395–408

Kambey JL, Farrell AP, Bendell-Young LI (2001) Influence of illegal gold mining on mercury levels in fish of North Sulawesi's Minahasa Peninsula, (Indonesia). Environ Poll 114:299–302

Karathanasis AD, Johnson CM (2003) Metal removal potential by three aquatic plants in an acid mine drainage wetland. Mine Water Environ 22:22–30

Kargbo DM, He J (2004) A simple accelerated rock weathering method to predict acid generation kinetics. Environ Geol 46:775–783

Kathren RL (1998) NORM sources and their origins. Appl Rad Isotop 49:149–168

Keith CN, Vaughan DJ (2000) Mechanisms and rates of sulphide oxidation in relation to the problems of acid rock (mine) drainage. In: Cotter-Howells JD, Campbell LS, Valsami-Jones E, Batchelder M (eds) Environmental mineralogy: microbial interactions, anthropogenic influences, contaminated land and waste management. Mineralogical Society, London (Mineralogical Society Series no 9, pp 117–139)

Keith DC, Runnells DD, Esposito KJ, Chermak JA, Hannula SR (1999) Efflorescent sulfate salts: chemistry, mineralogy, and effects on ARD streams. In: Tailings and mine waste '99. Balkema Publishers, Rotterdam, pp 573–579

Keith DC, Runnells DD, Esposito KJ, Chermak JA, Levy DB, Hannula SR, Watts M, Hall L (2001) Geochemical models of the impact of acidic groundwater and evaporative sulfate salts on Boulder Creek at Iron Mountain, California. Appl Geochem 16:947–961

Kelly DP, Wood AP (2000) Reclassification of some species of *Thiobacillus* to the newly designated genera *Acidithiobacillus* gen. nov., *Halothiobacillus* gen. nov. and *Thermithiobacillus* gen. nov. Int J Syst Evol Microbiol 50:511–516

Khalil NF, Alnuaimi NM, Mustafa MH (1990) Utilizations of phosphogypsum in Iraq. In: Chang WF (ed) Proceedings of the 3rd international symposium on phosphogypsum. Florida Institute of Phosphate Research (Publication no 01-060-083)

Kim H, Benson CH (2004) Contributions of advective and diffusive oxygen transport through multilayer composite caps over mine waste. J Contam Hydrol 71:193–218

Kim JJ, Kim SJ (2004) Seasonal factors controlling mineral precipitation in the acid mine drainage at Donghae coal mine, Korea. Sci Total Environ 325:181–191

Kim JJ, Kim SJ, Tazaki K (2002) Mineralogical characterization of microbial ferrihydrite and schwertmannite, and non-biogenic Al-sulfate precipitates from acid mine drainage in the Donghae mine area, Korea. Environ Geol 42:19–31

Kim CS, Rytuba JJ, Brown GE (2004) Geological and anthropogenic factors influencing mercury section in mine wastes: an EXAFS spectroscopy study. Appl Geochem 19:379–393

Kirby CS, Cravotta CA (2005a) Net alkalinity and net acidity 1: theoretical considerations. Appl Geochem 20:1920–1940

Kirby CS, Cravotta CA (2005b) Net alkalinity and net acidity 2: practical considerations. Appl Geochem 20:1941–1964

Kirby CS, Decker SM, Macander NK (1999) Comparison of color, chemical and mineralogical compositions of mine drainage sediments to pigment. Environ Geol 37:243–254

Kleinmann RLP (1989) Acid mine drainage in the United States: controlling the impact on streams and rivers. In: 4th world congress on the conservation of built and natural environments. University of Toronto, pp 1–10

Kleinmann RLP (1997) Mine drainage systems. In: Marcus JJ (ed) Mining environmental handbook: effects of mining on the environment and American environmental controls on mining. Imperial College Press, London, pp 237–244

Kleinmann RLP (1999) Bactericidal control of acidic drainage. In: Coal mine drainage prediction and pollution prevention in Pennsylvania. The Pennsylvania Department of Environmental Protection, Chapter 15, pp 15-1 to 15-6 (*http://www.dep.state.pa.us/dep/deputate/minres/districts/cmdp/main.htm*)

Kleinmann RLP, Hedin RS, Nairns RW (1998) Treatment of mine drainage by anoxic limestone drains and constructed wetlands. In: Geller W, Klapper H, Salomons W (eds) Acidic mining lakes. Springer, Berlin Heidelberg New York, pp 303–319

Koehnken L (1997) Final report Mount Lyell remediation research and demonstration program. Supervising Scientist report 126. Supervising Scientist, Canberra

Korcak RF (1998) Agricultural uses of phosphogypsum, gypsum and other industrial by-products. In: Wright RJ, Kemper PD, Millner JF, Power JF, Korcak RF (eds) Agricultural uses of municipal, animal, and industrial byproducts. US Department of Agriculture, Agricultural Research Service (Conservation research report no 44)

Kouloheris AP (1980) Chemical nature of phosphogypsum as produced by various wet process phosphoric acid processes. In: Proceedings of the international symposium on phosphogypsum. Florida Institute of Phosphate Research (Publication no 01-001-017)

Krause E, Ettel VA (1988) Solubility and stability of scorodite, $FeAsO_4 \cdot 2H_2O$: new data and further discussion. Amer Mineral 73:850–854

Kucha H , Martens A, Ottenburgs R, De Vos W, Viaene W (1996) Primary minerals of Zn-Pb mining and metallurgical dumps and their environmental behavior at Plombieres, Belgium. Environ Geol 27:1–15

Kuyucak N (2001) Acid mine drainage: treatment options for mining effluents. Mining Environmental Management 9:12–15

Kwong YTJ (1993) Mine site acid rock drainage assessment and prevention: a new challenge for a mining geologist. In: Proceedings of the international mining geology conference, Kalgoorlie, pp 213–217

Kwong YTJ, Roots CF, Roach P, Kettley W (1997) Post-mine metal transport and attenuation in the Keno Hill mining district, central Yukon, Canada. Environ Geol 30:98–107

Kwong YTJ, Swerhone GW, Lawrence JR (2003) Galvanic sulphide oxidation as a metal-leaching mechanism and its environmental implications. Geochem Explor Environ Anal 3:337–343

Lacerda de LD (2003) Updating global Hg emissions from small-scale gold mining and assessing its environmental impacts. Environ Geol 43:308–314

Lacerda de LD, Marins RV (1997) Anthropogenic mercury emissions to the atmosphere in Brazil: the impact of gold mining. J Geochem Explor 58:223–229

Lacerda de LD, Salomons W (1997) Mercury from gold and silver mining: a chemical time bomb? Springer, Berlin Heidelberg New York

Lachmar T, Burk NI, Kolesar PT (2006) Groundwater contribution of metals from an abandoned mine to the North Fork of the American Fork River, Utah. Water Air Soil Poll 173:103–120

Lamb HM, Dodds-Smith M, Gusek J (1998) Development of a long-term strategy for the treatment of acid mine drainage at Wheal Jane. In: Geller W, Klapper H, Salomons W (eds) Acidic mining lakes. Springer, Berlin Heidelberg New York, pp 335–346

Lambert JB (1997) Traces of the past. Unraveling the secrets of archaeology through chemistry. Addison Wesley, Reading

Lambert DC, McDonough KM, Dzombak DA (2004) Long-term changes in quality of discharge water from abandoned coal mines in Uniontown Syncline, Fayette County, PA, USA. Water Res 38:277–288

Landa ER (1999) Geochemical and biogeochemical controls on element mobility in and around uranium mill tailings. In: Filipek LH, Plumlee GS (eds) The environmental geochemistry of mineral deposits. Part B: Case studies and research topics. Society of Economic Geologists, Littleton (Reviews in economic geology, vol 6B, pp 527–538)

Landa ER, Gray JR (1995) US Geological Survey research on the environmental fate of uranium mining and milling wastes. Environ Geol 26:19–31

Langmuir D (1997) Aqueous environmental geochemistry. Prentice Hall, Upper Saddle River

Langmuir D, Mahoney J, MacDonald A, Rowson J (1999) Predicting the arsenic source-term from buried uranium mill tailings. In: Tailings and mine waste '99. Balkema, Rotterdam, pp 503–514

Lapakko K (2002) Metal mine rock and waste characterization tools: an overview. International Institute for Environment and Development (*http://www.iied.org/mmsd/*)

Laperdina TG (2002) Estimation of mercury and other heavy metal contamination in traditional gold-mining areas of Transbaikalia. Geochem Explor Environ Anal 2:219–223

Lapointe F, Fytas K, McConchie D (2006) Efficiency of bauxsol™ in permeable reactive barriers to treat acid rock drainage. Mine Water Environ 25:37–44

Larsen D, Mann R (2005) Origin of high manganese concentrations in coal mine drainage, eastern Tennessee. J Geochem Explor 86:143–163

Lasaga AC, Berner RA (1998) Fundamental aspects of quantitative models for geochemical cycles. Chem Geol 145:161–175

Lawrence RW, Scheske M (1997) A method to calculate the neutralization potential of mining wastes. Environ Geol 32:100–106

Lawrence EP, Erskine DW, Sealy CO (1997) Evaluation of groundwater remediation at a uranium mill site in Uravan, Colorado. In: Tailings and mine waste '97. Balkema, Rotterdam, pp 565–575

Lazareva EV, Shuvaeva OV, Tsimbalist VG (2002) Arsenic speciation in the tailings impoundment of a gold recovery plant in Siberia. Geochem Explor Environ Anal 2:263–268

Leblanc M, Achard B, Othman DB, Luck JM, Bertrand-Sarfati J, Personné JC (1996) Accumulation of arsenic from acidic mine waters by ferruginous bacterial accretions (stromatolites). Appl Geochem 11:541–554

Leblanc M, Morales JA, Borrego J, Elbaz-Poulichet F (2000) 4500-year-old mining pollution in southwestern Spain: long-term implications for modern mining pollution. Econ Geol 95:655–662

Ledin M, Pedersen K (1996) The environmental impact of mine wastes: roles of microorganisms and their significance in treatment of mine wastes. Earth-Sci Rev 41:67–108

Lee JS, Chon HT (2006) Hydrogeochemical characteristics of acid mine drainage in the vicinity of an abandoned mine, Daduk Creek, Korea. J Geochem Explor 88:37–40

Lee G, Bigham JM, Faure G (2002) Removal of trace metals by coprecipitation with Fe, Al and Mn from natural waters contaminated with acid mine drainage in the Ducktown mining district, Tennessee. Appl Geochem 17:569–581

Lee PK, Kang MJ, Choi SH, Touray JC (2005) Sulfide oxidation and the natural attenuation of arsenic and trace metals in the waste rocks of the abandoned Seobo tungsten mine, Korea. Appl Geochem 20:1687–1703

Lei L, Watkins R (2005) Acid drainage reassessment of mining tailings, Black Swan nickel mine, Kalgoorlie, Western Australia. Appl Geochem 20:661–667

Leigh DS (1997) Mercury-tainted overbank sediment from past gold mining in north Georgia, USA. Environ Geol 30:244–251

Lengke M, Tempel RN (2005) Geochemical modeling of arsenic sulfide oxidation kinetics in a mining environment. Geochim Cosmochim Acta 69:341–356

Levy DB, Custis KH, Casey WH, Rock PA (1996) Geochemistry and physical limnology of an acidic pit lake. In: Tailings and mine waste '96. Balkema, Rotterdam, pp 479–489

Levy DB, Custis KH, Casey WH, Rock PA (1997) A comparison of metal attentuation in mine residue and overburden material from an abandoned copper mine. Appl Geochem 12:203–211

Lewis-Russ A (1997) Ground water quality. In: Marcus JJ (ed) Mining environmental handbook: effects of mining on the environment and American environmental controls on mining. Imperial College Press, London, pp 162–165

Li MG, St-Arnoud L (1997) Hydrogeochemistry of secondary mineral dissolution: column leaching experiment using oxidized waste rock. In: Proceedings of the 4th international conference on acid rock drainage, vol 1. Vancouver, pp 465–477

Li MG, Jacob C, Comeau G (1996) Decommissioning of sulphuric acid-leached heap by rinsing. In: Tailings and mine waste '96. Balkema, Rotterdam, pp 295–304

Limbong D, Kumampung J, Rimper J, Arai T, Miyazaki N (2003) Emissions and environmental implications of mercury from artisanal gold mining in north Sulawesi, Indonesia. Sci Total Environ 302:227–236

Lin Z (1997) Mineralogical and chemical characterization of wastes from the sulfuric acid industry in Falun, Sweden. Environ Geol 30:152–162

Lin Z, Herbert Jr RB (1997) Heavy metal retention in secondary precipitates from a mine rock dump and underlying soil, Dalarna, Sweden. Environ Geol 33:1–12
Lind CJ, Creasey CL, Angeroth C (1998) In-situ alteration of minerals by acidic ground water resulting from mining activities: preliminary evaluation of method. J Geochem Explor 64:293–305
Lindahl LA (1998) The accident in Spain – a case study of Los Frailes. In: Proceedings of the workshop on risk assessment and contingency planning in the management of mine tailings. International Council on Metals and the Environment, Ottawa, United Nations Environment Programme, Paris, pp 67–90
Linklater CM, Sinclair DJ, Brown PL (2005) Coupled chemistry and transport modeling of sulphidic waste rock dumps at the Aitik mine site, Sweden. Appl Geochem 20:275–293
Ljungberg J, Öhlander B (2001) The geochemical dynamics of oxidising mine tailings at Laver, northern Sweden. J Geochem Explor 74:57–72
Loch R, Jasper D (eds) (2000) Soil research for mine rehabilitation. Austral J Soil Res 38(2)
Logsdon MJ, Hagelstein K, Mudder TI (1999) The management of cyanide in gold extraction. International Council on Metals and the Environment, Ottawa
Lopez-Archilla AI, Marin I, Amils R (1993) Bioleaching and interrelated acidophilic microorganisms from Rio Tinto, Spain. Geomicrobiol J 11:223–233
Loredo J, Ordonez A, Gallego JR, Baldo C, Garcia-Iglesias J (1999) Geochemical characterisation of mercury mining spoil heaps in the area of Mieres (Asturias, northern Spain). J Geochem Explor 67:377–390
Loring DH, Asmund G (1989) Heavy metal contamination of a Greenland Fjord system by mine wastes. Environ Geol Water Sci 14:61–71
Lottermoser BG (1995) Rare earth element mineralogy of the Olympic Dam Cu-U-Au-Ag deposit, South Australia: implications for ore genesis. Neues Jahrb Miner, Monatshefte 8:371–384
Lottermoser BG (2002) Mobilization of heavy metals from historical smelting slag dumps, north Queensland, Australia. Min Mag 66:475–490
Lottermoser BG (2005) Evaporative mineral precipitates from a historical smelting slag dump, Río Tinto, Spain. Neues Jb Miner Abh 181:183–190
Lottermoser BG, Ashley PM (2005a) Tailings dam seepage at the rehabilitated Mary Kathleen uranium mine, Australia. J Geochem Explor 85:119–137
Lottermoser BG, Ashley PM (2005b) Physical dispersion of radioactive mine waste at the rehabilitated radium Hill uranium mine site, South Australia. Austral J Earth Sci 53:485–499
Lottermoser BG, Ashley PM (2006) Mobility and retention of trace elements in hardpan-cemented cassiterite tailings, north Queensland, Australia. Environ Geol 50:835–846
Lottermoser BG, Schomberg S (1993) Thallium content of fertilizers: environmental implications. Fresenius Environ Bull 2:53–57
Lottermoser BG, Schütz U, Boenecke J, Oberhänsli R, Zolitschka B, Negendank JFW (1997a) Natural and anthropogenic influences on the geochemistry of Quaternary lake sediments from Holzmaar, Germany. Environ Geol 31:236–247
Lottermoser BG, Ashley PM, Muller M, Whistler BD (1997b) Metal contamination at the abandoned Halls Peak massive sulphide deposits, New South Wales. In: Ashley PM, Flood PG (eds) Tectonics and metallogenesis of the New England Orogen. Geological Society of Australia, Sydney (Special Publication no 19, pp 290–299)
Lottermoser BG, Ashley PM, Lawie DC (1999) Environmental geochemistry of the Gulf Creek copper mine area, northeastern New South Wales, Australia. Environ Geol 39:61–74
Lottermoser BG, Ashley PM, Costelloe MT (2005) Contaminant dispersion at the rehabilitated Mary Kathleen uranium mine, Australia. Environ Geol 48:748–761
Lu L, Wang R, Chen F, Xue J, Zhang P, Lu J (2005) Element mobility during pyrite weathering: implications for acid and heavy metal pollution at mining-impacted sites. Environ Geol 49:82–89
Lundgren T (2001) The dynamics of oxygen transport into soil covered mining waste deposits in Sweden. J Geochem Explor 74:163–173
Lussier C, Veiga V, Baldwin S (2003) The geochemistry of selenium associated with coal waste in the Elk River valley, Canada. Environ Geol 44:905-913
Luther GW (1987) Pyrite oxidation and reduction: molecular orbital theory considerations. Geochim Cosmochim Acta 51:3193–3199

Luther SM, Dudas MJ, Rutherford PM (1993) Radioactivity and chemical characteristics of Alberta phosphogypsum. Water Air Soil Poll 69:277–290

Maboeta MS, van Rensburg L (2003) Bioconversion of sewage sludge and industrially produced woodchips. Water Air Soil Poll 150:219–233

Macklin MG, Brewer PA, Balteanu D, Coulthard TJ, Driga B, Howard AJ, Zaharia S (2003) The long term fate and environmental significance of contaminant metals released by the January and March 2000 mining tailings dam failures in Maramures county, upper Tisa Basin, Romania. Appl Geochem 18:241–257

Macur RE, Wheeler, JT, McDermott TR, Inskeep WP (2001) Microbial populations associated with the reduction and enhanced mobilization of arsenic in mine tailings. Environ Sci Technol 35:3676–3682

Mager D, Vels B (1992) Wismut: an example for the uranium industry in eastern Europe? The Uranium Institute Annual Symposium 1992. The Uranium Institute, London, pp 1–7

Major R (2001) Natural radioactivity, hazards, wastes and the environment in Australia. In: Gostin VA (ed) Gondwana to greenhouse: Australian environmental geoscience. Geological Society of Australia, Sydney (Special Publication no 21, pp 111–124)

Manceau A (1995) The mechanism of anion adsorption on iron oxides: evidence for the bonding of arsenate tetrahedra on free $Fe(O,OH)_6$ edges. Geochim Cosmochim Acta 59:3647–3653

Maree JP, Hlabela P, Nengovhela R, Geldenhuys AJ, Mbhele N, Nevhulaudzi T, Waanders B (2004) Treatment of mine water for sulphate and metal removal using barium sulphide. Mine Water Environ 23:195–203

Marszalek H, Wasik M (2000) Influence of arsenic-bearing gold deposits on water quality in Zloty Stok mining area (SW Poland). Environ Geol 39:888–892

Martin P, Hancock GJ, Johnston A, Murray AS (1998) Natural-series radionuclides in traditional north Australian aboriginal foods. J Environ Rad 40:37–58

Martin JE, Garcia-Tenorio R, Respaldiza MA, Ontalba MA, Bolivar JP, da Silva MF (1999) TTPIXE analysis of phosphate rocks and phosphogypsum. Appl Rad Isotop 50:445–449

Martin AJ, Crusius J, McNee JJ, Yanful EK (2003) The mobility of radium-226 and trace metals in pre-oxidized subaqueous uranium mill tailings. Appl Geochem 18:1095–1110

Mascaro I, Benvenuti B, Corsini F, Costagliola P, Lattanzi P, Parrini P, Tanelli G (2001) Mine wastes at the polymetallic deposit of Fenice Capanne (southern Tuscany, Italy). Mineralogy, geochemistry, and environmental impact. Environ Geol 41:417–429

Mastrine JA, Bonzongo JC, Lyons WB (1999) Mercury concentrations in surface waters from fluvial systems draining historical precious metals mining areas in southeastern USA. Appl Geochem 14:147–158

Mathews WH, Bustin RM (1984) Why do the Smoking Hills smoke? Can J Earth Sci 21:737–742

Maxwell P, Govindarajalu S (1999) Do Australian mining companies pay too much? Reflections on the burden of meeting environmental standards in the late twentieth century. In: Azcue JM (ed) Environmental impacts of mining activities. Springer, Berlin Heidelberg New York, pp 7–17

McCarty DK, Moore JN, Marcus WA (1998) Mineralogy and trace element association in an acid mine drainage iron oxide precipitate: comparison of selective extractions. Appl Geochem 13:165–176

McConchie D, Clark M, Hanahan C (1998) The use of seawater neutralised red mud from bauxite refineries to control acid mine drainage and heavy metal leachates. Geological Society of Australia, Sydney (14th Australian Geological Convention, abstracts no 49, p 298)

McConchie D, Clark M, Davies-McConchie F, Hanahan C (2000) Treatment of acid mine waters using seawater-neutralised bauxite refinery residues. Geological Society of Australia, Sydney (15th Australian Geological Convention, abstracts no 59, p 320)

McCreadie H, Blowes DW, Ptacek CJ, Jambor JL (2000) Influence of reduction reactions and solid-phase composition on porewater concentrations of arsenic. Environ Sci Technol 34:3159–3166

McGregor RG, Blowes DW (2002) The physical, chemical and mineralogical properties of three cemented layers within sulfide-bearing mine tailings. J Geochem Explor 76:195–207

McGregor RG, Blowes DW, Jambor JL, Robertson WD (1998) Mobilization and attenuation of heavy metals within a nickel mine tailings impoundment near Sudbury, Ontario, Canada. Environ Geol 36:305–319

McKnight DM, Kimball BA, Bencala KE (1988) Iron photoreduction and oxidation in an acidic mountain stream. Science 240:637–640

McMahon B, Phillips JT, Hutton WA (1996) The evaluation of thickened tailings in a seismic area. In Tailings and mine waste '96. Balkema, Rotterdam, pp 133–143
McNulty T (2001) Alternatives to cyanide for processing precious metal ores. Mining Environmental Management, May 2001:35–37
McSweeney K, Madison FW (1988) Formation of a cemented subsurface horizon in sulfidic minewaste. J Environ Qual 17:256–262
MEND (1993a) Preventing AMD by disposing of reactive tailings in permafrost. Mine Environment Neutral Drainage Program, Ottawa (MEND report 6.1)
MEND (1993b) A case history of partially submerged pyritic uranium tailings under water. Mine Environment Neutral Drainage Program, Ottawa (MEND report 3.12.2)
MEND (1997a) Review and assessment of the roles of ice, in the water cover option, and permafrost in controlling acid generation from sulphidic tailings. Mine Environment Neutral Drainage Program, Ottawa (MEND report 1.61.1)
MEND (1997b) Characterization and stability of acid mine drainage treatment sludges. Mine Environment Neutral Drainage Program, Ottawa (MEND report 3.42.2)
Menzies NW, Mulligan DR (2000) Vegetation dieback on clay-capped pyritic mine wastes. J Environ Qual 29:437–442
Meyer WA, Williamson MA, Warren GC, Brown A (1997) Suppression of sulfide mineral oxidation in mine pit walls. Part II: Geochemical modeling. In: Tailings and mine waste '97. Balkema, Rotterdam, pp 433–441
Miedecke J, Miller S, Gowen M, Ritchie I, Johnston J, McBride P (1997) Remediation options (including copper recover by SX/EW) to reduce acid drainage from historical mining operations at Mount Lyell, western Tasmania, Australia. In: Proceedings from the 4th international conference on acid rock drainage, vol 4. Vancouver, pp 1451–1467
Miller SD (1995) Geochemical indicators of sulfide oxidation and acid generation in the field. In: Grundon NJ, Bell LC (eds) Proceedings of the 2nd Australian acid mine drainage workshop. Australian Centre for Minesite Rehabilitation Research, Brisbane, pp 117–120
Miller SD (1996) Advances in acid drainage: prediction and implications for risk management. In: Proceedings of 3rd international and 21st annual Minerals Council of Australia environmental workshop, vol 1. Minerals Council of Australia, Dickson, pp 149–157
Miller SD (1998a) Predicting acid drainage. Groundwork 2:8–9
Miller SD (1998b) Theory, design and operation of covers for controlling sulfide oxidation in waste rock dumps. In: McLean RW, Bell LC (eds) Proceedings of the 3rd Australian acid mine drainage workshop. Australian Centre for Minesite Rehabilitation Research, Brisbane, pp 115–126
Miller GC, Lyons WB, Davis A (1996) Understanding the water quality of pit lakes. Environ Sci Technol 30:118A–123A
Miller JR, Lechler PJ, Desilets M (1998) The role of geomorphic processes in the transport and fate of mercury in the Carson River basin, west-central Nevada. Environ Geol 33:249–262
Mills AL (1999) The role of bacteria in environmental geochemistry. In: Plumlee GS, Logsdon MS (eds) The environmental geochemistry of mineral deposits. Part A: Processes, techniques and health issues. Society of Economic Geologists, Littleton (Reviews in economic geology, vol 6A, pp 125–132)
Mining Journal Research Services (1996) Environmental and safety incidents concerning tailings dams at mines. The Mining Journal Ltd, London
Mitchell P (2000) Prediction, prevention, control and treatment of acid rock drainage. In: Warhurst A, Noronha L (eds) Environmental policy in mining: corporate strategy and planning for closure. Lewis Publishers, Boca Raton, pp 117–143
Mitsch WJ, Wise KM (1998) Water quality, fate of metals, and predictive model variation of a constructed wetland treating acid mine drainage. Water Res 32:1888–1900
Moberly C, Workman SR, Warner RC (2001) Calibration and evaluation of a hydrologic model for loose-dump mine spoil. Int J Surface Min Reclam Environ 15:147–162
Moncur MC, Ptacek CJ, Blowes DW, Jambor JL (2005) Release, transport and attenuation of metals from an old tailings impoundment. Appl Geochem 20:639–659
Monna F, Hamer K, Léveque J, Sauer M (2000) Pb isotopes as a reliable marker of early mining and smelting in the Northern Harz province (Lower Saxony, Germany). J Geochem Explor 68:201–210

Morin KA, Hutt NM (1994) An empirical technique for predicting the chemistry of water seeping from mine-rock piles. In: Proceedings of the international land reclamation and mine drainage conference and 3rd international conference on the abatement of acidic drainage, vol 1. United States Department of the Interior, Bureau of Mines (Special Publication SP06A-94, pp 12–19)

Morin KA, Hutt NM (1997) Environmental geochemistry of minesite drainage. MDAG Publication, Vancouver

Moses CO, Nordstrom DK, Herman JS, Mills AL (1987) Aqueous pyrite oxidation by dissolved oxygen and by ferric iron. Geochim Cosmochim Actan 51:1561–1571

Mosher JB, Figueroa L (1996) Biological oxidation of cyanide: a viable treatment option for the minerals processing industry? In: Tailings and mine waste '96. Balkema, Rotterdam, pp 525–533

Mudd GM (2001a) Critical review of acid in situ leach uranium mining: 1. USA and Australia. Environ Geol 41:390–403

Mudd GM (2001b) Critical review of acid in situ leach uranium mining: 2. Soviet Block and Asia. Environ Geol 41:404–416

Mudder T, Botz M (2001) A guide to cyanide. Mining Environmental Management 9:8–12

Mudder T, Botz M, Smith ACS (2001) Chemistry and treatment of cyanidation wastes. The Mining Journal Ltd, London

Mulligan DR (ed) (1996) Environmental management in the Australian minerals and energy industries. University of New South Wales Press, Sydney

Munk LA, Faure G, Pride DE, Bigham JM (2002) Sorption of trace metals to an aluminum precipitate in a stream receiving acid rock-drainage: Snake River, Summit County, Colorado. Appl Geochem 17:421–430

Munk LA, Faure G, Koski R (2006) Geochemical evolution of solutions derived from experimental weathering of sulfide-bearing rocks. Appl Geochem 21:1123–1134

Munro LD, Clark MW, McConchie D (2004) A Bauxsol™-based permeable reactive barrier for the treatment of acid rock drainage. Mine Water Environ 23:183–194

Munroe EA, McLemore VT, Kyle P (1999) Waste rock pile characterization, heterogeneity, and geochemical anomalies in the Hillsboro Mining District, Sierra County, New Mexico. J Geochem Explor 67:391–405

Murad E, Rojik P (2003) Iron-rich precipitates in a mine drainage environment: influence of pH on mineralogy. Amer Mineral 88:1915–1918

Murad E, Schwertmann U, Bigham JM, Carlson L (1994) Mineralogical characteristics of poorly crystallized precipitates formed by oxidation of Fe^{2+} in acid sulfate waters. In: Alpers CN, Blowes DW (eds) Environmental geochemistry of sulfide oxidation. American Chemical Society, Washington DC (Symposium Series 550, pp 190–200)

Murray L, Thompson M, Voigt K, Jeffery J (2000a) Mine waste management at Ok Tedi mine, Papua New Guinea. In: Tailings and mine waste '00. Balkema, Rotterdam, pp 507–515

Murray K, May P, Hefter G, Lotz P, Schulz R, Lye P (2000b) Cyanide waste management. Minerals & Energy Research Institute of Western Australia, Perth (Report no 214)

Naamoun T, Degering D, Hebert D, Merkel B (2000) Distribution of radionuclides in the tailings of Schneckenstein, Germany. In: Tailings and mine waste '00. Balkema, Rotterdam, pp 353–359

Nelson CH, Lamothe PJ (1993) Heavy metal anomalies in the Tinto and Odiel river and estuary system, Spain. Estuaries 16:496–511

Nicholson RV, Scharer JM (1994) Laboratory studies of pyrrhotite oxidation kinetics. In: Alpers CN, Blowes DW (eds) Environmental geochemistry of sulfide oxidation. American Chemical Society, Washington DC (Symposium Series 550, pp 14–30)

Nicholson RV, Gillham RW, Reardon EJ (1990) Pyrite oxidation in carbonate-buffered solution: 2. Rate control by oxide coatings. Geochim Cosmochim Acta 54:395–402

Nielson DL, Linpei C, Ward SH (1991) Gamma-ray spectrometry and radon emanometry in environmental geophysics. In: Ward SH (ed) Geotechnical and environmental geophysics, vol 1. Review and tutorial. Society of Exploration Geophysicists, Tulsa, pp 219–250

Nimick DA, Gammons CH, Cleasby TE, Madison JP, Skaar D, Brick CM (2003) Diel cycles in dissolved metal concentrations in streams: occurrence and possible causes. Water Resour Res 39:1247 (doi:10.1029/WR001571)

Noller BN, Hart BT (1993) Uranium in sediments from the Magela Creek catchment, Northern Territory, Australia. Environ Technol 14:649–656

Noller BN, Watters RA, Woods PH (1997) The role of biogeochemical processes in minimising uranium dispersion from a mine site. J Geochem Explor 58:37–50
Nordstrom DK, Alpers CN (1999a) Geochemistry of acid mine waters. In: Plumlee GS, Logsdon MS (eds) The environmental geochemistry of mineral deposits. Part A: Processes, techniques and health issues. Society of Economic Geologists, Littleton (Reviews in economic geology, vol 6A, pp 133–160)
Nordstrom DK, Alpers CN (1999b) Negative pH, efflorescent mineralogy, and consequences for environmental restoration at the Iron Mountain Superfund site, California. Proceedings of the National Academy of Sciences 96:3455–3462
Nordstrom DK, Alpers CN, Ptacek CJ, Blowes DW (2000) Negative pH and extremely acidic mine waters from Iron Mountain, California. Environ Sci Technol 34:254–258
Nriagu JO, Pacyna JM (1988) Quantitative assessment of world-wide contamination of air, water and soils by trace metals. Nature 333:134–139
Nuttall CA, Younger PL (2000) Zinc removal from hard, calcium-neutral mine waters using a novel closed-bed limestone reactor. Water Res 34:1262–1268
OECD (Organization for Economic Cooperation and Development) (1999) Environmental activities in uranium mining and milling. Joint report by the OECD Nuclear Energy Agency and the International Atomic Energy Agency. OECD Publications, Paris
Ogura T, Ramirez-Ortiz J, Arroyo-Villasenor ZM, Martinez SH, Palafox-Hernandez JP, Garcia de Alba LH, Fernando Q (2003) Zacatecas (Mexico) companies extract Hg from surface soil contaminated by ancient mining industries. Water Air Soil Poll 148:167–177
Ollier CD, Pain CF (1997) Regolith, soils and landforms. Wiley, New York
Ostergren JD, Brown GE, Parks GA, Tingle TN (1999) Quantitative speciation of lead in selected mine tailings from Leadville, CO. Environ Sci Technol 33:1627–1636
Oudjehani K, Zagury GJ, Deschenes L (2002) Natural attenuation of cyanide via microbial activity in mine tailings. Appl Microbiol Biotechnol 58:409–415
Paktunc AD (1998) Characterization of mine wastes for prediction of acid mine drainage. In: Azcue JM (ed) Environmental impacts of mining activities. Springer, Berlin Heidelberg New York, pp 19–40
Paktunc AD (1999) Mineralogical constraints on the determination of neutralization potential and prediction of acid mine drainage. Environ Geol 39:103–112
Parker G (1999) A critical review of acid generation resulting from sulfide oxidation: processes, treatment and control. In: Acid drainage. Australian Minerals & Energy Environment Foundation, Melbourne (Occasional paper no 11, pp 1–182)
Parker G, Noller BN, Woods PH (1996) Water quality and its origin in some mined-out open pits of the monsoonal zone of the Northern Territory, Australia. In: Proceedings of 3rd international and 21st annual Minerals Council of Australia environmental workshop. Minerals Council of Australia, Dickson, pp 50–63
Parsons MB, Bird DK, Einaudi MT, Alpers CN (2001) Geochemical and mineralogical controls on trace element release from the Penn Mine base-metal slag dump, California. Appl Geochem 16:1567–1593
Paschke SS, Harrison WJ, Walton-Day K (2001) Effects of acidic recharge on groundwater at the St. Kevin Gulch site, Leadville, Colorado. Geochem Explor Environ Anal 1:3–14
Peacey V, Yanful EK, Payne R (2002) Field study of geochemistry and solute fluxes in flooded uranium mine tailings. Can Geotech J 39:357–376
Pedersen TF, McNee JJ, Flather DH, Mueller B, Pelletier CA (1998) Geochemical behaviour of submerged pyrite-rich tailings in Canadian lakes. In: Geller W, Klapper H, Salomons W (eds) Acidic mining lakes. Springer, Berlin Heidelberg New York, pp 87–125
Pedersen HD, Postma D, Jakobsen R (2006) Release of arsenic associated with the reduction and transformation of iron oxides. Geochim Cosmochim Acta 70:4116–4129
Pellicori DA, Gammons CH, Poulson SR (2005) Geochemistry and stable isotope composition of the Berkeley pit lake and surrounding mine waters, Butte, Montana. Appl Geochem 20:2116–2137
Perkins EH, Gunter WD, Nesbitt HW, St-Arnaud LC (1997) Critical review of classes of geochemical computer models adaptable for prediction of acidic drainage from mine waste rock. In: Proceedings from the 4th international conference on acid rock drainage, vol 2. Vancouver, pp 587–601
Petruk W (ed) (1998) Waste characterization and treatment. Society for Mining, Metallurgy, and Exploration, Littleton
Petruk W (2000) Applied mineralogy in the mining industry. Elsevier, Amsterdam

Petrunic BM, Al TA (2005) Mineral/water interactions in tailings from a tungsten mine, Mount Pleasant, New Brunswick. Geochim Cosmochim Acta 69:2469–2483

Pettit CM, Scharer JM, Chambers DB, Halbert BE, Kirkaldy JL, Bolduc L (1999) Neutral mine drainage. In: Goldsack D, Belzile N, Yearwood P, Hall G (eds) Proceedings of Sudbury '99 Mining and the Environment, vol 2. Sudbury Environmental, Sudbury, Canada, pp 829–838

Pfeiffer WC, Lacerda LD, Salomons W, Malm O (1993) Environmental fate of mercury from gold mining in the Brazilian Amazon. Environ Rev 1:26–37

Piatak NM, Seal RR, Hammarstrom JM (2004) Mineralogical and geochemical controls on the release of trace elements from slag produced by base- and precious-metal smelting at abandoned mine sites. Appl Geochem 19:1039–1064

Pichler T, Hendry MJ, Hall GEM (2001) The mineralogy of arsenic in uranium mine tailings at the Rabbit Lake in-pit facility, northern Saskatchewan, Canada. Environ Geol 40:495–506

Pilon-Smits EAH, Zhu Y-L, Terry N (2000) Improved metal phytoremediation through plant biotechnology. In: Tailings and mine waste '00. Balkema Publishers, Rotterdam, pp 317–320

Pinto MMSC, Silva MMVG, Neiva AMR (2004) Pollution of water and stream sediments associated with the Vale de Abrutiga uranium mine, central Portugal. Mine Water Environ 23:66–75

Pirrie D, Camm GS, Sear LG, Hughes SH (1997) Mineralogical and geochemical signature of mine waste contamination, Tresillian River, Fal Estuary, Cornwall, UK. Environ Geol 29:58–65

Plumlee GS (1999) The environmental geology of mineral deposits. In: Plumlee GS, Logsdon MS (eds) The environmental geochemistry of mineral deposits. Part A: Processes, techniques and health issues. Society of Economic Geologists, Littleton (Reviews in economic geology, vol 6A, pp 71–116)

Plumlee GS, Logsdon MJ (1999) An Earth-system science toolkit for environmentally friendly mineral resource development. In: Plumlee GS, Logsdon MS (eds) The environmental geochemistry of mineral deposits. Part A: Processes, techniques and health issues. Society of Economic Geologists, Littleton (Reviews in economic geology, vol 6A, pp 1–27)

Plumlee GS, Smith KS, Montour MR, Ficklin WH, Mosier EL (1999) Geologic controls on the composition of natural waters and mine waters draining diverse mineral-deposit types. In: Filipek LH, Plumlee GS (eds) The environmental geochemistry of mineral deposits. Part B: Case studies and research topics. Society of Economic Geologists, Littleton (Reviews in economic geology, vol 6B, pp 373–432)

Pluta I (2001) Barium and radium discharged from coal mines in the Upper Silesia, Poland. Environ Geol 40:345–348

Poling GW, Ellis DV (1995) Importance of geochemistry: the Black Angel lead-zinc mine, Greenland. Marine Geores Geotechnol 13:101–118

Pond AP, White SA, Milczarek M, Thompson TL (2005) Accelerated weathering of biosolid-amended copper mine tailings. J Environ Qual 34:1293–1301

Posey HH, Renkin ML, Woodling J (2000) Natural acid drainage in the upper Alamosa River of Colorado. In: Proceedings from the 5th international conference on acid rock drainage, vol 1. Society for Mining, Metallurgy, and Exploration, Littleton, pp 485–498

Powell JH, Powell RE (2001) Trace elements in fish overlying subaqueous tailings in the tropical west Pacific. Water Air Soil Poll 125:81–104

Price WA, Morin K, Hutt N (1997) Guidelines for the prediction of acid rock drainage. Part II. Recommended procedures for static and kinetic testing. In: Proceedings from the 4th international conference on acid rock drainage, vol 1. Vancouver, pp 15–30

Ptacek CJ, Blowes DW (1994) Influence of siderite on the pore-water chemistry of inactive mine-tailings impoundments. In: Alpers CN, Blowes DW (eds) Environmental geochemistry of sulfide oxidation. American Chemical Society, Washington DC (Symposium Series 550, pp 172–189)

Puura E, Neretnieks I (2000) Atmospheric oxidation of the pyritic waste rock in Maardu, Estonia. 2: An assessment of aluminosilicate buffering potential. Environ Geol 39:560–566

Puura E, Neretnieks I, Kirsimäe K (1999) Atmospheric oxidation of the pyritic waste rock in Maardu, Estonia. 1: Field study and modelling. Environ Geol 39:1–19

Rager R, MacClanahan M, Thiers G, Grozescu S (1996) Disposal cell cover selection – riprap vs. grass. In: Tailings and mine waste '96. Balkema, Rotterdam, pp 219–228

Ragnarsdottir KV, Charlet L (2000) Uranium behaviour in natural environments. Cotter-Howells JD, Campbell LS, Valsami-Jones E, Batchelder M (eds) Environmental mineralogy: microbial interactions, anthropogenic influences, contaminated land and waste management. Mineralogical Society, London (Mineralogical Society Series no 9, pp 245–289)

Ramstedt M, Carlsson E, Lövgren L (2003) Aqueous geochemistry in the Udden pit lake, northern Sweden. Appl Geochem 18:97–108
Rankin M, Miller T, Petrovic S, Zapf-Gilje R, Davidson S, Drysdale K, Van Zyl D (1997) Submarine tailings discharge: optimizing the evaluation and monitoring process. In: Tailings and mine waste '97. Balkema, Rotterdam, pp 303–316
Ranville JF, Schmiermund RL (1999) General aspects of aquatic colloids in environmental geochemistry. In: Plumlee GS, Logsdon MS (eds) The environmental geochemistry of mineral deposits. Part A: Processes, techniques and health issues. Society of Economic Geologists, Littleton (Reviews in economic geology, vol 6A, pp 183–199)
Rauche HAM, Fulda D, Schwalm V (2001) Tailings and disposal brine reduction – design criteria for potash production in the 21st century. In: Tailings and mine waste '01. Balkema, Rotterdam, pp 85–91
Regenspurg S, Pfeiffer S (2005) Arsenate and chromate incorporation in schwertmannite. Appl Geochem 20:1226–1239
Rensburg van L, Maboeta MS, Morgenthal TL (2004) Rehabilitation of co-disposed diamond tailings: growth medium rectification procedures and indigenous grass establishment. Water Air Soil Poll 154:101–113
Richardson SG (1995) Establishing vegetation cover on phosphogypsum in Florida. Florida Institute of Phosphate Research (Publication no 01-086-116)
Rimstidt JD, Vaughan DJ (2003) Pyrite oxidation: a state-of-the-art assessment of the reaction mechanism. Geochim Cosmochim Acta 67:873–880
Rimstidt JD, Chermak JA, Gagen PM (1994) Rates of reaction of galena, sphalerite, chalcopyrite and arsenopyrite with Fe(III) in acidic solutions. In: Alpers CN, Blowes DW (eds) Environmental geochemistry of sulfide oxidation. American Chemical Society, Washington DC (Symposium Series 550, pp 2–13)
Ripley EA, Redmann RE, Crowder AA (1996) Environmental effects of mining. St Lucie Press, Delray Beach
Ritcey GM (1989) Tailings management. Problems and solutions in the mining industry. Elsevier, Amsterdam
Ritchie AIM (1994a) The waste-rock environment. In: Jambor JL, Blowes DW (eds) Environmental geochemistry of sulfide mine-wastes. Mineralogical Association of Canada, Nepean (Short course handbook, vol 22, pp 133–161)
Ritchie AIM (1994b) Sulfide oxidation mechanisms: controls and rates of oxygen transport. In: Jambor JL, Blowes DW (eds) Environmental geochemistry of sulfide mine-wastes. Mineralogical Association of Canada, Nepean (Short course handbook, vol 22, pp 201–245)
Ritchie AIM (1994c) Rates of mechanisms that govern pollutant generation from pyritic wastes. In: Alpers CN, Blowes DW (eds) Environmental geochemistry of sulfide oxidation. American Chemical Society, Washington DC (Symposium Series 550, pp 108–122)
Ritchie AIM (1995) Application of oxidation rates in rehabilitation design. In: Grundon NJ, Bell LC (eds) Proceedings of the 2nd Australian acid mine drainage workshop. Australian Centre for Minesite Rehabilitation Research, Brisbane, pp 101–116
Ritchie AIM (1998) The performance of covers. In: McLean RW, Bell LC (eds) Proceedings of the 3rd Australian acid mine drainage workshop. Australian Centre for Minesite Rehabilitation Research, Brisbane, pp 135–145
Ritchie AIM, Bennett JW (2003) The Rum Jungle mine – a case study. In: Jambor JL, Blowes DW, Ritchie AIM (eds), Environmental aspects of mine wastes. Mineralogical Association of Canada, Ottawa, pp 385–405
Robertson WD (1994) The physical hydrogeology of mill-tailings impoundments. In: Jambor JL, Blowes DW (eds) Environmental geochemistry of sulfide mine-wastes. Mineralogical Association of Canada, Nepean (Short course handbook, vol 22, pp 1–17)
Robinson BH, Chiarucci A, Brooks RR, Petit D, Kirkman JH, Gregg PEH, Dominicis de V (1997) The nickel hyperaccumulator plant Alyssum bertolonii as a potential agent for phytoremediation and phytomining of nickel. J Geochem Explor 59:75–86
Roddick-Lanzilotta AJ, McQuillan AJ, Craw D (2002) Infrared spectroscopic characterisation of arsenate (V) ion adsorption from mine waters, Macreas mine, New Zealand. Appl Geochem 17:445–454
Romano P, Blazquez ML, Alguacil FJ, Munoz JA, Ballester A, Gonzalez F (2001) Comparitive study on the selective chalcopyrite bioleaching of a molybdenite concentrate with mesophilic and thermophilic bacteria. FEMS Microbiol Lett 196:71–75

Romano CG, Mayer KU, Jones DR, Ellerbroek DA, Blowes DW (2003) Effectiveness of various cover scenarios on the rate of sulfide oxidation of mine tailings. J Hydrol 271:171–187
Romero A, Gonzalez I, Galan E (2006) Estimation of potential pollution of waste mining dumps at Peña del Hierro (Pyrite Belt, SW Spain) as a base for future mitigation actions. Appl Geochem 21:1093–1108
Rose AW, Cravotta CA (1999) Geochemistry of coal mine drainage. In: Coal Mine Drainage Prediction and Pollution Prevention in Pennsylvania. The Pennsylvania Department of Environmental Protection, Chapter 1, 1-1 to 1-22 (*http://www.dep.state.pa.us/dep/deputate/minres/districts/cmdp/main.htm*)
Rose S, Elliott WC (2000) The effects of pH regulation upon the release of sulfate from ferric precipitates formed in acid mine drainage. Appl Geochem 15:27–34
Rose S, Ghazi AM (1998) Experimental study of the stability of metals associated with iron oxyhydroxides precipitated in acid mine drainage. Environ Geol 36:364–370
Rosman KJR, Boutron CF, Chisholm W, Hong S, Candelone J-P (1997) Lead from Carthaginian and Roman Spanish mines isotopically identified in Greenland ice dated from 600 B.C. to 300 A.D. Environ Sci Technol 31:3413–3416
Rutherford PM, Dudas MJ, Samek RA (1994) Environmental impacts of phosphogypsum. Sci Total Environ 149:1–38
Rutherford PM, Dudas MJ, Arocena JM (1995) Radioactivity and elemental composition of phosphogypsum produced from three phosphate rock sources. Waste Management Res 13:407–423
Salomons W (1995) Environmental impact of metals derived from mining activities: processes, predictions, prevention. J Geochem Explor 52:5–23
Salomons W, Eagle AM (1990) Hydrology, sedimentology and the fate and distribution of copper in mine-related discharges in the Fly River system, Papua New Guinea. Sci Total Environ 97/98:315–334
Salzsauler KA, Sidenko NV, Sherriff BL (2005) Arsenic mobility in alteration products of sulfide-rich, arsenopyrite-beairng mine wastes, Snow Lake, Manitoba, Canada. Appl Geochem 20:2303–2314
Sams JI, Veloski GA (2003) Evaluation of airborne thermal infrared imagery for locating mine drainage sites in the Lower Kettle Creek and Cook Run Basins, Pennsylvania, USA. Mine Water Environ 22:85–93
Sams JI, Veloski GA, Ackman TE (2003) Evaluation of airborne thermal infrared imagery for locating mine drainage sites in the lower Youghiogheny River Basin, Pennsylvania, USA. Mine Water Environ 22:94–103
Sato M (1992) Persistency-field Eh-pH diagrams for sulfides and their application to supergene oxidation and enrichment of sulfide ore bodies. Geochim Cosmochim Acta 56:3133–3156
Savage KS, Tingle TN, O'Day PA, Waychunas GA, Bird DK (2000) Arsenic speciation in pyrite and secondary weathering phases, Mother Lode Gold District, Tuolumne County, California. Appl Geochem 15:1219–1244
Schafer WM (2000) Use of the net acid generation pH test for assessing risk of acid generation. In: Proceedings from the 5th international conference on acid rock drainage, vol 1. Society for Mining, Metallurgy, and Exploration, Littleton, pp 613–618
Scharer JM, Arthurs P, Knapp RA, Halbert BE (1999) Fate of cyanate and ammonia in gold tailings. In: Goldsack D, Belzile N, Yearwood P, Hall G (eds) Proceedings of Sudbury '99 Mining and the Environment, vol 2. Sudbury Environmental, Sudbury, Canada, pp 617–626
Scharer JM, Pettit CM, Kirkaldy JL, Bolduc L, Halbert BE, Chambers DB (2000) Leaching of metals from sulphide mine wastes at neutral pH. In: Proceedings from the 5th international conference on acid rock drainage, vol 1. Society for Mining, Metallurgy, and Exploration, Littleton, pp 191–201
Schemel LE, Kimball BA, Bencala KE (2000) Colloid formation and metal transport through two mixing zones affected by acid mine drainage near Silverton, Colorado. Appl Geochem 15:1003–1018
Schippers A, Sand W (1999) Bacterial leaching of metal sulfides proceeds by two indirect mechanisms via thiosulfate or via polysulfides and sulfur. Appl Environ Microbiol 65:319–321
Schmid S, Wiegand J (2003) Radionuclide contamination of surface waters, sediments, and soil caused by coal mining activities in the Ruhr district (Germany). Mine Water Environ 22:130–140
Schmiermund RL (1997) Acid mine drainage and other mining-influenced waters (MW). In: Marcus JJ (ed) Mining environmental handbook: effects of mining on the environment and American environmental controls on mining. Imperial College Press, London, pp 599–617
Schmiermund RL (2000) Non-acidic, sulfate-poor, copper-enriched drainage from a Precambrian stratabound chalcopyrite/pyrite deposit, Carbon County, Wyoming In: Proceedings from the 5th international conference on acid rock drainage, vol 2. Society for Mining, Metallurgy, and Exploration, Littleton, pp 1059–1070

Schramke JA, Murphy SF, Lewis RL, Medlock RL, Franko MJ (1997) Natural attenuation of ground water constituents at a uranium mill tailings site, Shirley Basin, Wyoming. In: Tailings and mine waste '97. Balkema, Rotterdam, pp 499–508

Schrenk MO, Edwards KJ, Goodman RM, Hamers RJ, Banfield JF (1998) Distribution of *Thiobacillus ferrooxidans* and *Leptospirillum ferrooxidans*: implications for generation of acid mine drainage. Science 279:1519–1522

Schuiling RD, van Gaans PFM (1997a) The waste sulfuric acid lake of the TiO_2-plant at Armyansk, Crimea, Ukraine. Part I. Self-sealing as an environmental protection mechanism. Appl Geochem 12:181–186

Schuiling RD, van Gaans PFM (1997b) The waste sulfuric acid lake of the TiO_2-plant at Armyansk, Crimea, Ukraine. Part II. Modeling the chemical evolution with PHRQPITZ. Appl Geochem 12:187–201

Schroth AW, Parnell RA (2005) Trace metal retention through the schwertmannite to goethite transformation as observed in a field setting, Alta mine, MT. Appl Geochem 20:907–917

Schwartz MO, Ploethner D (2000) Removal of heavy metals from mine water by carbonate precipitation in the Grootfontein-Omatako Canal, Namibia. Environ Geol 39:1117–1126

Seal RR, Hammarstrom JM (2003) Geoenvironmental models of mineral deposits: examples from massive sulfide and gold deposits. In: Jambor JL, Blowes DW, Ritchie AIM (eds) Environmental aspects of mine wastes. Mineralogical Association of Canada, Nepean (Short course handbook, vol 31, pp 11–50)

Seal RR, Hammarstrom JM, Foley NK, Alpers CN (2000) Geoenvironmental models for seafloor base- and precious-metal massive sulfide deposits. In: Proceedings from the 5th international conference on acid rock drainage, vol 1. Society for Mining, Metallurgy, and Exploration, Littleton, pp 115–123

Seidel H, Görsch K, Amstätter K, Mattusch J (2005) Immobilization of arsenic in a tailings material by ferrous iron treatment. Water Res 39:4073–4082

Sharma PV (1997) Environmental and engineering geophysics. Cambridge University Press, Cambridge

Shay DA, Cellan RR (2000) Use of a chemical cap to remediate acid rock conditions at Homestake's Santa Fe mine. In: Tailings and mine waste '00. Balkema, Rotterdam, pp 203–210

Shehong L, Baoshan Z, Jiangming Z, Xiaoying Y (2005) The distribution and natural degradation of cyanide in gold mine tailings and polluted soils in arid and semiarid areas. Environ Geol 47:1150–1154

Shelp GS, Chesworth W, Spiers G (1995) The amelioration of acid mine drainage by an in situ electrochemical method. Part 1. Employing scrap iron as the sacrificial anode. Appl Geochem 10:705–713

Sherlock EJ, Lawrence RW, Poulin R (1995) On the neutralization of acid rock drainage by carbonate and silicate minerals. Environ Geol 25:43–54

Shevenell LA (2000) Water quality in pit lakes in disseminated gold deposits compared to two natural, terminal lakes in Nevada. Environ Geol 39:807–815

Shevenell LA, Connors KA, Henry CD (1999) Controls on pit lake water quality at sixteen open-pit mines in Nevada. Appl Geochem 14:669–687

Shope CL, Xie Y, Gammons CH (2006) The influence of hydrous Mn-Zn oxides on diel cycling of Zn in an alkaline stream draining abandoned mine lands. Chem Geol 21:476–491

Shotyk W, Cheburkin AK, Appleby PG, Fankhauser A, Kramers JD (1996) Two thousand years of atmospheric arsenic, antimony, and lead deposition recorded in an ombrotrophic peat bog profile, Jura Mountains, Switzerland. Earth Planet Sci Lett 145:E1–E7

Shum M, Lavkulich L (1999) Speciation and solubility relationships of Al, Cu and Fe in solutions associated with sulfuric acid leached mine waste rock. Environ Geol 38:59–68

Sibrell PL, Watten BJ, Friedrich AE, Vinci BJ (2000) ARD remediation with limestone in a CO_2 pressurized reactor. In: Proceedings from the 5th international conference on acid rock drainage, vol 2. Society for Mining, Metallurgy, and Exploration, Littleton, pp 1017–1026

Sidenko NV, Sherriff BL (2005) The attenuation of Ni, Zn and Cu, by secondary Fe phases of different crystallinity from surface and ground water of two sulfide mine tailings in Manitoba, Canada. Appl Geochem 20:1180–1194

Sidenko NV, Gieré R, Bortnikova SB, Cottard F, Pal'chik NA (2001) Mobility of heavy metals in self-burning waste heaps of the zinc smelting plant in Belovo (Kemerovo Region, Russia). J Geochem Explor 74:109–125

Siegel J (1997) Ground water quantity. In: Marcus JJ (ed) Mining environmental handbook: effects of mining on the environment and American environmental controls on mining. Imperial College Press, London, pp 165–168

Simms PH, Yanful EK, St-Arnaud L, Aubé B (2001) Effectiveness of protective layers in reducing metal fluxes from flooded pre-oxidized mine tailings. Water Air Soil Poll 131:73–96
Singer PC, Stumm W (1970) Acid mine drainage: rate-determining step. Science 167:1121–1123
Skousen JG, Ziemkiewicz PF (compilers) (1996) Acid mine drainage control and prevention. West Virginia University, National Mine Land Rehabilitation Center, Morgantown
Skousen J, Renton J, Brown H, Evans P, Leavitt B, Brady K, Cohen L, Ziemkiewicz P (1997) Neutralization potential of overburden samples containing siderite. J Environ Qual 26:673–681
Skousen J, Simmons J, McDonald LM, Ziemkiewicz P (2002) Acid-base accounting to predict post-mining drainage quality on surface mines. J Environ Qual 31:2034–2044
Sladek C, Gustin MS (2003) Evaluation of sequential and selective extraction methods for determination of mercury speciation and mobility in mine waste. Appl Geochem 18:567–576
Sladek C, Gustin MS, Kim CS, Biester H (2002) Application of three methods for determining mercury speciation in mine waste. Geochem Explor Environ Anal 2:369–376
Smedley PL, Kinniburgh DG (2002) A review of the source, behaviour and distribution of arsenic in natural waters. Appl Geochem 17:517–568
Smit JP (2000) Potable water from sulphate polluted mine sources. Mining Environmental Management 8:7–9
Smith ACS (1994) The geochemistry of cyanide in mill tailings. In: Jambor JL, Blowes DW (eds) Environmental geochemistry of sulfide mine-wastes. Mineralogical Association of Canada, Nepean (Short course handbook, vol 22, pp 293–332)
Smith KS (1999) Metal sorption on mineral surfaces: an overview with examples relating to mineral deposits. In: Plumlee GS, Logsdon MS (eds) The environmental geochemistry of mineral deposits. Part A: Processes, techniques and health issues. Society of Economic Geologists, Littleton (Reviews in economic geology, vol 6A, pp 161–182)
Smith MW, Brady KBC (1999) Alkaline addition. In: Coal mine drainage prediction and pollution prevention in Pennsylvania. The Pennsylvania Department of Environmental Protection, Chapter 13, 13-1 to 13-13 (*http://www.dep.state.pa.us/dep/deputate/minres/districts/cmdp/main.htm*)
Smith KS, Huyck HLO (1999) An overview of the abundance, relative mobility, bioavailability, and human toxicity of metals. In: Plumlee GS, Logsdon MS (eds) The environmental geochemistry of mineral deposits. Part A: Processes, techniques and health issues. Society of Economic Geologists, Littleton (Reviews in economic geology, vol 6A, pp 29–70)
Smith ACS, Mudder TI (1991) The chemistry and treatment of cyanidation wastes. The Mining Journal, London
Smith ACS, Mudder TI (1999) The environmental geochemistry of cyanide. In: Plumlee GS, Logsdon MS (eds) The environmental geochemistry of mineral deposits. Part A: Processes, techniques and health issues. Society of Economic Geologists, Littleton (Reviews in economic geology, vol 6A, pp 229–248)
Smith A, Robertson A, Barton-Bridges J, Hutchison IPG (1992) Prediction of acid generation potential. In: Hutchison IPG, Ellison RD (eds) Mine waste management. Lewis Publishers, Boca Raton, pp 123–199
Smith KS, Ramsey CA, Hageman PL (2000) Sampling strategy for the rapid screening of mine-waste dumps on abandoned mine lands. In: Proceedings from the 5th international conference on acid rock drainage, vol 2. Society for Mining, Metallurgy, and Exploration, Littleton, pp 1453–1461
Smith AML, Hudson-Edwards KA, Dubbin WE, Wright K (2006) Dissolution of jarosite [$KFe_3(SO_4)_2(OH)_2$] at pH 2 and 8: insights from batch experiments and computational modelling. Geochim Cosmochim Acta 70:608–621
Smyth DJA, Blowes DW, Benner SG, Hulshof AM (2001) In situ treatment of groundwater impacted by acid mine drainage using permeable reactive materials. In: Tailings and mine waste '01. Balkema, Rotterdam, pp 313–322
SMME (Society for Mining, Metallurgy, and Exploration) (1998) Remediation of historical mine sites: technical summary and bibliography. Society for Mining, Metallurgy, and Exploration, Littleton
Sobek AA, Schuller WA, Freeman JR, Smith RM (1978) Field and laboratory methods applicable to overburdens and minesoils. US EPA 600/2-78-054, Washington DC
Somot S, Pagel M, Thiry J, Ruhlmann F (2000) Speciation of ^{226}Ra, uranium and metals in uranium mill tailings. In: Tailings and mine waste '00. Balkema, Rotterdam, pp 343–352

Soner Altundogan H, Attundogan S, Tumen F, Bildik M (2000) Arsenic removal by red mud. Waste Management 20:761–767
Stouraiti C, Xenidis A, Paspaliaris I (2002) Reduction of Pb, Zn and Cd availability from tailings and contaminated soils by the application of lignite fly ash. Water Air Soil Poll 137:247–265
Strömberg B, Banwart SA (1999) Experimental study of acidity-consuming processes in mine waste rock: some influences of mineralogy and particle size. Appl Geochem 14:1–16
Stumm W, Morgan JJ (1995) Aquatic chemistry, 3rd edn. Wiley, New York
Sullivan AB, Drever JI (2001) Geochemistry of suspended particles in a mine-affected mountain stream. Appl Geochem 16:1663–1676
Suzuki Y, Banfield JF (1999) Geomicrobiology of uranium. In: Burns PC, Finch R (eds) Uranium: mineralogy, geochemistry and the environment. Mineralogical Society of America, Washington (Reviews in mineralogy and geochemistry, vol 38, pp 393–432)
Sverdrup HU (1990) The kinetics of base cation release due to chemical weathering. Lund University Press, Lund
Swayze GA, Smith KS, Clark RN, Sutley SJ, Pearson RM, Vance JS, Hageman PL, Briggs PH, Meier AL, Singleton MJ, Roth S (2000) Using imaging spectroscopy to map acidic mine waste. Environ Sci Technol 34:47–54
Swedlund PJ, Webster JG (2001) Cu and Zn ternary surface complex formation with SO_4 on ferrihydrite and schwertmannite. Appl Geochem 16:503–511
Swedlund PJ, Webster JG, Miskelly GM (2003) The effect of SO_4 on the ferrihydrite adsorption of Co, Pb and Cd: ternary complexes and site heterogeneity. Appl Geochem 18:1671–1689
Szczepanska J, Twardowska I (1999) Distribution and environmental impact of coal-mining wastes in Upper Silesia, Poland. Environ Geol 38:249–258
Taschereau C, Fytas K (2000) Arsenic in mine tailings – the Chimo Mine case study. Int J Surface Min Reclam Environ 14:337–347
Taylor JR, Waring CL, Murphy NC, Leake MJ (1998) An overview of acid mine drainage control and treatment options, including recent advances. In: McLean RW, Bell LC (eds) Proceedings of the 3rd Australian acid mine drainage workshop. Australian Centre for Minesite Rehabilitation Research, Brisbane, pp 147–159
Tempel RN, Shevenell LA, Lechler P, Price J (2000) Geochemical modeling approach to predicting arsenic concentrations in a mine pit lake. Appl Geochem 15:475–492
Thomas MA, Conaway CH, Steding DJ, Marwin-DiPasquale M, Abu-Saba KE, Flegal AR (2002) Mercury contamination from historic mining in water and sediment, Guadalupe River and San Francisco Bay, California. Geochem Explor Environ Anal 2:211–217
Thornton I, Ramsey M, Atkinson N (1995) Metals in the global environment: facts and misconceptions. International Council on Metals and the Environment, Ottawa
Toyer GS (1981) The Woodsreef asbestos mine. Its effect on water quality. Mineral 24:5–12
Tyrrell W (1996) A review of wetlands for treating coal mine wastewater, particularly in low rainfall environments in Australia. Australian Minerals & Energy Environment Foundation, Melbourne (Occasional paper no 3)
UNEP/IFA (United Nations Environment Programme/International Fertilizer Industry Association) (2001) Environmental aspects of phosphate and potash mining. (*http://www.fertilizer.org*)
UNEP/OCHA (United Nations Environment Programme/Office for the Co-ordination of Humanitarian Affairs) (2000) Spill of liquid and suspended waste at the Aurul S.A. retreament plant in Baia Mare. (*http://www.mineralresourcesforum.org/*)
US DOE (United States Department of Energy) (1999) Integrated data base report – 1996: US spent nuclear fuel and radioactive waste inventories, projections, and characteristics. US Department of Energy Office of Environmental Management (*http://www.em.doe.gov/idb97/contents.html*)
US EPA (United States Environmental Protection Agency) (1997) Feasability analysis: a comparison of phosphogypsum and uranium mill tailing waste unit design. Office of Solid Waste US Environmental Protection Agency, Washington
USGS (United States Geological Survey) Mineral Resources Program (2001) Minerals yearbook. (*http://minerals.usgs.gov/minerals/pubs/myb.html*)
Uranium Information Centre (2001a) In situ leach mining of uranium. Uranium Information Centre, Melbourne (Nuclear Issues Briefing Paper 40, *http://www.uic.com.au/nip40.htm*)

Uranium Information Centre (2001b) Radiation and the nuclear fuel cycle. Uranium Information Centre, Melbourne (Nuclear Issues Briefing Paper 17, *http://www.uic.com.au/nip17.htm*)

Uranium Information Centre (2001c) Occupational safety in uranium mining. Uranium Information Centre, Melbourne (Nuclear Issues Briefing Paper 6, *http://www.uic.com.au/nip06.htm*)

Vance GF (2000) Problems associated with selenium leaching from waste shale. In: Daniels WL, Richardson SG (eds) Proceedings of the 17th annual meeting of the American Society for Surface Mining and Reclamation. American Society for Surface Mining and Reclamation, pp 71–82

Varnell CJ, Brahana van J, Steele K (2004) The influence of coal quality variation on utilization of water from abandoned coal mines as a municipal water source. Mine Water Environ 23:204–208

Vaughan DJ, Craig JR (1978) Mineral chemistry of metal sulfides. Cambridge Earth Science Series, Cambridge University Press, Cambridge

Velasco F, Alvaro A, Suarez S, Herrero JM, Yusta I (2005) Mapping Fe-bearing hydrated sulphate minerals with short wave infrared (SWIR) spectral analysis at San Miguel mine environment, Iberian Pyrite Belt (SW Spain). J Geochem Explor 87:45–72

Verlaan PA, Wiltshire JC (2000) Manganese tailings: a potential resource? Mining Environmental Management 8:21–22

Vick SG (1983) Planning, design and analysis of tailings dams. Wiley, New York

Wagener FM von, Craig HJ, Blight G, McPhail GI, Williams AAB, Strydom JH (1998) The Merriespruit tailings dam failure: a review. In: Tailings and mine waste '98. Balkema, Rotterdam, pp 925–952

Waggitt P (1994) A review of worldwide practices for disposal of uranium mill tailings. Technical Memorandum 48, Supervising Scientist for the Alligator Rivers Region, AGPS, Canberra

Walker FP, Schreiber ME, Rimstidt JD (2006) Kinetics of arsenopyrite oxidative dissolution by oxygen. Geochim Cosmochim Acta 70:1668–1676

Walton-Day K (1999) Geochemistry of the processes that attenuate acid mine drainage in wetlands. In: Plumlee GS, Logsdon MS (eds) The environmental geochemistry of mineral deposits. Part A: Processes, techniques and health Issues. Society of Economic Geologists, Littleton (Reviews in economic geology, vol 6A, pp 215–228)

Walton-Day K (2003) Passive and active treatment of mine drainage. In: Jambor JL, Blowes DW, Ritchie AIM (eds) Environmental aspects of mine wastes. Mineralogical Association of Canada, Nepean (Short course handbook, vol 31, pp 335–359)

Wang G, Fayram T, Begeman J (1997) Effluent treatment for removal of cyanide and nitrogen compounds. In: Tailings and mine waste '97. Balkema, Rotterdam, pp 625–631

Wanty RB, Miller WR, Briggs PH, McHugh JB (1999) Geochemical processes controlling uranium mobility in mine drainages. In: Plumlee GS, Logsdon MS (eds) The environmental geochemistry of mineral deposits. Part A: Processes, techniques and health Issues. Society of Economic Geologists, Littleton (Reviews in economic geology, vol 6A, pp 201–213)

Ward TA, Cox BJ (1985) Rehabilitation of the Mary Kathleen uranium mining and processing site. In: Proceedings of the 9th international symposium uranium and nuclear energy. Uranium Institute, London, pp 222–236

Ward TA, Flannagan JC, Hubery RW (1983) Rehabilitation of the Mary Kathleen uranium mine site after closure. In: Proceedings of the international specialist conference on water regime in relation to milling, mining and waste treatment including rehabilitation. Australian Water and Wastewater Association, Canberra, 32.1–32.9

Ward TA, Hart KP, Morton WH, Levins DM (1984) Factors affecting groundwater quality at the rehabilitated Mary Kathleen tailings dam, Australia. In: 6th symposium on uranium mill tailings management. Colorado State University, pp 319–328

Warhurst A (2000) Mining, mineral processing, and extractive metallurgy: an overview of the technologies and their impact on the physical environment. In: Warhurst A, Noronha L (eds) Environmental policy in mining: corporate strategy and planning for closure. Lewis Publishers, Boca Raton, pp 33–56

Warren GC, Brown A, Meyer WA, Williamson MA (1997) Suppression of sulfide mineral oxidation in mine pit walls. Part I: Hydrologic modeling. In: Tailings and mine waste '97. Balkema, Rotterdam, pp 425–431

Watten BJ, Sibrell PL, Schwartz MF (2005) Acid neutralization within limestone sand reactors receiving coal mine drainage. Environ Poll 137:295–304

Watzlaf GR, Ackman TE (2006) Underground mine water for heating and cooling using geothermal heat pump systems. Mine Water Environ 25:1–14

Weber PA, Stewart WA, Skinner WM, Weisener CG, Thomas JE, Smart RSC (2004) Geochemical effects of oxidation products and framboidal pyrite oxidation in acid mine drainage prediction techniques. Appl Geochem 19:1953–1974

Wendel W (1999) Laurion: Der antike Bergbau von Lavrion in Griechenland. Lapis 24:11–13

Wengel M, Kothe E, Schmidt CM, Heide K, Gleixner G (2006) Degradation of organic matter from black shales and charcoal by the wood-rotting fungus *Schizophyllum commune* and release of DOC and heavy metals in the aqueous phases. Sci Total Environ 367:383–393

Weterings K, Janssen J (1985) Recovery of uranium, vanadium, yttrium and rare earths from phosphoric acid by a precipitation method. Hydrometall 15:173–190

White WW, Lapakko KA, Cox RL (1999) Static-test methods most commonly used to predict acid-mine drainage: practical guidelines and interpretation. In: Plumlee GS, Logsdon MS (eds) The environmental geochemistry of mineral deposits. Part A: Processes, techniques and health issues. Society of Economic Geologists, Littleton (Reviews in economic geology, vol 6A, pp 325–338)

Whitehead PG, Prior H (2005) Bioremediation of acid mine drainage: an introduction to the Wheal Jane wetlands project. Scit Total Environ 338:15–21

Widerlund A, Shcherbakova E, Carlsson E, Holmström H, Öhlander B (2005) Laboratory study of calcite-gypsum sludge-water interactions in a flooded tailings impoundment at the Kristineberg Zn-Cu mine, northern Sweden. Appl Geochem 20:973–987

Willett IR, Bond WJ (1995) Sorption of manganese, uranium, and radium by highly weathered soils. J Environ Qual 24:834–845

Willett IR, Bond WJ (1998) Fate of manganese and radionuclides applied in uranium mine waste water to a highly weathered soil. Geoderma 84:195–211

Willett IR, Noller BN, Beech TA (1994) Mobility of radium and heavy metals from uranium mine tailings in acid sulfate soils. Austral J Soil Res 32:335–355

Williams PA (1990) Oxide zone geochemistry. Ellis Horwood, New York

Williams DJ (1997) Effectiveness of co-disposing coal washery wastes. In: Tailings and mine waste '97. Balkema, Rotterdam, pp 335–341

Williams M (2001) Arsenic in mine waters: an international study. Environ Geol 40:267–278

Williams MPA, Seddon KD (1999) Thickened tailings discharge: a review of Australian experience. In: Tailings and mine waste '99. Balkema, Rotterdam, pp 125–135

Williams TM, Smith B (2000) Hydrochemical characterization of acute acid mine drainage at Iron Duke mine, Mazowe, Zimbabwe. Environ Geol 39:272–278

Williams DJ, Wilson GW, Currey NA (1997) A cover system for a potentially acid forming waste rock dump in a dry climate. In: Tailings and mine waste '97. Balkema, Rotterdam, pp 231–236

Williams TM, Weeks JM, Apostol Jr AN, Miranda CR (1999) Assessment of mercury contamination and human exposure associated with coastal disposal of waste from a cinnabar mining operation, Palawan, Philippines. Environ Geol 39:51–60

Williams DJ, Bigham JM, Cravotta CA, Traina SJ, Anderson JE, Lyon JG (2002) Assessing mine drainage pH from the colour and spectral reflectance of chemical precipitates. Appl Geochem 17:1273–1286

Wilson WF (1994) A guide to naturally occurring radioactive material (NORM). PennWell Publishing, Tulsa

Wilson GW, Newman LL, Ferguson KD (2000) The co-disposal of waste rock and tailings. In: Proceedings from the 5th international conference on acid rock drainage, vol 2. Society for Mining, Metallurgy, and Exploration, Littleton, pp 789–796

Wilson NJ, Craw D, Hunter K (2004) Contributions of discharges from a historic antimony mine to metalloid content of river waters, Marlborough, New Zealand. J Geochem Explor 84:127–139

WISE Uranium Project (2006) Chronology of major tailings dam failures. (*www.wise-uranium.org*)

Wissa AEZ, Fuleihan NF (2000) Phosphogypsum stacks and ground-water protection. Phosphorus and Potassium 227:46–53

Wolf SF (1999) Analytical methods for determination of uranium in geological and environmental materials. In: Burns PC, Finch R (eds) Uranium: mineralogy, geochemistry and the environment. Mineralogical Society of America, Washington (Reviews in mineralogy and geochemistry, vol 38, pp 623–651)

Woodward A, Mylvaganam A (1993) Effects of uranium mining on the health of workers: findings from Radium Hill. In: Programme and abstracts of the international congress on applied mineralogy. Mineralogy in the Service of Mankind. Fremantle, pp 14–16

Woulds C, Ngwenya BT (2004) Geochemical processs governing the performance of a constructed wetland treating acid mine drainage, central Scotland. Appl Geochem 19:1773–1783

Xenidis A, Mylona E, Paspaliaris I (2002) Potential use of lignite fly ash for the control of acid generation from sulphidic wastes. Waste Manag 22:631–641

Yamauchi H, Fowler BA (1994) Toxicity and metabolism of inorganic and methylated arsenicals. In: Nriagu JO (ed) Arsenic in the environment. Part 1: Cycling and characterization. Wiley, New York

Yin G, Catalan LJJ (2003) Use of alkaline extraction to quantify sulfate concentration in oxidized mine tailings. J Environ Qual 32:2410–2413

Younger PL (1995) Hydrogeochemistry of minewaters flowing from abandoned coal workings in County Durham. Quart J Eng Geol 28:S101–S113

Younger PL (2000) Nature and practical implications of heterogeneities in the geochemistry of zinc-rich, alkaline mine waters in an underground F-Pb mine in the UK. Appl Geochem 15:1383–1397

Younger PL (2004) Environmental impacts of coal mining and associated wastes: a geochemical perspective. In: Gieré R, Stille P (eds) Energy, waste, and the environment: a geochemical perspective. Geological Society, London (Special Publications 236:169–209)

Younger PL, Robins NS (2002) Mine water hydrogeology and geochemistry. Geological Society, London

Younger PL, Banwart SA, Hedin RS (2002) Mine water: hydrology, pollution, remediation. Kluwer Academic Publishers, Dordrecht

Yunmei Y, Yongxuan Z, Williams-Jones AE, Zhenmin G, Dexian L (2004) A kinetic study of the oxidation of arsenopyrite in acidic solutions: implications for the environment. Appl Geochem 19:435–444

Yusof AM, Mahat MN, Omar N, Wood AKH (2001) Water quality studies in an aquatic environment of disused tin-mining pools and in drinking water. Ecol Eng 16:405–414

Zänker H, Moll H, Richter W, Brendler V, Hennig C, Reich T, Kluge A, Hüttig G (2002) The colloid chemistry of acid rock drainage solution from an abandoned Zn-Pb-Ag mine. Appl Geochem 17:633–648

Zagury GJ, Oudjehani K, Deschenes L (2004) Characterization and availability of cyanide in solid mine tailings from gold extraction plants. Sci Total Environ 320:211–224

Zhang YL, Evangelou VP (1998) Formation of ferric hydroxide-silica coatings on pyrite and its oxidation behaviour. Soil Science 163:53–62

Zielinski RA, Chafin DT, Banta ER, Szabo BJ (1997) Use of 234U and 238U isotopes to evaluate contamination of near-surface groundwater with uranium-mill effluent: a case study in south-central Colorado, USA. Environ Geol 32:124–136

Zinck JM (1997) Acid mine drainage treatment sludges in the Canadian mineral industry: physical, chemical, mineralogical and leaching characteristics. In: Proceedings from the 4th international conference on acid rock drainage, vol 4. Vancouver, pp 1691–1708

Zinck JM, Griffith WF (2000) An assessment of HDS-type lime treatment processes – efficiency and environmental impact. In: Proceedings from the 5th international conference on acid rock drainage, vol 2. Society for Mining, Metallurgy, and Exploration, Littleton, pp 1027–1034

Zouboulis AI, Kydros KA (1993) Use of red mud for toxic metals removal: the case of nickel. J Chem Technol 58:95–101

Zuddas P, Podda F (2005) Variations in physico-chemical properties of water associated with bio-precipitation of hydrozincite [$Zn_5(CO_3)_2(OH)_6$] in the waters of Rio Naracauli, Sardinia (Italy). Appl Geochem 20:507–517

Subject Index

A

B

C

D

Y

Z

Printing: Krips bv, Meppel
Binding: Stürtz, Würzburg

LaVergne, TN USA
14 January 2010
170016LV00003B/26/P

9 783540 486299